BIO-INSPIRED INNOVATION and NATIONAL SECURITY

edited by Robert E. Armstrong
Mark D. Drapeau
Cheryl A. Loeb
James J. Valdes

Published for the
Center for Technology and National Security Policy
by National Defense University Press
Washington, D.C.
2010

Opinions, conclusions, and recommendations expressed or implied within are solely those of the contributors and do not necessarily represent the views of the Defense Department or any other agency of the Federal Government. Cleared for public release; distribution unlimited.

This edition published by Books Express Publishing

ISBN 978-1-780390-40-6
To purchase copies please contact info@books-express.com

This book is dedicated to the memory of Dr. Robert Armstrong, visionary colleague, scholarly mentor, and good friend to all who knew him. Bob served the Nation as a Soldier and scientist and was passionate about the contributions that the biological sciences could make to national security. His provocative, often contrarian ideas challenged the status quo and were the genesis of this book.

CONTENTS

part five

IMPLICATIONS for the DEPARTMENT of DEFENSE

ILLUSTRATIONS

Figures

Tables

FOREWORD

Ever since man took to the battlefield, biology has played a significant role—both unintentionally and intentionally—in conflict. Prior to the discovery of the germ theory of disease, most combat deaths were the result of infections. And even before that understanding, biology was used in an offensive role. An early example was the Tatars' hurling of plague victims' bodies over the wall of the Crimean city of Kaffa in 1346, which probably helped spread the Black Death.

Despite various treaties and protocols, offensive biological weapons use has continued to this day, with the anthrax attacks of 2001 being the most recent incident. Such activity has led to a strong defensive program, with medical science developing numerous countermeasures that have benefited both civilian and military populations.

But that is the "old" biological warfare. Covert programs for the development of novel weapons will advance; likewise, the development of countermeasures will also continue. The present volume, however, does not address these issues. Rather, it reviews and analyzes current research and likely future developments in the life sciences and how they will significantly influence the biological material available to warfighters—not as weapons systems, but as augmentation to currently available equipment. This is the "new" face of biological warfare.

The editors of this volume have assembled experts in research, warfighting, and defense policy to describe biological applications from the smallest to the largest scale. In addition, they show how thinking in biological terms can improve our procurement cycle and enhance our development time and costs. Finally, no description of biotechnology would be complete without a consideration of ethical and legal issues related to such research and development.

This edited book is an important contribution to the literature and nicely captures a number of ongoing military basic science research projects with long-term implications for the Department of Defense. It does not purport to be an exhaustive accounting, but it is an excellent introduction for policymakers to garner an understanding of where biology is going to fit into 21st-century readiness and preparedness for our fighting force.

Andrew W. Marshall
Director of Net Assessment
U.S. Department of Defense

BIO-INSPIRED INNOVATION and NATIONAL SECURITY

INTRODUCTION

For the average person, the mainstream media, and perhaps even the typical warfighter, the application of biology to military operations probably conjures up negative thoughts of Cold War–era biological weapons programs, the use of plague or other diseases as weapons during war, the effects of the Spanish flu on forces in Europe during World War I, medical treatments for severed limbs caused by battlefield injuries, and mental anguish caused by post-traumatic stress disorder or traumatic brain injury. It is unfair to blame people for thinking of biological weapons or applications of biology to medicine when "biology in the military" is mentioned.[1] To some degree, the two are difficult to separate from each other.

Particularly outside national security circles, much biotechnology research focuses on medical or health issues, agriculture, and industrial processes. Yet while these are very important topics, they are a narrow and limited way of viewing biology and its potential applications to national security. This volume is about applications of the biological sciences, which we call "biologically inspired innovations," to the military. Rather than treating biology as a series of threats to be dealt with, such innovations generally approach the biological sciences as a set of opportunities for the military to gain strategic advantage over adversaries. These opportunities range from looking at tiny genes to large brains, from enhancing human performance to creating renewable energy, from sensing the environment around us to harnessing its power.

Many developments in the biological sciences have increasingly made an opportunities-based approach to biology and the military possible. During the past 20 years, advances ranging from DNA sequencing to various biotechnology manipulations to the intersection of biology with engineering, mathematics, and even aeronautical engineering have empowered the field to grow far more influential than it once was. Bioengineering, bio-inspired research, and biological metaphors have all become quite useful to the military and, indeed, society at large.

This trend shows no signs of abating and in fact is spreading globally and becoming less expensive. There are commercial DNA synthesis facilities in numerous countries, including Australia, Canada, China, Denmark, France, Germany, Iran, Japan, India, Russia, South Africa, Sweden, Switzerland, the United Kingdom, and the United States. One can buy high-quality DNA synthesizing machines on the auction site eBay and other places online for a few thousand dollars, not to mention less sophisticated yet incredibly useful biotech equipment such as PCR machines and centrifuges. Technology is increasingly powerful, reagents often can be readily acquired by trading, and data are freely available in repositories such as GenBank. It is an open market for innovative people who want to get in on biotechnology and all it has to offer. In response, a field known as synthetic biology—the goal of which is to engineer biology like so many building blocks of nature—is burgeoning.

In 1864, William Sellers introduced a standardized system of screw threads and nuts—connectors that were crucial for building things in the physical world. What will be the equivalent standard set of connectors in the biological engineering world? Perhaps the "BioBricks" project from the Massachusetts Institute of Technology (MIT), which seeks to create "Lego" block-like biological parts with standard ends that plug into each other simply and easily. These "bricks" are freely available from the MIT Web site, which keeps track of a Registry of Standard Biological Parts.[2] In association with this, the annual International Genetically Engineered Machine (iGEM) competition challenges high school students from around the world to develop new parts and products. U.S. students have competed against teams from Africa, Canada, India, Japan, Scotland, Slovenia, South Korea, Spain, Switzerland, and the United Kingdom.

It is difficult to bridge the gap between what is happening in the world of iGEM and the research and operations inside the Department of Defense (DOD). There are very few "bio types" in the military, the DOD, or in national security jobs more generally. Getting an advanced degree in biology is not a typical route to becoming a senior military officer, a Foreign Service Officer, or even a DOD scientist. Biology is not emphasized at military academies, where students are much more likely to be trained in traditional engineering, computer science, or military history. Conversely, there are very few "military types" in biology jobs in academia, biotech companies, and the like. Despite many genuine advances in medicine inspired by war[3] that have helped humanity, the stigma attached to the intersection of biology and the military for the aforementioned reasons widens the gap still further.

Yet there are many revolutionary opportunities for the military stemming from biologically inspired innovations (BII) beyond traditional "threat-based" medicine and weapons applications of the technology. Biotechnologies can:

- impact logistics by reducing the size and weight of objects carried by warfighters
- provide novel portable power sources
- potentially influence manufacturing via rapid prototyping or other engineering advances
- provide opportunities for new kinds of battlefield sensors, soldier monitoring and therapeutics, human performance enhancement, health monitoring, foods that can provide vaccines and nutrients and other chemicals in novel ways, data analysis, combat identification involving biology, and various forms of camouflage and concealment inspired by the natural world.

Biological systems—from flocks of geese to complete ecosystems—are complex adaptive systems that affect human existence in fundamental ways. From the food supply to global climate change, humans are part of a global ecosystem, and biology impacts us at the core. Biological science, combined with engineering principles, can bridge the gap between today's standards and systems and desired long-term capabilities important for national security.

In the flat, global economy that is the backdrop of the military-industrial complex, this is not just about America. Many other countries have invested heavily in biology, realizing that it is the future. As one example, Singapore has set up a plan called Intelligent National 2015 to help the country meet the demands of its national security and economic growth structures and the population. The plan will use sensors, biocomputing, and nanotechnology.[4] Does the United States have such a roadmap—a national strategy for biology?

This volume is designed as an overview of the many applications of biology to the military and national security writ large (with the exception of bioweapons and biomedicine). Policy issues are covered, and original research is presented. The diversity of the authors' backgrounds reflects the breadth of applications and opportunities of biology to modern problems. The reader should walk away enlightened as to the many possible ways in which biology is influencing and will continue to influence national security. Authors were asked to provide foresight on trends and indicators in their areas of interest (but not forecasts, which would describe a single future state) in order to better understand their implications for the next 20 years or so.

The chapters are organized into five sections. Part one, "Perspectives on Biological Warfare," contains four chapters that review some traditional threat-based perspectives on biology and the military and place them in the opportunity-based framework of this volume. We asked warfighters, ethicists, and bioweapons experts to put emerging biologically inspired innovations into context. How do they change bioethics considerations? How will biotechnology impact the average warfighter? Is a spider silk net bazooka a bioweapon in an international legal context? Could it ever be? Should it? There are many issues to be considered against the backdrop of the more advanced research in later chapters.

The contributors to part two, "Biomolecular Engineering," are all active laboratory scientists who have worked on engineering in the small world of living molecules. Despite the scale of biomolecular work being tiny, generally at the level of genetic material, proteins, or cells, the implications range from large to small—from custom-designing molecules for specific tasks such as new materials that soldiers can wear, to creating large-scale renewable energy resources.

The authors in part three, "Bio-inspired Machines," paint a broader picture. Here, the natural world's behavior and design become an inspiration for building nonliving machines that perform desired tasks. In one case, researchers have built autonomous machines that can learn using an artificial brain modeled after neurobiological principles. In another, animals such as dogs and octopi become the design of robots that can walk and pick up objects in unconventional ways. Still other authors write about the human-robot interface, an important issue for warfighters as autonomous machines become more prevalent on the battlefield.

In part four, "Human Applications," authors examine how to better understand human beings for the purpose of enhancing their performance in some way. A good deal of the work being done in this field is in the general area of neuroscience, including neural plasticity, learning, resilience, and brain imaging. In one chapter, the authors look at bio-inspired network science, an attempt to comprehend how human networks are related at a fundamental level to genomic, hive, traffic, and other networks so we can better understand how humans interact in groups and how those groups interact with other kinds of nonhuman groups.

Finally, in part five, "Implications for the Department of Defense," authors discuss thorny issues surrounding BII policy that DOD and other parts of the Government currently (and will continue to) face. A military ethicist looks at potential ethical pitfalls related to the kind of research outlined in earlier chapters. A military lawyer considers the legality of biological inventions and their use in warfare. Finally, two lifelong defense scientists take a quantitative point of view on where the future BII workforce will come from, and the results are not encouraging; change will have to occur in order to ensure that DOD is prepared to take strategic advantage of BII opportunities during the next 20 years.

All of these biological advances are part of the larger picture of a "global technology revolution" in which we are engulfed.[5] The United States is no longer the sole proprietor of world-class research and development (R&D) programs, and emerging technologies in numerous areas combined with niche specialty companies will ensure that this trend continues. Moreover, as the world changes, technologies in other parts of the world will evolve in response to their own novel environments, and not necessarily to ones familiar to those looking only at how technologies are used within the United States. Intersecting developments on numerous scientific and technological fronts (in addition to biology), including nanotechnology, materials science, information technology, and human factors research, are creating new subfields and inventions of possible strategic importance. World events that involve emerging technologies could evolve in predictable—or unpredictable—ways that create disruptive game-changers affecting U.S. grand strategy and international security and stability.

Advances in other areas of science and technology will also bear on BII as they converge. The fields of nanotechnology, biotechnology, information technology, and cognitive science (collectively termed NBIC) increasingly interact in the laboratory and the industrial space. It is difficult to predict accurately the long-range effects this interaction will have, but clearly the trends in each of these areas separately, and more so together, are important to monitor. To some degree, advances in BII outlined in this volume, and the ethical, legal, education and training, and budgetary issues associated with them, must be seen in this greater NBIC context.

Analysts in the Office of Net Assessment in the Pentagon as well as thinkers in other institutions have been talking for about a decade about the emerging area of biotechnology and how it affects the military. They have previously recommended that DOD exploit the power of BII to address military challenges, that leaders inside DOD should formally advocate for rapid BII integration into DOD processes and programs (for example, by establishing a Department of Biotechnology and Military Life Sciences), and that DOD should establish a cooperative framework with the commercial biotechnology industry to support BII R&D. Yet there is no definitive book on the topic written for the defense policymaker, the military student, and the private-sector bioscientist interested in the "emerging opportunities market" of national security. We hope that this edited volume helps to close this gap and provides the necessary backdrop for thinking strategically about biology in defense planning and policymaking. DOD biology writ large needs to be better understood, should be a higher budget priority, needs a comprehensive human resource plan, and should collaborate with and be wary of other countries' biological efforts, as appropriate.

As DOD and other Government agencies move from study and research on BII toward product development and deployment on a large scale,[6] we hope that this book helps educate current and future leaders in the national security realm to take advantage of opportunities that biotechnology has created and to develop sustained U.S. technological dominance.

NOTES

1 Guo Ji-wei and Xue-sen Yang, "Ultramicro, Nonlethal, and Reversible: Looking Ahead to Military Biotechnology," *Military Review* (July-August 2005), 75–78.

2 For more information on the Registry of Standard Biological Parts, see **http://parts.mit.edu**.

3 The history of biology and warfare is reviewed by Robert E. Armstrong and Jerry B. Warner, *Biology and the Battlefield*, Defense Horizons 25 (Washington, DC: National Defense University Press, 2003), and "History of Biological Warfare," available at **www.gulfwarvets.com/biowar.htm**.

4 Comtex News Network, "Singapore Announces Latest 10-year IT Roadmap," *Xinhua*, March 9, 2005.

5 Richard Silberglitt et al., *The Global Technology Revolution* 2020 (Santa Monica, CA: RAND, 2006). See also **www.dni.gov/nic/NIC_2025_project.html**.

6 The Office of Net Assessment has published a number of reports that were some inspiration for this volume, including "Exploring Biotechnology: Opportunities for the Department of Defense" (January 2002), "Biotechnology II: Capitalizing on Technical Opportunities" (October 2002), and "An Investment Strategy for Non-Traditional Applications of Biotechnology for the Department of Defense" (October 2005).

part one

PERSPECTIVES on BIOLOGICAL WARFARE

chapter 1

BIOTECH IMPACT on the WARFIGHTER

Thomas X. Hammes

The past few decades have been shaped by the remarkable evolution of information technology (IT), which has allowed humans to connect in ways we never considered possible in the past. Yet in its fascination with these new technologies, the United States has made major mistakes in how it integrated them into its strategic concepts. We assumed that technology, as represented by network-centric warfare, would allow us to dominate war. As a result, we focused investments, force structure, and training on ways to establish dominance by replacing people with technology. Unfortunately, despite claims by its proponents that IT fundamentally changed the nature of war, it proved incapable of overcoming 4,000 years of recorded history. The enemy continued to have a vote. Today, America's enemies have voted to use irregular/fourth-generation warfare to counter U.S. dominance in high-technology conventional weaponry. Only the painful lessons of Iraq, Afghanistan, and the global war on terror forced the Pentagon to accept the fact that humans remain the dominant factor in war.

The next few decades will see a similar explosion in both biotechnology and nanotechnology. Today, these sciences are about where computers were in the late 1970s. Considering that it took us less than three decades to go from floppy disks with eight kilobytes of random access memory to terabyte hard drives, we have to expect the next few decades will bring progress as unforeseen as the Internet was in 1975. The United States must not make the same mistake in trying to apply the new technologies to replace humans. Rather, it should explore how to use the technologies to enhance and support humans. Given the nature of biological advances, this should be a commonsense approach to integrating biotechnology advances into our strategic concepts.

Short-term Developments

Although we are on the verge of the biotechnology explosion, the uncomfortable reality is that the early products of this biological revolution have much more potential to empower terrorists than to provide state-based warfighters with effective new tools.

This is not to say that modern biotechnology is not delivering some enhancements for today's warfighter—just that the current enhancements are less impressive than the perceived threat. Looking first at the advantages that biotechnology is bringing to the warfighter, we can see some progress in pharmaceuticals, significant advances in combat medical care, solid progress in sensors and new materials, and even some "green" manufacturing for propellants and other energetics.

For instance, pharmaceuticals that manipulate circadian rhythms are being used to enhance pilot alertness on long missions. More advanced tests indicate it might be possible to chemically sustain a person's alertness as long as several days without adverse effect, then allow the person to fully recover in one 10- to 12-hour stretch of sleep.

In a similar vein, sports medicine is constantly seeking ways, both legal and illegal, to enhance athlete performance. While legality is not a critical issue for combat soldiers, safety is. Current strength enhancers that cause long-term health problems obviously cannot be employed by the military. However, one area with promise is blood doping—that is, simply providing the soldier with more of his own red blood cells to increase his oxygen carrying capacity. In an extension of this concept, experiments with artificial blood cells that carry oxygen present another possibility of enhancing human performance. These are fairly minor improvements in performance with limited applications. While there is no doubt that biotechnology will bring dramatic changes in the capabilities of humans, it will take time, and improvements will be incremental in nature.

In contrast to human enhancement, major progress has been made in trauma care for wounded personnel. Not only have survival rates increased dramatically, in part due to new biological products, but also postinjury care is greatly improving the quality of life of those treated. In particular, prosthetic devices have made enormous strides. We have developed devices that can be directly controlled by a patient's nervous system. As progress continues, artificial limbs will improve in strength and agility until they exceed natural capabilities. Despite the obvious ethical questions, we can assume some individuals, groups, and perhaps even nations will experiment with fighters using artificial limbs.

In a more benign development, the Defense Advanced Research Projects Agency has produced a cooling glove that vastly enhances human performance in hot or cold environments. Slipped over the hand, the glove cools or heats the subject's blood as it passes under the glove and has resulted in considerable improvements in physical capability in the lab. In some hot environment tests, strength is increased 50 percent for short duration tasks, while endurance increases dramatically. Although the glove is a mechanical device, it clearly capitalizes on biological research into how the body functions under stress.

Other forms of technology have already enhanced individual human performance. We take night vision and thermal imagery for granted as normal capabilities for an individual soldier. We are working hard on exoskeletons that will allow individuals to carry heavy loads without fatigue or physical injury, although providing sufficient power will probably restrict operations to base camps with adequate power generation capability. As an interim step, we are exploring "super" Segways that allow an individual soldier to control the vehicle while remaining able to fire a machinegun and transport over 1,000 pounds. With its narrow wheelbase, this "super" Segway functions as a first-generation exoskeleton that allows soldiers to move quickly, with light armor protection, into small spaces.

The current state of biotechnology indicates that short-term enhancements will most likely come primarily from the use of technology to mechanically enhance human performance rather than directly from biological improvements.

Short-term Dangers

While there is great potential for long-term benefits to the warfighter and society as a whole from biotechnology, the primary short-term impact of biotechnology is to put biological weapons within reach of even small, poorly funded groups.

A primary short-term risk lies in synthetic biology, which differs from genetic engineering primarily in that the latter is the manipulation of existing living systems by modifying the genome (by either adding or subtracting genetic material), whereas synthetic biology is the creation of new genomes and organisms. The objective of the genetic engineering is simply to modify the genome of an organism or to bring the beneficial traits of one organism to another; synthetic biology seeks to reengineer entire organisms, create existing organisms using only their genetic base pairs, and even create new ones that have never been seen in nature.

Synthetic biology obviously has enormous potential for changing any element of our world—from creating plants that will produce energetic compounds or inexpensive medicines in theater, to producing organisms that neutralize industrial waste. Synthetic biology's potential is limited only by imagination and funding. Unfortunately, the same may (and I emphasize *may*, since this is easier said than done for a number of technical reasons) also be true of its potential as a weapon.

The following disconnected but very important developments foreshadow the potential for even small terrorist groups to successfully create biological weapons in the future:

- The nucleotides to make smallpox can be purchased from a variety of suppliers without identity verification.
- Smallpox has about 200,000 base pairs. DNA with up to 300,000 base pairs has already been successfully synthesized.
- An Australian research team enhanced the pathogenicity of mousepox virus by activating a single gene. The modification increased its lethality, killing all 10 mice in the experiment. It was even lethal to 60 percent of an immunized population. The team posted its results on the Internet.[1]
- Biohackers are following in the footsteps of their computer software hacker predecessors. They are setting up labs in their garages and creating products. A young researcher recently invested $20,000 in equipment and produced two new biological products before selling the company for $22 million. We can assume that others will try to replicate her actions.

Smallpox offers an example of the potential devastation a terrorist release of a biological weapon could cause. The disease is well known for its high mortality rate and propensity to leave survivors with disfiguring facial and body scars. Smallpox is communicable from person to person and has an incubation period of 7 to 17 days. While a vaccine for smallpox does exist, there is no completely effective medical treatment once a person displays overt symptoms of the disease. Smallpox is considered a potential terrorist threat because the virus was declared eradicated in 1980 and countries halted their mandatory smallpox vaccination campaigns, thus leaving the current population highly susceptible to the virus if it were to be released by a terrorist group.

DARK WINTER, an exercise conducted in 2001, simulated a smallpox attack on three U.S. cities. In a period of 13 days, smallpox spread to 25 states and 15 countries in several epidemiological waves, after which one-third of the hundreds of thousands of Americans who contracted the disease died. It was estimated that a fourth generation of the disease would leave 3 million infected and 1 million dead, although there are questions about the assumptions underpinning the epidemiological model. The good news is that this exercise was one of the key drivers behind the U.S. Government purchase of enough smallpox vaccine to inoculate every American in an emergency. Unfortunately, most of the world does not have access to adequate supplies of the vaccine.

It is essential to remember that not only could a terrorist release of smallpox cause an exceptional number of deaths, but it could also shut down a large portion of global trade until the outbreak is controlled or burns itself out. Given that the 2002 West Coast longshoremen's strike cost the U.S. economy $1 billion per day, the cost of a complete shutdown of all transportation would be catastrophic.

Biological weapons have the potential to kill many more people than a nuclear attack. Further, unlike nuclear weapons, which are both difficult and relatively expensive to build, biological weapons are quickly becoming inexpensive to produce as life sciences and related technologies advance. While smallpox was selected as an example for this chapter, other agents could also be selected. An attack using a contagious human disease represents the worst case scenario. However, there are numerous less devastating and correspondingly less difficult biological attacks that possibly could occur. For example, a terrorist could use a series of anthrax attacks similar to those perpetrated in 2001 to cause massive disruption but few deaths among the population (anthrax is not contagious). By routing anthrax letters through every major mail and package distribution center in the Nation, he could force the various carriers to shut down while they decontaminate their facilities and devise safeguards to prevent further contamination. A terrorist could also simply introduce foot and mouth disease into feedlots in America, then announce the presence of the disease on the Internet or with video to the news media. Repeated attacks could cause massive financial losses and security costs for the beef industry. Similar attacks using mad cow disease could also be high impact.

Longer Term Developments

It has already become a cliché that the 20th century was the century of physics and the 21st will be the century of biology. The fact that it is a cliché does not make it any less true. In the longer term, we can see a number of intriguing possibilities coming from biotechnology. We are already extracting spider silk from transgenic goats. Five times stronger than steel, spider silk offers some novel potential for new materials ranging from clothing with body armor capabilities to lightweight but incredibly strong ropes. While we are still learning to weave the material, we have proven that industrial quantities can be produced. This should be one of the first of a series of developments that make use of the natural evolution of materials and structures that have evolved over millions of years.

In the same way that we are learning to make materials as effectively as a spider does, we are studying many aspects of other natural designs to improve how we build and process material and even deal with complexity. Thus, the study of biological systems is improving our manmade ones, as is pointed out in chapter 19 of this book, where network science—understanding one kind of network, such as bird flocking or fish schooling behavior—has larger implications for other networks, such as unmanned micro-aerial vehicles that can swarm.

Beyond the exploitation of biology to improve technology lies the frontier of actual genetic manipulation of humans. While there are huge ethical questions about even experimenting in this field, we have to accept that in a society where parents give their children human growth hormones to make them more athletically competitive, some individuals, organizations, or nations will seek genetic enhancement. Ethical issues such as these are discussed in more detail by other authors in this volume.

Because we are on the knee of the change curve, we cannot begin to predict the impacts biology will have on the warfighter in the next 30 years—only that biotechnology, like IT before it, will revolutionize not just security but all aspects of our lives as well.

Implications for the Warfighter

While the longer term exceeds our ability to predict, the short-term impacts of biology are already becoming clear. We know that the use of biological weapons will present unprecedented political and strategic issues to our senior leadership. While those considerations exceed the purview of the warfighter, the impact will be no less intensive. In any major biological attack on the United States, the military, in particular the National Guard and the Reserve Components, will be called upon to execute an extensive and perhaps unanticipated range of tasks. These tasks could easily range from providing emergency medical treatment to establishing refugee centers to enforcing quarantines. While the first two are fairly routine, the third has enormous implications. Beyond the legal ramifications, we have to consider the practical aspects. Will U.S. troops fire on apparently healthy civilians if ordered? What tactics, training, procedures, and equipment would allow us to enforce a quarantine without using deadly force? What other tasks might we be called upon to execute? Distribution of food into contaminated zones? Protection of vaccines or medical facilities? Closing U.S. borders? Operating critical infrastructure? How do we operate when the families of our Servicepeople are exposed to the same threat? Clearly, we are unprepared to assist our nation in the event of a major domestic biological attack.

We also have to consider the use of biological weapons by various players on U.S. forces overseas. Such attacks could be designed to either create large numbers of casualties or render the force ineffective. What force protection measures, equipment, tactics, and training are necessary to protect a force against a potential terrorist attack with biological agents? How will we even determine whether we are under attack? Might the disease be a new outbreak of a naturally occurring disease? What are the procedures for identifying, verifying, and acting against such an attack? What is the impact of biological protective equipment? For many diseases, simple high efficiency particulate air filter masks to protect respiration will suffice. Others will travel by other vectors. How will an enemy introduce such weapons into our forces? Are there implications for how we provide contractor support in a combat zone? Would a terrorist go after the contracted logistics to cripple our forces?

The questions above barely scratch the surface of the variables in a biological environment. Obviously, we must work on systems to protect our personnel from a wide variety of agents and to rapidly identify a particular agent and on processes for rapid development of treatments and vaccines for newly identified agents. It is not a matter of too few resources but rather of priorities. Currently, we spend billions in the hope of shooting down an inbound missile that might carry a small nuclear warhead—a weapon very few nations in the world will ever develop. Even those that do develop the system know such an attack would bring a devastating if not annihilating response. Yet we spend comparatively minuscule amounts on countering weapons that we know could be available to a range of nonstate actors in the near future. Some will simply not be deterrable. It is time to rethink our priorities. Our current commitments to Iraq and Afghanistan provide us with a unique opportunity to do so.

Reset for What?

The extensive wear and tear on our forces in Iraq and Afghanistan has altered the debate about the future military force. Many have come to the conclusion we will have to focus our efforts on resetting rather than transforming the force. Of course, the key question is, "Reset for what?" Do we rebuild our forces to fight a near-peer competitor? Or to fight a smaller state? Or to fight a prolonged insurgency? Whatever scenarios we prepare for, there is a high probability they will include some form of biological weapons. We have to use the reset requirement to incorporate biological defense in our future force structure.

NOTE

1 Federation of American Scientists, "Mousepox Case Study," available at **www.fas.org/biosecurity/education/dualuse/FAS_Jackson/2-8.html**.

chapter 2

NEW BIOLOGICAL ADVANCES and MILITARY MEDICAL ETHICS

EDMUND G. HOWE

Rapid advances in biotechnology and the changing nature of modern warfare have raised ethical questions for the military and for military physicians. Issues such as the treatment of prisoners of war, the use of psychotropic drugs on soldiers on the battlefield, and the requirement of certain protective measures against chemical and biological weapons are all difficult ethical issues to be considered by the military physician.[1]

In this chapter, I present some examples of ethical concerns for the physician on the battlefield. New developments in biotechnology, nanoscience, and neurobiology further complicate these ethical concerns by opening the possibility of manipulating the fundamental nature of human physiology, but these developments can contribute greatly to combat success. Whether and how such technologies should be employed remain uncertain and must be weighed with respect to both ethics and national security. Overall, I suggest that now, as before, the overriding priority must be the military's missions.[2] If it is not, countless lives could be lost. However, at the same time that the military's missions are given highest priority, limits still must be established regarding acceptable means with which to achieve victory. Where these lines should be drawn is the crux of many of the ethical questions that will be posed.

Military Medical Triage

An ethical dilemma that has been posed for some time in military medicine is how triage should be carried out during combat.[3] In civilian settings, physicians are generally required to treat dying patients first. In some combat situations, however, this may not always be the case, and military physicians may be faced with difficult decisions when the military priority is to meet the needs of the military mission before meeting the needs of the individual patient.

In optimal conditions, when the resources are available (such as time, personnel, equipment, and supplies), and the combat situation allows for it, the most seriously ill patients would be treated first. In nonoptimal conditions, the military medical triage principle holds that military physicians can give priority to less severely ill or injured soldiers in two instances: if it would save a substantially greater number of soldiers' lives, and if it is necessary to treat soldiers so they can return to combat. In these circumstances, it is likely that the injured soldiers are needed on the front lines in order to prevail against enemy forces or in order for the mission to be completed.[4] When military medical physicians make these difficult decisions, they recognize that by doing so, some patients may die who would have survived under different circumstances and with adequate resources.

One example of this difficult decision process occurred during World War II in North Africa. Despite a shortage of penicillin, the decision was made to use it to treat soldiers infected with venereal disease in order to ensure that the soldiers would recover quickly and return to the front lines. Because of this policy, however, soldiers infected with other illnesses such as pneumonia did not receive the penicillin they needed, and, in some cases, they died as a result of their injuries/illnesses.[5]

Military physicians share their commitment to saving patients' lives with their civilian counterparts.[6] Civilian physicians performing medical triage after a natural disaster such as an earthquake or flood, in fact, adopt a practice comparable to that of military physicians. The ethical justification of sacrificing individuals' lives to save greater numbers of soldiers or to further the mission has not changed in recent years.[7] What may have changed is the way in which these principles should be applied.

Rescuing Soldiers on the Battlefield

Recent advances in the life sciences, such as in biotechnology, nanoscience, and neurobiology, have introduced new questions about when soldiers, such as medics, should risk their lives to try to save the lives of other soldiers wounded on the battlefield.[8] An ethical question posed now, as in the past, is how many soldiers should risk their lives to try to rescue one fallen colleague.

New technology has altered how this question might be answered. Thanks to new biosensors that can, for example, monitor vital signs such as the heartbeat of a wounded soldier, military physicians and their commanders can better monitor how the soldier is faring on the battlefield.[9] If the physician observes that the soldier's pulse is falling due to blood loss, for example, he or she may feel a heightened sense of urgency to carry out a rescue attempt. These new observations raise the question of what medical criteria should be used to determine if a rescue attempt should be made and how much weight each criterion should have on the decision. During the heat of battle, commanders can best decide what risks to both fallen soldiers and rescuers are warranted. However, military physicians can best judge the clinical implications of a soldier's vital signs. Thus, new biotechnology developments may require commanders and physicians to find ways to work together more closely.[10]

In addition to preserving the maximum number of soldiers' lives, an important ethical value is to respect soldiers' dignity by trying to save them.[11] This value may justify a second or third soldier attempting a rescue, even when the life of a previous rescuer has been lost. If rescue attempts are not made, soldiers' morale may be undermined. Higher morale will, of course, result in soldiers feeling more motivated to fight. The overriding ethical considerations here must be the same ones underlying military medical triage. The battle must be won. If this can be accomplished and a substantially greater number of soldiers' lives will not be lost, the next most important value to be furthered is respecting soldiers' dignity.

Paradoxically, the choices that most promote soldiers' dignity may be the same ones that would result in the most sacrifice of their individual interests. Soldiers know that their lives will be sacrificed if and when it is necessary for the mission. When soldiers enter the military, they promise implicitly to give their lives if they must. Military physicians, analogously, implicitly promise both the soldiers who will be their patients and the military that they will abide by military triage principles when their commanders indicate that they must.

Both soldiers and military physicians want to fulfill these prior promises, which are their reasons for serving in the first place. Thus, to fully respect soldiers' dignity, military physicians must allow them to sacrifice their individual interests when necessary for the mission.

Treating Soldiers for Combat Fatigue

Another dilemma involving conflicting values arises when military doctors must treat soldiers temporarily immobilized by combat fatigue.[12] Combat fatigue, also known as battle fatigue and shell shock, is a neurotic disorder cause by the stress involved in war. Combat fatigue has a number of characteristics including hypersensitivity to noises, movement, and lights, easy irritability leading in extreme cases to violence, and sleep disturbances including nightmares, restlessness, and insomnia.[13]

Under optimal conditions, one potential option for treating soldiers experiencing combat fatigue could be to remove them permanently from the front lines. This might increase the severity and permanence of their psychiatric symptoms but eliminate the possibility that they would die during battle. If asked, many might choose to be permanently removed from combat rather than risk death by reentering it. In civilian contexts, doctors generally respect what patients want, even when they are significantly emotionally impaired.

In combat, however, removing the soldier from the battlefield might result in extremely negative consequences. First, other soldiers might also want to be relieved from the front lines and the attendant risk of death. If they had learned that soldiers experiencing combat fatigue could be removed permanently from further duty, they might develop symptoms of combat fatigue themselves, consciously or unconsciously, in order to be removed from combat. These actions could decimate the war effort.[14]

Military physicians, therefore, are taught to treat such soldiers near the front lines if possible and give them food, sleep, and emotional support, with the clear expectation that they will rejoin their unit in battle within days, rather than treating them far from the front lines.[15] Treating soldiers for combat fatigue may require giving priority to the needs of the military mission to avoid unintentionally opening the floodgates to soldiers wanting to escape front line duty.

This practice has been successful in the past. In general, soldiers temporarily removed from duty and treated in this manner have done well emotionally after rejoining their units. Depending on their closeness and loyalty to their unit, they will likely receive emotional support from others in their unit. The powerful emotional influence of unit cohesion is believed to be a reason that returning soldiers to their unit helps them to overcome combat fatigue.[16] On the other hand, permanent removal from the front lines could further enhance their combat fatigue symptoms, partly as a result of survivor guilt—feeling guilty that other soldiers in their unit stayed at the front and died.

Further, if soldiers are returned to serve in a more isolated environment, the benefits of unit cohesion could be reduced, and they could experience relapses of combat fatigue. This situation could occur in the future due to the ongoing counterinsurgency that U.S. forces are fighting in Iraq and Afghanistan. Soldiers in unprotected environments may fear that their lives are always in danger and may be traumatized more than once, either by being wounded themselves or by seeing a fellow soldier wounded or killed. Repeated trauma may result in their combat fatigue becoming worse, particularly if they remain in this situation. It may be, then, that in this kind of environment, the criteria for removing soldiers from the front lines versus returning them to duty should change.

The ethical priority is again to do whatever is necessary to maximize the success of the mission. Military physicians treating soldiers with combat fatigue as they traditionally have done—with the expectation that the soldiers will return to the front lines—may therefore be more necessary than it has been in the past. As a result of this practice, soldiers may pay a dearer price.[17]

This suggests in turn the possibility of a greater need to try to prevent combat fatigue before it occurs. One example of the ethical problems this uncertainty brings about is whether to place mental health personnel at or near the front lines. If such personnel are stationed with soldiers, they may better gain their trust and thus be able to treat them more quickly. Yet this would also place these personnel, who are limited in number, at higher risk. Such a deployment of mental health personnel to the front lines raises the same kinds of questions discussed previously in regard to rescue attempts. How much risk to other soldiers, who in this instance are mental health personnel, is justifiable to possibly benefit soldiers at increased risk for combat fatigue?

Preventive Measures

In the future, new approaches may be developed to prevent soldiers from experiencing combat fatigue.[18] One possibility is the use of medications.[19] Another is genetic screening, which would eliminate soldiers who are more genetically vulnerable to stress from serving in combat conditions in which they would be more likely to experience combat fatigue.

Medication might decrease soldiers' risk of developing combat fatigue by reducing physiological responses to stress, such as heart palpitations, trembling, and sweating. This, could, in turn, reduce their subsequent emotional morbidity and could potentially be accomplished with so-called beta-blocking medications such as propranolol. This medication was first developed to treat hypertension and is often used to treat chest pain, high blood pressure, irregular heartbeats, migraine headaches, and tremors.[20] Other uses of propranolol have included the treatment of anxiety and stress, and it has been used by performers and others to prevent stage fright. However, because neither this nor any other approach has been shown to be successful in reducing occurrences of combat fatigue, further research needs to be conducted. It may be that the fear and stress some soldiers experience during combat could be so profound that it would overcome preventive efforts.[21]

The use of a medication such as propranolol could potentially have serious negative effects. Such medication could potentially cause soldiers to be more callous during combat and/or to remember less of the horrors of war. Empirically, it is questionable whether these negative effects would actually occur, and more research is necessary before a true analysis can be made. Persons under the influence of hypnosis, for example, will not carry out acts that violate their personal conscience, whereas the influence of some drugs or alcohol may cause people to behave in ways out of character for them that could even be harmful to others. Ethically, it may not be the case that good and bad effects (if both occurred) should be balanced in some way to produce the "best overall result." To do this would subject soldiers to unnecessary risk—a disrespectful and possibly ethically impermissible course of action.

Further, research into the use of psychotropic medications, including some beta-blocking medications, and psychostimulants such as amphetamines, which could be used to help soldiers stay awake and fight fatigue, must be measured against the adverse risk of addiction and potentially serious side effects such as psychosis.[22]

Genetic Screening

Alternatively, the risk of soldiers experiencing combat fatigue might be reduced by excluding from combat those who are genetically susceptible to being psychologically harmed due to stress—for example, if they had exceptional vulnerability based on genetic testing, if they had prior responses of exceptional anxiety, or even if they had close relatives with anxiety disorders.[23]

A study published in the November 2009 edition of the *Archives of General Psychiatry*, for example, has shown that a gene variant makes people who experienced trauma as children or adults more susceptible to psychiatric disease, including post-traumatic stress disorder (PTSD). The Yale researchers who undertook the study "looked at a particular variant, or allele, of the serotonin transporter protein gene, which was previously linked to an increased risk for depression in individuals who had experienced adverse events. People with this allele were more likely to develop PTSD if they had experienced childhood adversity or adult trauma."[24]

Performing genetic screening on soldiers and excluding some from combat on the basis of their results could, however, violate soldiers' privacy and equality by requiring other soldiers to take disproportionate risks.[25] Both concerns would also be more ethically problematic because the extent to which more genetically vulnerable soldiers would be likely to experience combat fatigue is uncertain.

This challenging nature of discriminating on a genetic-related basis was manifested previously when soldiers with the sickle cell trait were barred from being military pilots, even if they had flown previously without having negative symptoms.[26] These soldiers felt that their dignity had been violated, particularly because there was only a slightly increased likelihood, if any, that those with this trait would develop more symptoms.

There are numerous other examples in which genetic screening could also be used to try to reduce risks,[27] such as screening out soldiers with certain genes who may be more vulnerable to heat effects if serving in desert environments.[28] Key considerations in determining whether this kind of screening is justifiable are the magnitude and probability of harm to soldiers genetically at risk. Unless this harm is substantial and/or likely, the inherent violation of soldiers' privacy and dignity would tend to preclude this screening from being ethically justifiable.

A still more problematic potential use of screening is to exclude soldiers from specific roles on the basis of conditions that will affect them later in their lives. An example of such a condition is Huntington's disease. Screening can determine that a person carries this gene and thus will, with some degree of certainty, be affected in the future. Such individuals could be excluded from certain military specialties.

In summary, excluding soldiers on the basis of their genes violates their privacy and the principle of equality. Research into genetic screening must take these factors into account. However, the importance of the military mission should prevail if this discrimination was necessary to heighten the chance of victory.

Preventive Measures against Biological and Chemical Weapons

Policy

During the first Gulf War (1990–1991), soldiers were required to take certain protective measures against possible biological and chemical weapons attack, even though these protective agents had not been fully tested for these purposes.[29] The courts determined that such measures were permissible because there was a need to give absolute priority to saving the maximum number of soldiers' lives and to support the military's mission. It held specifically that under the circumstances then existing, soldiers could be vaccinated to protect them from chemical and biological agents even against their will because of the belief that Iraq would use weapons of mass destruction against American forces. After the war, it was discovered that Iraq did indeed have chemical and biological weapons and the delivery apparatuses, such as missiles, for them.

The ethical argument in these instances is straightforward. While these protective agents may not have provided total protection from these types of weapons, and the enemy may not have even deployed them, requiring soldiers to get vaccinated against suspected threats was not only the best but also the only means of protecting U.S. troops in the Persian Gulf if Iraq chose to use this weaponry.

In the future, there may be new threats and the need to balance the risks of potentially protective measures against these threats. These agents cannot be tested on humans to determine their actual efficacy because testing could seriously harm or even kill them.[30] Furthermore, in the future it is possible that novel biological and chemical weapons will be developed at a pace that far outstrips a nation's capacity to develop prophylactic vaccines or other protective measures in response to these new threats. In this case, the obligatory use of preventive measures might no longer be ethically justifiable.[31]

Because events and circumstances change often, vaccination policies should be continually reevaluated. The scientific conclusions regarding these agents should be made by those who are most knowledgeable, and the policy decisions should be made by persons who most understand the threats at stake. If the risks seem small, but the likelihood that an enemy would use this weaponry is great, the ethical choice is to require protection, just as the military requires soldiers to wear helmets.

Honesty in Communication

In the first Gulf War, soldiers were initially told that they would have booster shots (in this case, three injections of the vaccine over a specific time frame) of the vaccine against botulism. When supplies later ran out, they were told that one dose of the botulinum toxoid vaccine would actually be sufficient in order for them to obtain protection against the toxin. Some soldiers felt angry and others afraid when they heard of this shift in immunization policy. This experience exemplifies the importance of the degree to which soldiers should be accurately informed and the importance of honest communication.

The military acknowledges and accepts the notion that information may be withheld from soldiers when necessary for the mission. Should soldiers be given information when such knowledge could result in their feeling unnecessarily angry or afraid? How this question should be answered is difficult to determine. What should be done, for example, if there is an attack using radiological weapons? What if the dosage soldiers receive will ultimately be fatal? If military physicians told soldiers that they had received a lethal dose of radiation, the possible effects could be quite different. Soldiers who knew they were going to die might feel that they had nothing to lose and fight harder. On the other hand, they might lose their will to continue to fight.

What should be done, then, if new biological or chemical weapons of known or suspect lethality are used?[32] What if the presence of these weapons is detected on the battlefield but the prophylactics required to protect against these agents are unknown? What should military physicians say? Should they tell soldiers under these circumstances that they have been or may have been lethally affected? Or should they withhold this information, at least for a time, so that they would still fight? Soldiers implicitly agree, when volunteering to serve, to forego information and to undergo any other sacrifice, including loss of life, when necessary for the success of the mission. The military's need for soldiers' sacrifice underlies the justification of military physicians giving them protective agents such as vaccines against botulism without their consent.

Protecting Others

Should military physicians offer the protective agents they would give to soldiers to prisoners of war (POWs) and detainees? These prisoners may also be at risk of being subjected to biological or chemical warfare. The Geneva Convention holds that since a POW is no longer a threat, he or she should be given equal treatment.[33] Ethically, the same equity could be extended to apply to detainees. This would suggest that all protective measures offered to soldiers, and especially those that they are required to take, should be offered to POWs and detainees as well.[34]

The same question can be raised regarding civilians. For example, if Saddam Hussein had used biological weapons during the first Gulf War, civilians in Kuwait would have needed the same protection as our own soldiers, Allied soldiers, and POWs. Providing such protective measures, however, raises greater problems regarding adequate supplies. Unlike many countries, the United States has not agreed to the section in the Geneva Convention that would require it to provide civilians with the same medical supplies it must give its own soldiers and POWs.[35] The U.S. refusal occurred because the United States believed that in some situations, it would be unable to provide the potentially vast amount of medical resources to soldiers, POWs, and civilians that equal treatment might require.

Undoubtedly, as the war on terror and the wars in Afghanistan and Iraq continue, military physicians will face ethical dilemmas regarding the treatment of detainees and civilians if chemical or biological weapons are used.

Conclusion

The increased risks and heightened concerns posed by technological advances and the changing nature of modern warfare have raised a number of ethical questions for the military physician. Indeed, the pace of biotechnological advances and their availability for use by lay persons have outstripped society's ability to understand their full ethical implications. Prior to this new "age of biology," moral priorities had been carefully worked out in regard to conflicts in military medicine.[36] However, new advances in technology and the battlefield require that we reexamine traditional ethical assumptions, including principles of military medical triage, procedures for treating soldiers with combat fatigue, and grounds for providing soldiers with protection from biological and chemical warfare without their consent.

NOTES

1 E.G. Howe, "Dilemmas in Military Medical Ethics Since 9/11," *Kennedy Institute of Ethics Journal* 13, no. 2 (2003), 175–188.

2 R.J. Ursano and D.R. Jones, "The Individual's Versus the Organization's Doctor? Value Conflict in Psychiatric Aeromedical Evacuation," *Aviation, Space, and Environmental Medicine* 52, no. 11 (1981), 704–706.

3 U.S. Department of Defense, *Emergency War Surgery*, 1st rev. ed. (Washington, DC: Department of Defense, 1975).

4 T.E. Beam, "Medical Ethics on the Battlefield: The Crucible of Military Medical Ethics," in *Military Medical Ethics*, vol. 2 (Washington, DC: Office of the Surgeon General, Department of the Army, 2003).

5 E.G. Howe, "Mixed Agency in Military Medicine: Ethical Roles in Conflict," in *Military Medical Ethics*, vol. 2.

6 J.T. Janousek et al., "Mass Casualty Triage Knowledge of Military Medical Personnel," *Military Medicine* 164, no. 5 (May 1999), 332–335; E.G. Howe, "Ethical Issues in Military Medicine: Mixed Agency," in *Principles and Practice of Military Forensic Psychiatry*, ed. R.G. Lande and D.T. Armitage (Springfield, IL: Charles C. Thomas, 1997), 469–514.

7 M.C. Brickness, "The Evolution of Casualty Evacuation in the 20th Century (Part 5)—Into the Future," *Journal of the Royal Army Medical Corps* 149, no. 4 (December 2003), 357–363.

8 R.F. Bellamy, "Conserving the Fighting Strength," *Military Medicine* 153, no. 4 (1988), 185–187; R.G. Malish, "The Medical Preparation of a Special Forces Company for Pilot Recovery," *Military Medicine* 164, no. 12 (December 1999), 881–884.

9 R.W. Hoyt et al., "Combat Medical Informatics: Present and Future," *Proceedings of the American Medical Informatics Association Symposium* (2002), 335–339.

10 M. Pueschel, "Pilot FBI Unit Links Medicine and Intelligence in Combating Bioterrorism," *U.S. Medicine*, May 14, 2004.

11 E.G. Howe, "Ethical Issues Regarding Mixed Agency of Military Physicians," *Social Science and Medicine* 23, no. 8 (1986), 803–815.

12 H.T. Engelhardt, "Fear of Flying: The Psychiatrist's Role in War," *Hastings Century Reprint* 6, no. 1 (1976), 21.

13 Encyclopedia Britannica Online, "Combat Fatigue," 2009, available at **www.britannica.com/EBchecked/topic/127295/combat-fatigue**.

14 S. Hazen and C. Llewellyn, "Battle Fatigue Identification and Management for Military Medical Students," *Military Medicine* 156, no. 6 (June 1991), 263–267.

15 J. Pearn, "Traumatic Stress Disorders: A Classification with Implications for Prevention and Management," *Military Medicine* 165, no. 6 (June 2000), 434–440; S.M. Gerardi, "The Management of Battle-Fatigued Soldiers: An Occupational Therapy Model," *Military Medicine* 16, no. 8 (August 1996), 483–438.

16 A. Fontana, R. Rosenheck, and T. Horvath, "Social Support and Psychopathy in the War Zone," *Journal of Nervous and Mental Disease* 185, no. 11 (November 1997), 675–681; F. Armfield, "Preventing Post-Traumatic Stress Disorder Resulting From Military Operations," *Military Medicine* 159, no. 12 (December 1994), 739–746.

17 C.W. Hogue et al., "Combat Duty in Iraq and Afghanistan, Mental Health Problems, and Barriers to Care," *New England Journal of Medicine* 251, no. 1 (July 1, 2004), 13–22.

18 D.W. Foy et al., "Trauma Focus Group Therapy for Combat-Related PTSD: An Update," *Journal of Clinical Psychology* 58, no. 8 (August 2002), 907–918; Z. Kaplan, I. Iancu, and F.

Bodner, "A Review of Psychological Debriefing After Extreme Stress," *Psychiatric Services* 52, no. 6 (June 2001), 824–827.

19 G. Miller, "Learning to Forget," *Science* 304, no. 5667 (April 2004), 4–36.

20 MedicineNet.com, "Medications and Drugs: Propranolol," available at **www.medicinenet.com/propranolol-oral/article.htm**.

21 D. Grossman, "On Killing II: The Psychological Cost of Learning," *International Journal of Emergency Mental Health* 3, no. 3 (Summer 2001), 137–144.

22 T.E. Beam and E.G. Howe, "A Look Toward the Future," in *Military Medical Ethics*, vol. 2.

23 K.C. Koenen et al., "A Twin Registry Study of Familial and Individual Risk Factors for Trauma Exposure and Posttraumatic Stress Disorder," *Journal of Nervous Mental Disorders* 190, no. 4 (April 2002), 209–218; D.S. McLeod et al., "Genetic and Environmental Influences on the Relationship Among Combat Exposure, Posttraumatic Stress Disorder Symptoms, and Alcohol Use," *Journal of Traumatic Stress* 14, no. 2 (April 2001), 259–275.

24 PhysOrg.com, "Gene Increases Susceptibility to Post-Traumatic Stress, Researchers Find," available at **www.physorg.com/news176408321.html**.

25 N.A. Lewis, "Two Marines Who Refused to Comply with Genetic-Testing Order Face Court-Martial," *The New York Times*, April 13, 1996, A7.

26 E.G. Howe, J.A. Kark, and D.G. Wright, "Studying Sickle Cell Trait in Healthy Army Recruits: Should the Research Be Done?" *Clinical Research* 31, no. 2 (1983), 119–125.

27 E.H. Hanson et al., "An Aviator with Cardiomyopathy and Genetic Susceptibility to Hereditary Hemochromatosis: A Case Report," *Aviation, Space, and Environmental Medicine* 72, no. 10 (October 2001), 924–927; W.C. Fox and W. Lockette, "Unexpected Syncope and Death During Intense Physical Training: Evolving Role of Molecular Genetics," *Aviation, Space, and Environmental Medicine* 74, no. 12 (December 2003), 1223–1230.

28 D. Woods et al., "Elite Swimmers and the D allele of the ACE I/D Polymorphism," *Human Genetics* 108, no. 3 (March 2001), 230–232.

29 E.G. Howe and E.D. Martin, "Treating the Troops/The Use of Investigational Drugs Without Obtaining Servicepersons' Consent," *Hastings Century Reprint* 21, no. 20 (1991), 21–24.

30 Howe, "Mixed Agency in Military Medicine," 337.

31 Ibid., 338.

32 A.L. Brooks, "Biomarkers of Exposure and Dose: State of the Art," *Radiation Protection Dosimetry* 97, no. 1 (2001), 39–46.

33 "Geneva Conventions of 1949," in *Human Rights Documents: Compilations of Documents Pertaining to Human Rights* (Washington, DC: U.S. Government Printing Office, 1983).

34 J.A. Singh, "American Physicians and Dual Loyalty Obligations in the 'War on Terror,'" *BMC Medical Ethics* 4, no. 4 (2003).

35 W.H. Parks, Memorandum for the Surgeon General, Article 10, 1977; Protocols Additional to the Geneva Convention, August 12, 1949.

36 E.M. Spiers, "The Use of the Dum-Dum Bullet in Colonial Warfare," *Journal of Imperial Commonwealth History* 4 (1975), 3–14.

chapter 3

THE LIFE SCIENCES, SECURITY, and the CHALLENGE of BIOLOGICAL WEAPONS: AN OVERVIEW

Michael Moodie

Since the tragedies of September 11, 2001, and the mailings later that year of letters containing anthrax that resulted in the deaths of five people, the prospect of a biological weapons (BW) attack against the United States or its interests overseas has come to be considered, along with the possibility of a nuclear attack, as a paramount challenge to U.S. national security. Since that time, the United States has pursued a wide-ranging agenda both at home and abroad to address the BW risk. To that end, it has spent tens of billions of dollars to prevent and protect itself against such a contingency.

Despite all the attention the issue of biological weapons has received, not everyone agrees that it deserves to be made such a high priority. Indeed, since perceptions of heightened biological-related security risks came to the fore in the mid- to late 1990s, a debate has raged over the extent to which biological weapons—whether used by states or terrorists—represent a significant national security challenge. The public debate has often swung like a pendulum between hype and complacency, and at times it has been hard to hear the voices calling for an approach that provides calibrated responses to this multifaceted problem based on a sophisticated understanding of all its dimensions. The challenge of getting the approach to biological weapons "right" is exacerbated by the rapidly changing context within which the BW risk is shaped and manifested. Major developments such as rapid advances in relevant sciences and related technology, sweeping trends such as globalization, and shifts in the style, location, and form of conflict in today's world are all fostering a strategic environment far different from the one in which modern concerns about BW emerged almost a century ago.

An appropriate, cost-effective, and, most importantly, successful response to meeting the BW challenge demands an appreciation of its complexity and an awareness of those changes that will shape its form and expression in the future. It will call for a policy and strategy that are as multifaceted as the issue is multidimensional. It will also require the active involvement of a wider range of players—both government and nongovernment—than has been the case traditionally in addressing national security issues. Clearly, a significant challenge lies ahead.

Biological Weapons: Background

Biological weapons can be defined as living organisms or their inanimate byproducts that are used as weapons against humans, animals, and plants. They can take several forms including bacteria, or small, single-celled organisms; viruses, or organisms requiring living cells in which to replicate; rickettsia, or microorganisms with characteristics similar to both bacteria and viruses; and toxins, or poisonous chemical substances produced by living things. Examples of traditional biological weapons that fall into these categories include the bacteria anthrax, brucellosis, and plague; the smallpox virus as well as viruses that produce dengue fever or hemorrhagic fevers such as Ebola and Marburg; and rickettsia such as Q fever or Rocky Mountain spotted fever. Examples of toxins include snake venom, ricin (a byproduct of processing castor beans), and saxitoxin, a poison produced by shellfish.

Biological weapons have their roots in antiquity.[1] Most discussions of biological weapons, however, begin with the 1346 siege of Kaffa (now Feodosiya) on the Crimean coast during which Tatars catapulted plague-infested bodies into the city. Other early uses included the distribution of plague-infested blankets by the British among Native Americans during the French and Indian War, and the possible spreading by British troops of smallpox in Boston following their evacuation of the city during the American Revolution.[2] In more modern times, the Germans initiated sabotage programs using BW in World War I, such as the use of glanders in an attempt to sicken U.S. Cavalry horses that were to be shipped to Europe.[3] Virtually all of the major participants in World War II had active BW programs, although Japan is the only major power to have attempted to use them. The infamous Unit 731 conducted a number of experiments and attacks, for example, in Manchuria during Japan's occupation of that territory in the late 1930s.[4]

The 50 years following the end of World War II witnessed virtually no BW use, although a number of states pursued BW programs in the decades following the war. The search for effective offensive biological weapons was supposed to end, however, after President Richard Nixon announced the unilateral termination of the U.S. offensive BW program in 1969. President Nixon's decision also prompted negotiations with the Soviet Union and other countries that eventually produced the Biological and Toxin Weapons Convention (BWC), which was concluded in 1972 and entered into force in 1975. The BWC was a historic milestone because it was the first arms control agreement in history to ban totally an entire class of weapons, making it illegal to develop, produce, otherwise acquire, store, deploy, or transfer them.[5]

The BWC's entry into force did not fully eliminate concerns about the weapons, however, particularly as the treaty had no provisions to verify a state party's compliance with its treaty obligations or to investigate allegations that states were in noncompliance. For the next two decades, the United States continued to allege that at various times, between a dozen and two dozen countries were violating their treaty commitments and pursuing offensive BW programs.[6]

Washington was especially concerned about illicit BW activities in the Soviet Union. Although it had been concerned about the Soviet program from the time of the BWC's 1975 entry into force, U.S. allegations became much more public following an anthrax outbreak in the Siberian city of Sverdlovsk in 1979, which Washington claimed originated in an illegal BW research facility but which Soviet authorities dismissed as the product of people eating tainted meat. U.S. concerns intensified substantially following the defection to Britain in the late 1980s of a Soviet scientist who had been deeply involved in the illegal program. Vladimir Pasechnik provided significant details about the so-called *Biopreparat* program, which had been created in the Soviet Union's civil sector shortly after Moscow signed the BWC as a parallel effort to an existing military program. *Biopreparat*, which was funded by the defense ministry but which operated under the auspices of the Main Administration of Microbiology Industry and had close ties to the health and agriculture ministries, involved hundreds of research facilities and production plants, employed tens of thousands of people, and operated on a budget of 100 to 200 billion rubles per year.[7] Western intelligence services knew nothing about it. It was not until Pasechnik's defection that they began to appreciate the enormity of the Soviet effort. As one British report noted, "The information was stunning: a whole ministry exposed; billions of rubles spent; a complete organization shown to be a front; and there was the clear involvement of Gorbachev, this friend of the West. It just went on and on."[8] It was the defection of Pasechnik that set in motion major U.S. and U.K. efforts to bring the illegal Soviet effort to a halt, including meetings "at the highest levels." Following a public accusation by the United States and Britain of Soviet noncompliance at the 1991 Third BWC Review Conference, Boris Yeltsin, who had become President of Russia following the collapse of the Soviet Union, acknowledged in April 1992 that the claims about the illegal Soviet program were correct and ordered the program to be shut down.[9]

Apprehensions about BW were reinforced during this same general time period by developments in Iraq. After Operation *Desert Storm* reversed the Iraqi invasion of Kuwait in 1991, discoveries by the United Nations Special Commission on Iraq were especially distressing. In particular, Iraq was discovered to have had an extensive offensive BW program on the eve of *Desert Storm*. The program included research and development (R&D) on a wide spectrum of agents, large-scale weaponization of botulinum toxin, anthrax, and the carcinogen aflatoxin, and potential use of a variety of delivery systems including aerial bombs and al-Hussein missiles. Several facilities, some of which were also engaged in civilian efforts, were involved in the program. Many Iraqi BW scientists were found to have received their scientific training at universities and other facilities in Europe and the United States. What was never made clear, however, was Iraq's concept of operations for use of its BW or the conditions under which such weapons would be used.

A third major development in this period that moved biological weapons up on the national security agenda even before 9/11 and the "Amerithrax" letters was the emergence of terrorist interest in BW capabilities. Most notable in this regard were the activities of the Aum Shinrikyo, the Japanese cult responsible for the March 1995 sarin chemical weapons attacks in the Tokyo subway. Following that attack, investigations revealed that the group's activities were not limited to production and use of chemical weapons; indeed, biological weapons may have been their weapon of choice. The Aum performed research, for example, on both anthrax and botulinum toxin, and reportedly it attempted to conduct at least nine separate attacks against both Japanese targets and U.S. military bases in Japan, all of which failed. Fortunately, the cult had acquired an attenuated strain of anthrax that rendered it useless. Reports also indicated that the Aum tried unsuccessfully to secure a sample of Ebola under the guise of sending a "mercy mission" to what was then Zaire.[10]

The revelations regarding Aum Shinrikyo prompted attention to other instances in which groups rather than states had tried to use biological materials, including in the United States. One such case was the use in 1986 of salmonella on restaurant salad bars in The Dalles, Oregon, by the Rajneeshi cult in order to influence the outcome of a local election and referendum. Another was the attempted use of ricin by the Minnesota Patriots Council, an American militia group, to kill jurists who were engaged in legal proceedings against them.[11]

This coming to light of nonstate actor interest in BW heightened perceptions that such capabilities were insufficiently appreciated as a risk to national security. The fact that the Japanese subway attack and the bombing of the Murrah Federal Building in Oklahoma City were juxtaposed closely in time fostered the sense among both policymakers and the public that the United States was now vulnerable to terrorist attacks, that those attacks could include unconventional weapons, and that the United States was not prepared to handle such attacks or their consequences. Efforts to address these unconventional threats, then, including BW, began in the mid-1990s, before the events of 9/11 and the autumn of 2001. The challenges posed by biological weapons have been at or near the top of the national security agenda ever since.

The Current Challenge

Today's BW challenge has two dimensions, and debate continues over which one is the more important. To some, the major risk continues to emanate from states. For the last two decades, concern has existed that up to perhaps two dozen countries have been pursuing offensive BW programs. Countries that have admitted to programs include the United States, United Kingdom, France, the Soviet Union, South Africa, and Iraq. Other countries with past or current programs include Iran, Syria, Israel, China, North Korea, and Taiwan. Little information is available in the public record about any of their programs, with the exception of those that have admitted to past offensive BW efforts (the United States, United Kingdom, and France).

For much of the Cold War, the United States indicated that it believed that as many as a dozen or more states were engaged in illegal BW efforts. In recent years, however, it has moved to a somewhat different formulation. At the Sixth BWC Review Conference in late 2006, for example, John Rood, Assistant Secretary of State for International Security and Nonproliferation, alleged only that three states—North Korea, Iran, and Syria—were engaged in activities in violation of the BWC.[12] Although the most recent State Department assessment of nations' compliance with their arms control obligations (published in 2005) noted that the U.S. Government had "concerns regarding the activities of other countries," the report itself focused only on alleged noncompliance with the BWC by Russia, China, and Cuba, in addition to the three countries mentioned by Secretary Rood.[13]

Those who argue that state-based BW programs should be the primary concern do so because they see states as the entities with the technical and financial resources to develop such capabilities. There are other reasons, however, why states might be interested in these weapons: they are relatively simple and cheap to produce, especially compared to either nuclear weapons or large quantities of chemical weapons; the necessary agents and equipment are widely available because they are also used in legitimate scientific and commercial enterprises; the entry costs for beginning a program are relatively low; covert programs are reasonably easy to conceal because neither a large space nor specialized equipment is needed to produce a significant amount of agent; and gains in such scientific and technological fields as genetic engineering and biotechnology are diminishing the barriers that traditionally have made BW development, production, and weaponization difficult.

A further reason that states might be interested in biological weapons is that, used in the right conditions, they could be strategic in their impact. A well-regarded study by the Office of Technology Assessment in 1993 estimated, for example, that, used under ideal meteorological conditions (that is, a clear, calm night), an aerosol version of anthrax could kill between 1 million and 3 million people, a total equivalent to the impact of a small nuclear weapon.[14]

Some of these attractions are also of concern to those who argue that the more severe risks related to BW stem from their potential use by terrorists rather than states. As noted, the autumn 2001 "Amerithrax" letters represent the latest in a series of events in which nonstate actors have sought to exploit biological weapons. Moreover, they highlighted a few important points that perhaps had not been well understood until those attacks. First, biological weapons do not have to produce catastrophic levels of casualties in order to have a mass effect. Although only five people died, the impact on the public's psyche was enormous, exemplified by the fear people had of retrieving and opening their daily mail. Underscoring the disruption that BW can cause was the fact that because of necessary decontamination procedures, it took months for Congressional offices to return to their normal operations, and the Brentwood postal facility did not reopen for more than a year. Second, because of the nature of biological weapons, it will be difficult to attribute an attack.

The question of technical skill is important because some analysts downplay, if not dismiss, the risks of terrorist BW use in the belief that terrorist groups have neither the inclination nor the capability to conduct such attacks.[15] This may, however, be changing. It is clear, for example, that al Qaeda has developed an elaborate rationale justifying the use of chemical, biological, nuclear, or radiological weapons, and materials found in caves in Afghanistan document the group's interest in and pursuit of such capabilities.

Heavy emphasis on the technical capabilities of a terrorist group skews to some extent the analysis of potential terrorist BW use. Such single-factor analyses oversimplify the issues and imply a precision in the ability to identify bioterrorism risks that does not exist. The risk of biological terrorism is in fact the product of a complex interaction among several categories of factors—actors, agents, targets, and operational considerations—each of which includes many variables. Taken together, these variables produce a matrix that offers a large set of possible combinations and permutations, each of which constitutes a pathway to a particular outcome. Some of these pathways produce catastrophic consequences; others result in significant casualties; still others yield no consequences at all. As a terrorist seeks to achieve the more demanding goals of higher casualties or more widespread disruption, fewer pathways are available to reach them, and those that remain are more difficult. The degree of risk declines, therefore, as the level of desired casualties/disruption increases, insofar as it becomes less likely. This is not to say, however, that the degree of risk is zero, or even negligible.

Looking to the Future

The key question regarding the prospects of both state and terrorist BW use is whether the future will resemble the past. Will the factors that have shaped the BW challenge for the last 50 years remain the dominant influences, or are changes under way that could reshape it?

Three factors are converging that probably will promote considerable change in both the security landscape and the form that the biological challenge is likely to take in the future. Adapting to those changes represents perhaps the most significant demand confronting policymakers with responsibility for meeting the biological challenge.

Rapidly Advancing Science and Technology

The speed at which the science and technology related to the biological challenge are moving is hard to exaggerate; indeed, it is happening at a rate faster than the well-known Moore's Law in information technology.[16] The rate at which the life sciences and its associated technologies are progressing and knowledge of the processes of life at the molecular level is advancing has led some commentators to suggest that the life sciences are undergoing a profound "revolution." They contend that the 21st century could well be "the century of biology." The expansion of knowledge in the life sciences, however, is not only rapid, but it is also deep, broad in that it encompasses a wide range of disciplines, and widely available.

Moreover, the knowledge that is being gained is being put to use; that is, it is not just remaining in the realm of basic science. More and more of that knowledge is applied in forms that are increasingly making it a commodity. This "commoditization" has several significant implications. Most importantly, it demonstrates that the civilian sector—both science and business—is the key driver in pushing this scientific and technological advance. The corollary is that governments are less and less in control of scientific and technological progress and "spinning on" those commercially driven developments into their products, including in the security realm.

An Expanding Risk Spectrum

The second major factor shaping the future perspective of the biological challenge is the growing practice of placing the biological weapons problem along a wider biological risk spectrum (see figure 3–1).

At one end of this risk spectrum stand naturally occurring developments such as chronic disease and emerging or reemerging infectious diseases. The spectrum then runs through human-induced developments, including accident or misadventure (the unintended consequences of otherwise benign activities), to the other end at which stands deliberate misuse, whether for political power, ideology, or greed.

The two elements of this risk spectrum that have received the greatest attention in recent years are naturally occurring infectious diseases and deliberate misuse. Over the last few years, the relationship between these two issues has undergone a distinct change. Following 9/11, bioterrorism was the overwhelming focus. The argument was made that if effective bioterrorism preparations could be developed, they would also have utility in dealing with a naturally occurring disease outbreak. Recently, however, perhaps prompted by the SARS experience, the threat of an influenza pandemic such as the recent H1N1 pandemic, and the continuing scourge of HIV/AIDS, the priority given to these two elements has been completely reversed. Today, more attention seems to be given to infectious disease; deliberate misuse in such forms as bioterrorism has become the "lesser included case." Some analysts might suggest that this assertion is an exaggeration and that health and bioterrorism threats are addressed at the same level. There is no question that these issues are often considered together because many of the requirements for dealing with either contingency are the same.

This shifting priority is not just an interesting intellectual point. It has repercussions for policy priorities and the allocation of limited resources, particularly the balance of investment in efforts to address infectious disease and those that are directed toward more narrow bioterrorism and biodefense concerns. Moreover, the relative priority given to each of these components along this spectrum could be an issue of some disagreement between the United States and those it wants and needs to work with in combating proliferation.

figure 3–1. THE BIOLOGICAL RISK SPECTRUM

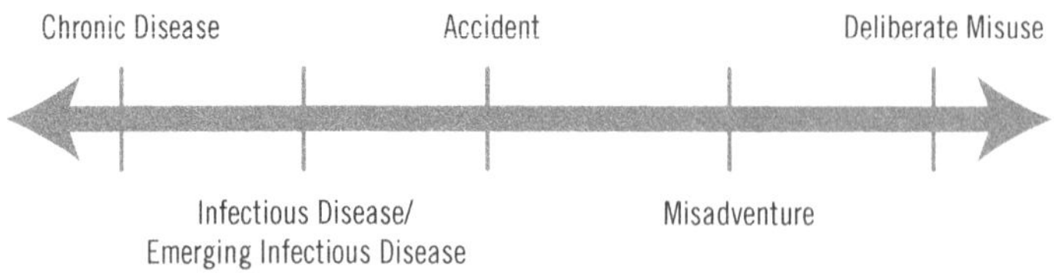

At the same time, it is important to appreciate this spectrum because, in fact, segments are not readily divided but blend into one another. This has important implications for policy because measures designed to deal with one aspect of the problem will have important implications—and sometimes important tradeoffs—for others. Development of medical treatments for infectious disease or biological attacks is only the most obvious of many potential examples.

Globalization

The third key factor promoting the complexity of the future biological challenge is globalization. It is a trend marked by multiplying, deepening, intensifying, accelerating, and more complex interactions, transactions, and interrelationships among a growing number of players on a worldwide scale. It is further characterized by the exploitation of the information revolution, innovation, and an increasingly prominent role of nonstate actors.

The importance of globalization for the biological challenge is difficult to underestimate. First, globalization has fostered a rapid diffusion of the life sciences and related technologies across the world, driven largely by the creation of small- to medium-sized biotechnology firms.

Biotechnology growth, however, is not occurring only in the developed world. China and India in particular have been identified as developing countries that are likely to be biotechnology leaders in years to come. Many other countries outside the "developed world," including Singapore, Malaysia, South Korea, Cuba, Brazil, and South Africa, are also investing heavily in biotechnology as a driver of their future economic growth. As a result of these developments, today a critical scientific or commercial breakthrough could come from almost anywhere in the world.

Second, globalization has transformed patterns of industrial production at both national and international levels, including in life sciences industries. Together with new production processes, agile manufacturing, miniaturization, lower technology costs, and increased productivity of a global talent pool, these trends are restructuring business enterprises in fundamental ways.

Third, globalization has spurred the emergence of nongovernmental entities operating on an international basis and in greater numbers. More and more, these new and increasingly empowered nonstate actors are able to express their singular interests through the tools and channels globalization provides, allowing them to operate beyond the control of any single government. The result is that even relatively weak actors can have disproportionate impact both positively and negatively.

This changing scientific and technological landscape has critical implications for security. The accelerating pace of discovery in the life sciences and its widespread diffusion around the globe have fundamentally altered the threat spectrum, which is now exceptionally broad and continually evolving:

- The number of regions of the world where people can be found with the requisite ability to exploit knowledge that can do harm has obviously grown significantly. That people might know how to apply this knowledge for malign purposes does not, of course, mean automatically that they will. But an increase in the number of people with the requisite knowledge does imply an increase in the burden of potential risk that must be managed.
- These trends have created lower entry costs for joining the proliferation process as well as the potential to enter that process at a higher point on the learning curve.
- The new structures of commercial and scientific enterprises will provide a wider range and more diverse array of legitimate dual-use covers for malign activities. They could also create multiple, parallel, possibly nontraditional pathways to the development of critical biological or chemical weapons–related capabilities.
- These trends will make it even harder to determine noncompliance with global norms, leading to uncertain enforcement and response at both the international and national levels.

In such an environment, serious problems exist with respect to governments' abilities to meet the challenges that lie ahead. Government bureaucracies are notoriously slow to adapt, as are international institutions. How, then, can governments and international institutions keep pace with the speed at which science and technology are moving, especially in terms of understanding both the negative and positive implications of those changes for security?

Treaty regimes could be especially affected. The BWC, for example, is now almost 35 years old, and the need to adapt it to this new set of circumstances is perhaps the single greatest challenge confronting BWC states parties. The inability to do so will have an extremely corrosive effect. But adaptation will be unsuccessful if those responsible fall back on business as usual.

What Constitutes "Proliferation"?

The traditional view of proliferation is that a government, for a variety of reasons, makes a commitment to achieve a nuclear, chemical, or biological weapons capability, and then moves through a series of programmatic steps to the eventual deployment of full military systems. In light of the trends discussed above, however, the process of proliferation is moving away from this classic paradigm for several reasons.

First, the decision to seek a BW capability is not likely to be primarily driven by the perceived efforts of an adversary, but propelled by ongoing global advances in microbiology, biomedical research, and biotechnology. Rather, as a report by the Institute of Medicine of the U.S. National Academies of Science put it, "Tomorrow's world of proliferating threats . . . may be quite different from the twentieth-century agents and indeed may not even exist at this time."[17]

Second, if scientific and technological advances are becoming the drivers of actors' interest in BW capabilities, the process of proliferation may become increasingly characterized by a move down a path toward those capabilities without necessarily a commitment to the deployment of a specific weapons system or the implementation of a well-developed concept of use. Advancing science and technology make "dabbling" much more possible, and they increase the potential for the creation of a "warm capabilities base" or a latent capability on which an actor could rapidly build if perceived changes in the security environment or other factors propel it to do so.

Third, the shift in technological innovation from the military to the civilian sector suggests that much of the R&D phase of a BW program need not be conducted in a dedicated military facility or even in the context of an offensive military program. Since many questions relating to development and performance of agents and delivery devices are the same for civilian and military applications, understanding of fundamentals can be achieved without engaging in activities that would be considered violations of international treaty obligations. Drug companies, for example, for perfectly legitimate reasons are strenuously exploring improvements in aerosol technology, which bear directly on what the single most difficult technical challenge is in developing an effective weapons capability.

Fourth, proliferation is not just the province of governments. The problem now encompasses a wider range of potential actors. It is not only the number of possible participants in the process of proliferation, however, that has changed, but also the nature of their interaction. Traditionally, proliferation was seen as the product of a hierarchical, top-down process that began with a government's commitment to seek a given capability. Today, the process is much more free form. The operation of formal or informal networks is especially important. The emergence of networks is one of the key factors that have empowered the wide array of players now active in the security landscape, and it has facilitated the flow and exchange of not only critical materials and equipment, but also the increasingly important element of scientific and technological knowledge that now resides at the core of the proliferation process.

Proliferation today, then, is not so much about classic military programs. Rather, it is about the security implications of key trends, including:

- expansion of knowledge and the capability provided by technology
- growth in the number of legitimate applications of advancing life sciences and technology

- the increase in the number and variety of players able to exploit that knowledge base
- the geographic expansion of regions of the world where such expertise can be found.[18]

The argument is not that latent capabilities become realized with no effort, but that the process of that realization will take a very different form than the traditional notion of proliferation and that the time needed for such a transformation will be shorter and shorter. Moreover, the issue is about an interaction between intent and capability that is much less straightforward than the standard proliferation model suggests.

New Responses for an Evolving Challenge

The changing nature of the biological challenge demands as much new thinking about the response as it does about the nature of the problem. The first requirement is a change in thinking. Those who have responsibility for meeting the biological challenge must jettison the comfortable reliance on old concepts and past practice to address the future.

The United States has already demonstrated such a conceptual shift and has provided considerable leadership in promoting new measures that seek to respond to the changes mentioned above. For some people, U.S. efforts in this regard reflect its discarding of traditional approaches, and multilateralism in particular, in favor of either a "go-it-alone" model or efforts of "coalitions of the willing." In particular, they lament the U.S. rejection of the draft protocol to the BWC designed to enhance confidence in compliance with the treaty. British analyst Jez Littlewood points out, however, that "[a]ny approach to dealing with BW that is based only on the BWC or, alternatively, based on *anything but* the BWC is politically simplistic and fails to grasp the scale of the BW problem."[19] Rather, what is needed is a strategy that combines old and new into a policy framework that is as multidimensional as the problem is multifaceted and that can accommodate contributions from the full range of players—government and nongovernment—that will be needed to be effective.

Nor is the issue one of choice between multilateral and unilateral approaches. Posing the issue in such terms is overly simplistic; it is not an "either/or" choice. But the answer of "both" multilateralism and unilateralism is also inadequate, although both national and multinational efforts are, indeed, required. So, too, are local efforts within countries as well as regional measures involving several neighboring states or ad hoc arrangements among geographically disparate states.[20] Again, some of these measures, such as regional cooperative arrangements to counter proliferation (in the former Soviet Union, for example) are already in place, at least in a basic form. But more is needed.

Finally, governments cannot do the job of managing biological risks alone. As has already been suggested, governments will find it hard to keep up. To do so, they must benefit from the inputs of those "working at the coal face" of scientific and technological change every day. This means in the first instance that governments must reach out to the academic research community and life sciences industries. Moreover, many of the assets on which governments must rely for managing biological risks lie outside government control, in the private sector.

Managing the risks associated with developments in the life sciences, therefore, will require a range of new partners, especially in the private sector. However, creating enduring partnerships also demands new thinking and new approaches on the part of all of those who must be engaged.

The foundation for effectively promoting new partnerships is a shared conceptual approach to the problem among all the stakeholders, on both the national and international level. This is not to argue for a unanimously agreed threat or risk assessment. It is unlikely that a consensus could or ever will be achieved. But neither is it necessary. What is important is finding enough common ground to create sufficient shared space for working together.

Developing that common ground might begin with agreement on the following assumptions:

- First, "biological threats to security" do not represent problems to be solved but rather risks to be managed. Work in the life sciences should and will continue for important legitimate reasons. This means, however, that the potential for misuse of those sciences remains a permanent reality. The challenge inherent in managing the risks of that misuse—whether by states or nonstate actors—is not to prevent international actors from acquiring the capabilities to exploit biology for malign purposes. That is not possible. Rather, the challenge is, as British Ministry of Defence official Paul Schulte put it, "to keep it out of their behavioral repertoires."
- Second, even if it cannot be totally eliminated, the risk of deliberate misuse can be reduced. But risk reduction is a multifaceted, complex challenge that includes deterrence, norm building, prevention, defense, preparedness, and consequence management, mitigation, and amelioration. No single instrument will be sufficient.
- Third, resources to meet the biological challenge are not endless, even if many countries are involved. In order to maximize the impact of limited resources (in terms of time, money, and people), smart choices must be made.

- Finally, the task of dealing effectively with the biological challenge cannot be met by trying to predict precisely what form an attack using biological agents will take. Rather, the objective must be preparedness for as wide a range of plausible contingencies as possible. That can only be accomplished if a country has a robust set of critical response capabilities that are integrated into a genuinely strategic approach that stresses flexibility and adaptability.

Developing those capabilities is not easy. It entails identifying requirements, establishing criteria to determine the appropriate level of capabilities to meet those requirements, balancing a wide set of competing interests, and involving the right set of players. One set of required capabilities that emphasizes prevention includes measures in such critical areas as law enforcement, intelligence, pathogen security, export controls, and cooperative threat reduction. A second set of requirements relates to preparedness and includes such elements as effective disease surveillance and reporting, health monitoring, good epidemiology, robust laboratory-based analysis, appropriate diagnostics and medical countermeasures, and sufficient medical stockpiles, among others. Both sets of capabilities must be supported by a dynamic R&D agenda and facilitated by effective communications strategies for policymakers and responders as well as for the general public. Finally, a capabilities-based strategy should be implemented in a context of international cooperation.

Creating New Partnerships

These demands will be met successfully only if all of the critical stakeholders are involved and their cooperation is fully integrated. In particular, strong partnerships are vital between government and the private sector. Past approaches to dealing with the biological challenge failed to generate badly needed inputs from these important players. The life sciences communities, in particular, were not well integrated, despite the fact that they stand at the cutting edge of both the remarkable scientific advances and their applications that are shaping the environment within which the biological challenge must be addressed.

Those involved in life sciences research and commerce emphasize their enormous contributions to improving the quality of life for people around the world. They are right to do so. But what has been missing from their thinking is recognition of their role and responsibility in helping to manage the risks of misuse of the science and technology with which they are involved.

The attitudes of those working in the life sciences contrast sharply with the nuclear community. From the beginning of the nuclear age, with the active participation of Albert Einstein himself, physicists understood that they had to not only think about the negative implications of atomic power, but also to participate actively in managing the risks they had helped to create. That recognition has been lacking in life sciences communities. Their single-minded focus on the good they are trying to do for humanity or on scientific discovery for its own sake has too often blinded life scientists to the risks that stand alongside the benefits they seek. At best, security issues have been kept at arm's length.

Such an attitude is an unaffordable luxury. As the drivers of critical science and technology, industry and the academic scientific community must understand their stakes in the challenge and be fully integrated into the response. The direct contribution that this community can make to dealing with the problem is obvious. It should be the source of the technical tools that must be deployed to help manage biological risks, including sensors for detection and identification, new medical treatments, and improvements in passive and active protective gear.

The indirect contribution of the life sciences community, however, is also crucial. Even scientists and companies that have no direct relationship to the security sector engage in activities with risks attached. Insufficient safety and security measures at laboratories and other facilities pose clear and obvious risks. But what about risks that are not so obvious? A device for the application of medical treatments without the use of needles—for example, involving absorption of a drug through the skin, or a flu vaccine in an aerosol form—could be of great medical value. But in some contexts, such devices might also be useful for terrorists. This does not mean that these beneficial products should not be made. But they should go forward with the relevant communities being fully aware of the risks that might be involved.

While attitudes within life sciences communities appear to be changing, they still have a long way to go before they can provide the requisite strong leadership and sustained engagement. Life scientists should now work with governments to help strengthen the norms against biological weapons research, acquisition, and use. Codes of conduct, peer reviews and panels, and self-regulation that define appropriate restrictions in scientific research are all ways in which the scientific community can contribute to the development of a culture of responsibility with respect to risk management.

For its part, governments must encourage this development further and engage the life sciences community in a way that takes into account legitimate security concerns without harming the innovation and creativity that are such a key feature of work in the life sciences. Governments must exercise caution, therefore, to ensure that they do not, in fact, drive scientists away from such engagement. Onerous requirements that force scientists or companies to conclude that potential benefits from particular research or commercial ventures are not worth the hassle will lead them to forego efforts that could prove to be valuable from both a scientific and a security standpoint.

Anecdotal evidence suggests that just such things are happening. Therefore, governments must focus on developing means for managing security-related risks that achieve an acceptable balance with the requirements of the scientific process and good business practices.

International cooperation is especially important in reconciling such tensions. The world's life science and technology base is thoroughly globalized, with capabilities available worldwide that can contribute to domestic security for many nations. Success in exploiting these capabilities will be possible, however, only if nations' efforts are informed by awareness of developments elsewhere. Information must be exchanged on an ongoing basis. No country has the resources to fund all of its demands for security investments, and maximizing the return on whatever investment is made requires a coordinated international approach.

NOTES

1 See, for example, Adrienne Mayor, *Greek Fire, Poison Arrows, and Scorpion Bombs: Biological and Chemical Warfare in the Ancient World* (New York: Overlook Duckworth, 2003).

2 For more detail, see Mark Wheelis, "Biological Warfare Before 1914," in *Biological and Toxin Weapons: Research, Development, and Use from the Middle Ages to 1945*, ed. Erhard Geissler and John Ellis van Courtland Moon (London: Oxford University Press for the Stockholm International Peace Research Institute, Chemical and Biological Warfare Studies no. 18, 1999), 8–34.

3 See Mark Wheelis, "Biological Sabotage in World War I," in ibid., 35–62.

4 See Sheldon H. Harris, *Factories of Death: Japanese Biological Warfare, 1932–1945, and the American Cover-Up* (London and New York: Routledge, 1994). For details on other countries' biological weapons (BW) programs prior to 1945, see the respective chapters on Poland, France, Germany, the Soviet Union, Britain, Canada, and the United States in Geissler and Moon.

5 Thus, the Biological Weapons Convention (BWC) complemented the 1925 Geneva Protocol, which outlawed the use of both chemical and biological weapons. The protocol, however, had not been ratified by a number of countries, including the United States, at the time that the BWC entered into force. Moreover, several countries that had ratified the protocol had done so with reservations, making it little more than a no-first-use agreement.

6 The BWC is especially difficult to verify because it does not ban any specific activity, but rather any action that is in service of an offensive BW program. Therefore, many of the same activities (for example, certain forms of research) can be either legal or illegal depending on the intent of the country taking that action.

7 For a detailed discussion of the operations of *Biopreparat*, see Ken Alibek with Stephen Handelman, *Biohazard* (New York: Random House, 1999). Alibek defected following Pasechnik. Alibek had an M.D. and Ph.D. in microbiology and served, at the rank of colonel, as first deputy chief of the *Biopreparat* program.

8 Cited in James Adams, *The New Spies: Exploring the Frontiers of Espionage* (London: Hutchinson, 1999), 270–283.

9 For more detail regarding these efforts and the U.S. approach in general to Soviet noncompliance with the BWC, see Michael Moodie, "The Soviet Union, Russia, and the Biological and Toxin Weapons Convention," *Nonproliferation Review* 8, no. 1 (Spring 2001).

10 For more details on the Aum Shinrikyo, see David A. Kaplan and Andrew Marshall, *The Cult at the End of the World* (New York: Crown, 1996).

11 For the most authoritative public source on such events, see W. Seth Carus, *Bioterror and Biocrimes: Illicit Use of Biological Agents Since 1900* (Washington, DC: Center for the Study of Weapons of Mass Destruction, National Defense University, February 2001).

12 John C. Rood, "Remarks to the Sixth BWC Review Conference," U.S. Department of State, November 20, 2006.

13 The report also took note of the past illicit programs of Libya and Iraq. "Adherence to and Compliance with Arms Control, Nonproliferation, and Disarmament Agreements and Commitments," U.S. Department of State, August 2005.

14 U.S. Congress, Office of Technology Assessment, *Proliferation of Weapons of Mass Destruction: Assessing the Risks* (Washington, DC: U.S. Government Printing Office, August 1993), 54.

15 See, for example, Milton Leitenberg, *Assessing the Biological Weapons and Bioterrorism Threat* (Carlisle, PA: U.S. Army War College, Strategic Studies Institute, December 2005).

16 See Rob Carlson, "The Pace and Proliferation of Biological Technology," *Biosecurity and Bioterrorism: Biodefense Strategy, Practice, and Science* 1, no. 3 (August 2003).

17 Institute of Medicine, *Globalization, Biosecurity, and the Future of the Life Sciences* (Washington, DC: National Academies of Science, 2006), 59.

18 Gerald L. Epstein, *Global Evolution of Dual-Use Biotechnology: A Report of the Project on Technology Futures and Global Power, Wealth, and Conflict* (Washington, DC: Center for Strategic and International Studies, April 2005), 3–4.

19 Jez Littlewood, *Managing the Biological Weapons Proliferation Problem: From the Individual to the International*, Weapons of Mass Destruction Commission Paper no. 14 (Stockholm: Weapons of Mass Destruction Commission, August 2004).

20 Ibid., 9. Littlewood points out, for example, that the ad hoc groupings that have emerged in recent years did not do so because of a desire on the part of some states to restrict certain activities "to a 'club' of states or some desire to push forward the boundaries of international law or control mechanisms per se. They were created because of the perception that they were required to perform necessary tasks."

chapter 4

BIOLOGICAL WARFARE: A WARFIGHTING PERSPECTIVE

John B. Alexander

Biological warfare is insidious, repulsive, frightening, effective, and marginal. Social, legal, military, and practical constraints have limited, but not eliminated, the employment of biological weapons (BW) in past conflicts. Currently at issue is whether weaponization of the agents resulting from the rapid advances in biology and related fields can possibly be restrained. Given the likely asymmetric nature of future conflicts and the difficulty of holding nonstate adversaries at risk, it seems probable that BW will be employed at some point in the future. While adding an order of complexity to the traditional battlefield, the most likely applications will be attacks against supporting elements and independent civilian targets. Despite low probability of a major attack, the extremely high-risk nature of the BW threat demands that greater attention be paid to this problem.

Role of the Fear Factor

Reduced to the lowest common denominator, the psychological effects that are generated by the potential for BW far outweigh the physical effects in most cases. Paradoxically, the threat is both overstated and understated. While the threat from disease is real and must be addressed, it has taken on near-mythical properties in the minds of many observers and certainly in the news media. To function in an environment in which biological weapons are available, realism in assessing the situation is essential. As expert Milton Leitenberg concludes, "Bioterrorism may or may not develop into a serious concern in the future, but it is *not* 'one of the most pressing problems we have on the planet today.'"[1] Still, there are viruses such as smallpox that, if reintroduced into an urban population, could ignite massive outbreaks of disease. Because smallpox was declared eradicated in 1980, vaccinating the general population was stopped, leaving billions of people vulnerable to the virus. In addition, and as will be described later, there are categories of BW agents that can attack other than living targets that should be of concern to military planners.

While biological weapons are categorized as weapons of mass destruction (WMDs), this categorization could be considered a misnomer. Unlike nuclear weapons that inflict massive infrastructural damage, even small amounts of BW can have large-scale impact without doing much physical damage. Therefore, it can be stated that they are *weapons of mass effect*, although that term is not in the accepted canon.

Recent history has persuasively demonstrated the mass effectiveness of BW in disrupting civil operations, if not in causing massive casualties. The most glaring example of disproportional effects was seen in the 2001 anthrax attacks that took place along the East Coast of the United States. In September and October of that year, a perpetrator launched two waves of attacks against high-profile people and institutions from postal facilities in New Jersey. The targets initially included news media, but later two U.S. Senators, Tom Daschle of South Dakota and Patrick Leahy of Vermont, also received letters. As a result of those mailings, 22 people developed anthrax infections, resulting in 5 deaths. While some of the victims included U.S. Postal Service personnel or people receiving the letters, others had no known established relationship to the chain of events.

In response to these small-scale but highly publicized attacks, near-panic set in across the country. The fact that these attacks occurred right after 9/11 intensified the psychological effect, and many people who came in contact with a powder they could not immediately identify notified law enforcement authorities of possible terrorist incidents. The investigation was massive, and numerous Congressional hearings were held. By September 2006, the Federal Bureau of Investigation reported having participated in over 9,100 interviews, issued over 6,000 grand jury subpoenas, and conducted many extensive searches.[2] The case was recently closed following the apparent suicide of the primary suspect, a government scientist who worked with biological agents.

There were other disproportionate direct and indirect costs incurred by taxpayers. More than $200 million was spent to decontaminate buildings, and some estimates place the total damage at over $1 billion. Due to public pressure, Congressional funding of BW defense, including the Project Bioshield Act of 2004, which provides $5.6 billion over 10 years, rose dramatically.[3]

While the five deaths were terrible, the casualty level was small when compared with other far worse statistics that evoked little response. As examples, each day in 2001 in the United States, approximately 104 people died in traffic accidents, and another 44 were murdered. Even the deaths of over 1,500 people *per day* from cancer did not get the same level of publicity as the BW events. Clearly, the response was disproportionate to the scale of the attack, highlighting the severe fear and psychological responses biological weapons invoke in individuals.

Fielded Military Implications

The psychological impact of BW is as apparent in military forces as it is in civilian circles. During Operation *Iraqi Freedom*, there was constant concern that Iraqi forces would deploy BW and chemical weapons (CW) against U.S. and Coalition forces. As part of the predeployment activities, anthrax vaccinations became mandatory for all U.S. troops going into the Gulf region. During ground operations, when initially approaching Baghdad, troops were ordered into protective gear as some commanders thought such an attack was imminent.

Even though no BW attack was ever launched, the mere threat was sufficient to take the precautionary measures even while understanding their potential negative consequences for operational performance. While the protective gear works relatively well, it can dramatically degrade the physical capability of the fighter, the degree of degradation increasing with the level of protection required. Some of the generic impacts are problems in communication, impaired hearing, loss of visual acuity, dehydration, heat injuries, combat stress, and impaired thinking and judgment.[4]

Selected examples from specific studies on the effects of wearing protective gear reported the following observations:

- task degradation of between 20 and 50 percent occurred
- leader performance declined as they become exhausted and disoriented
- sensory awareness was reduced, making it harder to stay awake
- soldiers required 1½ to 3 times longer to perform manual dexterity tasks
- individual rate of weapons fire dropped 20 to 40 percent
- calls for indirect fire increased 3 times
- communication was about half as effective, and total time on radio doubled
- cognitive tasks declined up to 23 percent
- land navigation was seriously impaired, especially at night
- units bunched up to maintain control.

After-action reports from Operation *Iraqi Freedom* outlined a number of problems inherent in deploying to an environment in which a CW/BW attack is anticipated. Highlighted was the need for involvement of leaders at all levels as well as enhancing individual and collective skills required to respond to these threats. The report also noted that the new protective suits were lighter and better suited for operations in warm temperatures. However, while allowing better performance, they tended to rip and caused problems with many activities. Of specific concern was damage to gear worn by infantry Soldiers.[5]

The recommendation for increased emphasis on training should not come as a surprise. That call has been sounded for decades, but in general CW/BW has been put in the *too tough to handle* bin. Like nuclear weapons before it, while the BW threat is acknowledged, it is often ignored in training and wargames as it can overwhelm the players and the outcome. Therefore, the rationale is to limit or eliminate the BW possibility so that other aspects of the game can produce useful results.

This is not to say there has not been training conducted; there has. The problem is that the amount of training required to adequately prepare to fight in a contaminated environment is too extensive for the total training time allotted for either individual skills or unit training. The conundrum that emerges is that adequate training is essential for survival on the CW/BW battlefield, yet it must compete with realistic time constraints that will always fall short of optimal, or even acceptable, standards.

This leads to a strategic observation that indicates that with CW/BW, the advantage *always* goes to an enemy force *willing* to employ them. In addition, armed forces tend to consider the threat from a military perspective—that is, how will they fight on a potentially contaminated battlefield. Thus, in preparing for the potential battle, they become a somewhat hardened target and reduce the threat to themselves in the direct fight.

Three Legs of Biological Warfare

The problem is very complex, and it is the author's opinion that even the approved Department of Defense (DOD) definition of BW is incomplete. The official definitions reads, "Biological operations: employment of biological agents to produce casualties in personnel or animals, or damage to plants."[6] There is no mention made of the potential of BW agents that attack nonliving targets. There are three legs to the BW threat, and only two—antipersonnel agents and those that attack agricultural targets such as animals or crops—are covered by this definition. The third leg, and the one that is unaccounted for, is antimateriel attacks by biological organisms that degrade inanimate material objects.

The majority of DOD efforts in countering BW have rightfully focused on prevention of infectious diseases. Many Americans do not realize that naturally occurring diseases have caused more casualties than enemy combatants in most wars prior to World War II. Even so, disease has been a major concern for military medical personnel. Therefore, it is a logical extrapolation that controlling specific diseases and employing them on the battlefield is a method to be considered in destroying the enemy force.

The use of disease as a weapon dates back many centuries. Because of the catastrophic consequences, the siege at the Crimean peninsula city of Kaffa, now known as Feodosiya, Ukraine, is often cited as an example. In 1346, invading Tatar forces experienced a deadly outbreak of bubonic plague and began to catapult infected bodies over the city walls to infect the defending forces within. This ultimately resulted in an outbreak of plague in the defending city. Shortly thereafter, the surviving defending forces sailed for Constantinople, carrying with them what became known as the Black Death, which would ultimately cause approximately 25 million deaths throughout Europe.[7]

In the last century, BW received considerable attention. Based on the dreadful experience with chemical weapons in World War I, the Geneva Protocol was adopted in 1925 (officially known as the Protocol for the Prohibition of the Use in War of Asphyxiating, Poisonous or Other Gases, and of Bacteriological Methods of Warfare), which banned the use of both biological and chemical weapons in war. While many countries continued to conduct research on BW agents, the only known military use of biological weapons during this time was by Japan against the Chinese during World War II.

In 1972, the Biological and Toxin Weapons Convention was signed by all major parties involved in the Cold War. While the United States honored the treaty and halted all offensive BW research and development, the Soviet Union initiated the largest and most comprehensive program of its kind. Ken Alibek, a former senior official in the Soviet BW program, alleged that this program moved production of selected agents, such as anthrax, to an industrial scale, explored a wide range of antipersonnel agents, and developed the essential distribution means by which they could be delivered.[8] The program continued undetected until a defector provided partial information about it in 1989.

Anthrax offers a good case study for the potential damage that could be caused by a successful large-scale BW attack, especially against a civilian population. If early diagnosis and immediate treatment do not occur, the mortality rate could be greater than 85 percent for inhalational anthrax (other forms of anthrax, such as cutaneous or gastrointestinal, have lower mortality rates).[9] In 1993, the U.S. Congressional Office of Technology Assessment estimated that 100 kilograms of agent dispersed from a low flying aircraft in a densely populated area, while a low probability event, could kill between 1 million and 3 million people within 5 to 7 days.[10] It should be noted, however, that these numbers are not universally agreed upon in the scientific community, and the successful delivery of anthrax requires a terrorist group to overcome numerous technical challenges.

When a specific BW agent is known, prophylaxes (for example, vaccines) or therapeutics (for example, antibiotics) can be employed to prevent or treat the disease. However, when confronted by multiple possibilities, it is nearly impossible to defend against all of the agents. After defecting from the Soviet Union, Alibek indicated that the Soviet program did include a mix-and-match capability, thus increasing the difficulty in countering such an attack. Even if vaccines were available for each agent, the combinations would likely have devastating health effects.

Since the breakup of the Soviet Union in 1992, numerous programs, such as the Nunn-Lugar Cooperative Threat Reduction Program, have strived to keep former weapons scientists gainfully employed in their fields of expertise. The Soviet BW program, however, employed thousands of people, and only a small subset of the researchers have been hired via the Cooperative Threat Reduction Program to prevent the spread of this technical knowledge. There have been cases where former Soviet BW scientists, with sufficient information to be quite dangerous, have either emigrated to suspected proliferant countries or attempted to sell their expertise on the black market. It is also known that terrorists groups such as al Qaeda have actively sought access to both BW information and weapons.

The second leg of the BW triad involves attacks against agriculture, both plants and animals. Despite decades of known research by many countries into biological weapons targeting agriculture, it has only recently been recognized as a national security threat. The potential impact of agroterrorism is huge. In the United States, farming employs only 2 percent of the population, but a total of 16 percent work in food- and fiber-related industries. This sector accounts for 11 percent of America's gross domestic product and 23 percent of wheat, 54 percent of corn, and 43 percent of soybean exports.[11]

There is an extensive range of possible agents that could be used against crops and livestock. Some of them are zoonotic, meaning that the disease could be transmitted to humans via animal vectors. Among the most studied agents are anthrax, Rinderpest, African swine fever, foot and mouth disease, wheat blast fungus, fowl plague, and many others. In addition, there are many nonindigenous insects and rodents that could be used to spread disease, and these would be difficult to detect until a major outbreak has begun.

Agricultural production is vital to our economic interests. Unlike many industrial facilities, by its nature, agricultural production is geographically dispersed and thus nearly impossible to secure. Experience has shown that even a single case of a debilitating illness can shut down international trade for extended periods of time.

The potential impact that a limited BW attack could have on American products is seen in response to a case of bovine spongiform encephalopathy (BSE), popularly known as mad cow disease. In December 2003, a single case of BSE was detected in the American food chain. This led to full or partial bans on beef imports by 65 countries at an estimated loss of over $2 billion each of the following years. Once import bans are established, it often takes years to lift them. While the introduction of BSE was not intentional, this example does point to the large economic vulnerabilities that exist in the agriculture sector.[12]

There are other examples, including intentional terrorist acts that have adversely affected international trade. Some suggest that in 1989, an ecoterrorist group called the Breeders released Mediterranean fruit flies into California groves. The estimated cost in lost revenue and other damage was nearly $900 million.[13] Estimates for such an insect infestation happening now would be the demise of 132,000 jobs and $13.4 billion in lost revenue.[14] In an article in the *Boston Globe*, Jeffery Lockwood described insects as "the box cutters of biological weapons—cheap, simple, and wickedly effective."

In addition to direct fiscal costs, there are other indirect vulnerabilities. Some analysts are concerned that a successful agroterrorism attack could degrade public confidence in the existing government. In the United States, we have established a carefully balanced food production and distribution system that relies on the smooth operation of many parts.[15] The public is quick to respond to small outbreaks such as the incident of contaminated spinach in September 2006, which caused 205 confirmed illnesses and 3 deaths.[16] If confronted with severe disruption in availability of food, public confidence in developed countries would be tested. However, such interruption in developing countries could be sufficient to destabilize governments.[17]

Militarily, agroterrorism is a strategic versus a tactical issue. Fielded forces have sufficient reserves of prepackaged food to be able to conduct immediate missions even in the event of substantial interruption of supplies. More at risk are popular support for the conflict and confidence that the government can protect the society.

The definition of biological warfare must be revised to include antimateriel agents as the third leg of the BW triad. The potential threat has enormous implications and is growing at a rate that should not be ignored. In addition, advances in genetically engineered designer microorganisms will increase the potential threat dramatically.

The issues of antimateriel BW can be understood easily in nontechnical terms. Trash and environmental contamination are global problems. The volume of waste material produced by developed countries greatly exceeds their proportionate population and their ability to dispose of it. Trash is an arbitrary designation assigned when an item is no longer deemed useful to the owner. The trash items contain the same physical properties they had when they were still useful. To deal with this mounting problem, bioremediation and microbial biodegradation are often used. That means that biological organisms are applied that intentionally degrade the physical material. There is almost nothing that some biological organism will not attack. Therefore, research is rapidly advancing to identify organisms that can be used to reduce the volume of waste material and to eliminate contaminating materials from the environment. This is considered positive and encouraged as a peaceful application of biological development.

However, from a military perspective, the same research and development can be employed as an offensive biological weapon. The only change is the intent for which the biological agent is introduced to the target. Therefore, offensive and defensive antimateriel BW research is virtually indistinguishable. The development of these agents will continue to accelerate and has strategic implications.

In my book *Future War*, I posited the application of nonlethal systems, including biological agents, to create strategic paralysis of a complex target set. In that case, it was a nation-state that was vulnerable to attack, albeit over time. The reality is that microorganisms could be developed that could accomplish that mission.[18] Most military analysts are aware that bioremediation is the treatment of choice when oil spills occur on the seas. It only takes a small step in logic to understand the implications of introduction of those same microorganisms into fuel or lubricant supplies at any stage from national petroleum reserves to vehicle fuel tanks. Once vehicle engines are befouled, they are extremely difficult to cleanse and repair short of depot level maintenance.

There are many other substances that are vulnerable to attacks of microorganisms. With the possible exception of titanium, all common metals and alloys are susceptible to biocorrosion.[19] Possibly most surprising to military planners will be that even concrete can be degraded, thus rendering many bridges, dams, and buildings as potential targets. Widely used polyurethane plastics, while resistant to oxidation, are also susceptible to biodegradation, especially when employed as thin films. That means it is not necessary to destroy bulk quantities of plastic to successfully damage key components of sensitive equipment. Relatively small amounts applied at critical locations would be sufficient to degrade operational capabilities of the item. Even fiberglass and polymeric composites can be targeted so as to degrade the integrity of the item.

Except by using the most advanced techniques, it would be difficult to detect and monitor an antimateriel BW program. The equipment necessary to initiate the project is easy to both obtain and conceal. A simple approach to acquiring microorganisms of interest is to go to a dump where the desired target material has been buried for some time. Due to their inherent ability to mutate and adapt to environmental changes, it is highly likely that the microorganisms of interest have already begun to degrade that material. It is then a matter of harvesting those agents and initiating a breeding program to develop the desired characteristics of the BW agent. As an example, introduction of a biocide will kill most, but not all, of the organisms. The surviving organisms are then bred until a biocide-resistant strain is produced.

A major constraint to application of antimateriel BW agents is that they are very slow to produce the degradation of the target. For that reason, many experts have discounted the probability of them being employed. There is no doubt that American military leaders want to know that when they pull the trigger, an immediate and predictable effect will occur. This is a shortsighted view and some potential adversaries are willing to move on timelines that are quite different from ours, and we must guard against discounting this threat because it does not mirror how we would prefer to accomplish the task.

Full-spectrum Analysis

Threat analysis must include the full spectrum of vulnerabilities, and it is insufficient to divide complex systems into small elements and establish protection for each part independently. Synergistically, the threat is greater than the sum of the parts, and seamless defensive system architectures are required. Combat systems must be evaluated in their entirety, and it is then essential that the analysis be broadened to a national or even an international level. Most threats to the military will not be constrained within the DOD organizational structures.

As previously indicated, troops in the field, while vulnerable, represent a somewhat hardened target set. However, the recent propensity for contracting services in a variety of areas dramatically complicates BW defensive measures. For operations in the Gulf theater, the ratio of military to contractors reached about 1:1. Contractors were deeply involved in all aspects of the logistics and maintenance systems as well as many other areas such as security. From the beginning, this was problematic from a BW perspective. In fact, one of the findings in the Third Infantry Division after-action report was that their supporting civilian contractors arrived without any protective equipment. This placed a burden on the deploying forces to supply them with the necessary equipment.[20] Not even addressed was whether or not those contractors had any training on how to respond in the event of a BW attack.

When evaluating military support systems, consideration must be given to the nationality of those performing services and what access they have at every stage. In Operation *Iraqi Freedom*, it was common to have third-party, and even host-nation, workers employed at fairly sensitive locations. The threat is twofold. Distribution of BW agents is one of the most difficult issues for offensive operations. If cooperation can be obtained from a worker already inside the chain of operations, a BW agent could be introduced directly to the targeted personnel. As an example, consider who works in consolidated mess facilities, where food is procured, how it is transported, where is it stored, and what level of protection is afforded throughout the process. It would only take one operative at a critical location to inject the BW agent, with deadly consequences.

The second issue is the impact a BW attack, or threat of an attack, might have on the workforce. As contractors, these workers cannot be forced to remain in a hostile environment. Rumors are likely to spread rapidly in the event of unexplained illnesses. Fear of further contamination would likely lead to large defections of foreign contractors. Strategically, threats of BW attacks could be used to drive a wedge between coalition partners, similar to what was accomplished by selective targeting of suicide bombings. While tolerance for violent casualties is already relatively low, threats of BW casualties would have serious repercussions in all countries and at all levels of the population.

Recommendations

Most of the BW threats have not gone unnoticed. In some areas, there have been Congressional hearings and funding increases to boost protection. These are extremely difficult threats against which to defend but there are tasks to be undertaken. Specifically, we must improve intelligence collection, develop and deploy near-ubiquitous sensors, establish perception management, and conduct training programs that extend to all sectors.

Intelligence

Probably the most important countermeasure is to expand intelligence collection aimed at the entire spectrum of the BW threat. It is vitally important that as much information as possible is collected about what specific biological agents are being researched, developed, and manufactured. Given the rapid advances that are occurring in biology, it can be assumed that new technologies and agents will be adapted as potential weapons. The Intelligence Community needs to stay current and be able to predict future innovations.

While human intelligence activities targeted against tribal adversaries are difficult, every effort must be made to ascertain the capabilities and intentions of potential adversaries. Every means available should be dedicated to this effort. Also useful is development of advanced computer models that can provide threat warnings and be predictive about the direction BW research, development, and applications will take.

Sensors

Sensor technologies are advancing at rates that parallel those of the BW field. Given the extremely broad vulnerabilities of both military units and society in general, there is a need for ubiquitous sensors that can quickly detect the presence of chemical or biological agents and report them immediately. State-of-the-art nanoscale sensors should be placed at every anticipated site of vulnerability. Lab-on-a-chip technology already provides the capability to constantly observe, and simultaneously test for, a wide range of agents. This effort should be perpetually technically upgraded and distribution expanded to the maximum extent conceivable.

Perception Management

As indicated, psychological factors play an extremely important role in the effectiveness of BW attacks. Unfortunately, news media often proclaim worst-case scenarios even without supporting facts. The best countermeasure is to provide honest and accurate information directly to the public as quickly as possible, as success depends on the public having a high degree of trust in the reporting agencies. There is little doubt that the American people currently have diminished confidence in the accuracy and honesty of the information provided by government sources. It is therefore essential that all information that is provided during noncrisis situations be completely open and honest. While difficult to accomplish, there should be a long-term comprehensive program to educate news media personnel about the realities of BW and provide them with an understanding about how their doomsday prognostications directly support the goals of the adversaries.

Training

Training for BW events must continue at all levels of the military. Realistic scenarios should be included in high-level wargames. They should be played out even if they disrupt the anticipated flow of the exercise. Individual and unit BW training should be conducted continuously. There can never be too much.

In addition, military commanders must ensure that nonmilitary support elements can function in a BW environment by providing adequate equipment and training. Support contracts should include specific requirements that contractors have minimum established levels of training for operations in contaminated environments. There will certainly be additional costs associated with including BW/CW capabilities requirements, but if the units must rely on contractor support, the latter must be capable of functioning in the same threat situations.

Hold at Risk

Worth considering is development of policy that might be called Biological Assured Destruction. The policy would be aimed at nation-states that might be tempted to make a large-scale BW attack or support other groups that might carry out such an act. The message would be clear: Any nation that attempts or supports a major BW act against the United States or its interests will risk conventional but lethal countermeasures.

NOTES

1 Milton Leitenberg, *Assessing the Biological Weapons and Bioterrorism Threat* (Carlisle, PA: U.S. Army War College, Strategic Studies Institute, December 2005).

2 Federal Bureau of Investigation, "Amerithrax Fact Sheet," September 2006, available at **www.fbi.gov/anthrax/amerithrax_factsheet.htm.**

3 Susan Wright, "Taking Biodefense too Far," *Bulletin of the Atomic Scientists* 60, no. 6 (November-December 2004).

4 Office of the Special Assistant for Gulf War Illnesses, "MOPP Information Paper," available at **www.gulflink.osd.mil/mopp/mopp_s03.htm**.

5 *Third Infantry Division (Mech) After Action Report—Operation* Iraqi Freedom, available at **www.globalsecurity.org/military/library/report/2003/3id-aar-jul03.pdf**.

6 Joint Publication 1–02, *Department of Defense Dictionary of Defense and Associated Terms* (Washington, DC: Department of Defense, amended October 17, 2007).

7 John B. Alexander, *Future War: Non-Lethal Weapons in Twenty-First Century Warfare* (New York: St. Martin's Press, 1999).

8 Ken Alibek, *Biohazard: The Chilling True Story of the Largest Covert Biological Weapons Program in the World—Told from the Inside by the Man Who Ran It* (New York: Random House, 1999).

9 According to the U.S. Army Medical Research Institute of Infectious Diseases, the mortality rate for cases of inhalational anthrax in the 2001 anthrax attacks was only 45 percent. It is suggested that this decrease in mortality rates could be due to advancements in intensive care medicine.

10 U.S. Congress, Office of Technology Assessment, *Proliferation of Weapons of Mass Destruction: Assessing the Risks* (Washington, DC: U.S. Government Printing Office, August 1993).

11 Jim Monke, *Agroterrorism: Threats and Preparedness*, RL32521 (Washington, DC: Congressional Research Service, August 13, 2004).

12 Colm A. Kelleher, *Brain Trust: The Hidden Connection Between Mad Cow and Misdiagnosed Alzheimer's Disease* (New York: Paraview Pocket Books, 2004).

13 Robert P. Kadlec, "Biological Weapons for Waging Economic Warfare," in *Battlefield of the Future: 21st-Century Warfare Issues* (Maxwell Air Force Base, AL: U.S. Air Force Air University, 1995), available at **www.airpower.maxwell.af.mil/airchronicles/battle/chp10.html**.

14 Jeffery A. Lockwood, "Bug Bomb: Why Our Next Terrorist Attack Could Come on Six Legs," *The Boston Globe*, October 21, 2007.

15 Monke.

16 U.S. Food and Drug Administration press release, "FDA Finalizes Report on 2006 Spinach Outbreak," March 23, 2007.

17 Mark Wheelis, "Agriculture and Bioterrorism: An Analytical Framework and Recommendations for the Fifth BTWC Review Conference," presented at annual meeting of the Association of Politics and the Life Sciences in Atlanta, Georgia, September 1999, available at **www.fas.org/bwc/agr/agwhole.htm**.

18 Alexander.

19 Margaret E. Kosal, "Anti-Materials Agents," *Encyclopedia of Bioterrorism Defense* (New York: John Wiley and Sons, 2005).

20 *Third Infantry Division (Mech) After Action Report.*

part two

BIOMOLECULAR ENGINEERING

chapter 5

ABIOTIC SENSING

James J. Valdes, Andrew Ellington, William E. Bentley, Roy G. Thompson, and James P. Chambers

We all feel humble in the face of the ingenuity and complexity of nature. Yet it need not prevent us from learning, exploring, and even attempting to construct useful artificial systems having a few of the simpler properties of their natural component.

—T.M.S. Chang

The most obvious quality of living cells is that they are complicated and highly organized. However, these complex structures and functions are made from simple lifeless molecules that, in appropriate combinations and number, can now be hypothetically put together in such a manner as to render "lifelike" functional circuits and networks exhibiting distinct and extraordinary properties.

The most sensitive biodetector known today is the olfactory system of animal species such as dogs and pigs,[1] capable of detecting a few molecules of C4 escaping from explosives hidden underwater, from cocaine frozen inside ice blocks, or from truffles buried underground. It was clear more than two decades ago that biological detection systems that capitalize on an understanding of the molecular physiology of ligand-receptor interaction, as well as the integrative properties of neuronal systems, would offer the best prospects for developing true biosensors. An early 1980s concept for threat agent detection involved coupling neuroreceptors with microsensors to create biosensors capable of detecting classes of threat agents. For example, using the acetylcholine receptor as the biological recognition element (BRE) would enable the detection of more than 60 cholinergic neurotoxins, while using the enzyme acetylcholinesterase would permit detection of all organophosphorus nerve agents and pesticides.

Such broad-spectrum, almost generic, biosensors are desirable for a number of reasons. From a military point of view, threat agents can be broadly considered to be xenobiotic entities that interact with human physiology and in some way interrupt normal function. This interaction is mediated by BREs known generically as "receptors." Receptors are the key to communication among the body's many networks as well as with the external environment, and recently a vast body of literature has identified nodes at the organ, cellular, subcellular, and genomic levels that are critical to proper function of these networks.

The nature of the military—and, by extension, public health—threat is now such that it cannot be predicted a priori. Broadly defined, nontraditional agents include genetically engineered pathogens, midspectrum agents such as toxins and biological regulators, lentiviral and other vectors capable of introducing genetic material into host genomes, and new classes of chemical agents, to name but a few. The Cold War–era approach of threat-oriented detection must necessarily give way to a target-oriented approach, the targets being BREs. While this shift was postulated two decades ago, the technology to bring it to fruition simply did not exist.

The technological bottlenecks in the 1980s were many. A number of receptors had been purified and studied but most were membrane bound and could not function outside of their native membranes, which were necessary to maintain proper conformation. Some receptors were cloned and produced in small quantities but manufacturing technology was essentially nonexistent. Many studies therefore relied on isolating acetylcholine receptors from rich sources such as the *torpedo* electric organ, or isolating synaptosomes from animal brains. Many receptors have since been cloned and efficiently expressed. G-protein–coupled receptors (GPCRs) are now known to be important targets for drug discovery, and assay kits for high throughput screening have been developed.[2]

A second technological barrier was lack of knowledge about the interactions between BREs and nonbiological supports. Specifically, techniques required to keep BREs and whole cells stable and functional on sensor platforms were very crude. Driven by the development of prosthetics and artificial organ systems, the interface between synthetic and biological materials has been an object of intense study in recent years.[3]

Third, there was a very limited ability to design novel BREs tailored for specific affinities or selectivity; nor was there much success with artificial receptors such as molecular imprints or engineered antibodies such as Fabs. Defining the minimum requirements for biomolecular interaction is critical for the design of stable artificial BREs. Two such approaches are the minimization of known protein binding motifs or, conversely, a biomimetic approach in which binding pockets are designed de novo.[4] Engineering novel antibodies is another successful approach, and these include recombinant antibody fragments (for example, Fab, scFv) and variants such as diabodies and triabodies.[5] Finally, molecular imprints made from synthetic materials have been successfully applied to both analyte detection and drug release.[6] More importantly, the genomic tools required to identify all proteins required for both normal and abnormal functioning, the precise role of the proteins in key metabolic pathways, or even determining a receptor's natural ligand did not exist. Using modern genomics mapping techniques, one can now identify all relevant proteins, but it is still problematic to definitively elucidate function. Nevertheless, enormous strides have been made in understanding GPCR cascades, signal amplification, and translocation of cellular proteins to specific intracellular compartments, knowledge that will result in design of high-content and high-throughput screens for drug discovery.[7]

Finally, the materials and micromachining techniques required to assemble an integrated biosensor were inadequate. Nanotechnology and its offshoots provide unprecedented control over material properties via structural control using directed and self-assembly. Advances in microcontact printing, soft photolithography, microfluidics, and laser desorption techniques have proceeded rapidly in recent years. Micropatterning techniques, which include laminar flow patterning and microchannels, have resulted in the design of numerous functional biosensors.[8]

Synthetic Biology

The term *synthetic biology* was coined in the literature over 20 years ago. The first use of the term referred to the genetic engineering of bacteria using recombinant DNA technology,[9] in which natural biological systems are altered by scientists to produce a biological system that performs unnatural function(s)—an approach that initiated the emerging field of bioengineering.

The term was somewhat usurped in 2000[10] to describe the field of chemistry striving to design the synthesis of unnatural molecules that could function in living systems. In this use, synthetic biology became an extension of biomimetic chemistry, in which artificial molecules are synthesized that can functionally replace natural biological molecules, such as artificial enzymes, aptamers, and peptide mimics.[11, 12, 13, 14] The extension of biomimetic chemistry to fulfill the larger scope of synthetic biology requires the assembly of individual chemical constructs into a larger chemical system that displays the natural, functional properties of a living system such as sustained metabolic activity, cell synthesis, and genetic evolution. Most recently, the term *synthetic biology* has been adopted by an engineering cohort to define the process by which natural biological molecules (enzymes, DNA, proteins, and so forth) are extracted from living systems and defined as basic building blocks to be reassembled in unnatural order and environments to create novel "devices or machines" that perform specific, predictable functions that may or may not be found in natural biological systems. This engineering approach differs significantly from *systems biology* in that the individual biological constructs most suited to constructing a device are those units that act independently in contributing to the whole: the whole can be predicted from the sum of its individual parts.

Artificial Cells

The unit of life is the cell, the smallest entity capable of displaying attributes associated with the living state: growth, metabolism, stimulus response, and replication. Even the simplest of cells (for example, prokaryotes) possess and utilize highly integrated networks of chemical reactions. Despite great diversity in form and function, all cells share extensive biochemistry. How many genes does it take to make a cell or, beyond that, a network of different cells? Important insight can be gleaned from the smallest known genome for an independently replicating organism, that of *Mycoplasma genitalium. M. genitalium* DNA consist of 517 genes, in contrast to 30,000 genes in the human genome. Thus, there exists a broad hierarchy of cell complexity as well as component processes to borrow and engineer into artificial systems.

A number of advances in fields such as evolutionary biology, synthetic biology, and nanotechnology have made it possible to contemplate the development of acellular networks of synthetic genetic circuits (Xcells). In conception, Xcells would be nonreplicating compartments genetically programmed to process information, make decisions, and carry out reactions, and would network with one another to form artificial ecologies (Xcologies) that would in turn perform higher order parallel and even emergent tasks.

Some of the advances that make possible Xcell and Xcology development include the explosion of knowledge about systems and synthetic biology, and the robust development of acellular biotechnologies. Furthermore, there is much precedent for formation of acellular supramolecular complexes of nucleic acid, either DNA or RNA, encapsulated in a protein coat and, in some instances, surrounded by a membrane envelope (that is, the viruses). Although viruses are not alive, they represent supramolecular assemblies that act as parasites of cells, underscoring the functional "culling out" of specific cellular processes, albeit within the confines of living cells.

Systems biology is the integration of the many different levels of knowledge (genomics, proteomics, metabolomics) about cells and organisms to gain a global understanding of function. The informatics tools for systems biology, including databases such as Genbank and the Kyoto Encyclopedia of Genes and Genomes and analytical tools such as Cytoscape, now make possible the ready visualization and analysis of virtually any set of pathways, natural or contrived.

As a counterpart, genes, proteins, and functionalities are becoming increasingly fungible real-world entities that are being engineered as synthetic biology "parts," such as BioBricks, that can be readily put together in novel synthetic genetic circuits. The modularity of synthetic biology parts is useful not only for creating circuits in vivo but also for generating tools and pathways in acellular systems. The modularity of many BREs has become increasingly apparent from basic research into signal transduction and biotechnology explorations of directed evolution. Many cellular receptors and signal transduction enzymes can already be readily "rewired" to take on new functions. For example, Wendell Lim and coworkers have retasked the yeast mating pathway to generate osmoresistance, as opposed to mating type changes,[15] and new receptors can be created almost at will by directed evolution techniques such as phage display, cell surface display, and Systematic Evolution of Ligands by Exponential Enrichment.

Other processes necessary to support BREs and signal transduction pathways can now be brought together in an acellular milieu. As a result of years of product development in molecular biology, complex biological reactions such as transcription and translation can now routinely be carried out efficiently in vitro. Moreover, in vitro transcription, translation, and other cellular machines can be ensconced within in vitro compartments that resemble cells.[16, 17]

While it may be possible to standardize synthetic circuits in the context of a minimal biological cell, another more visionary approach is to generate a cell-like compartment that could reproducibly and programmably support Xcells. Several elements might be considered in Xcell approaches:

- Beyond the in vitro compartmentalization methods so far cited, it may be possible to design and evolve self-amplifying or self-selecting genetic circuits completely within an Xcell[18] and to harness these to robust extracellular energy generating and harvesting systems for biology that have become available as a result of years of bioprocess development in industry.
- The boundaries of the Xcell can potentially be much more robust and more functional than conventional membrane-based vesicles or liposomes, due to intense efforts in nanotechnology and drug delivery that have led to the generation of "smart" materials and particles that are interactive with their environments. For example, it has now proven possible to create "double emulsions" in which dispersed water droplets are further compartmentalized and moved to aqueous environments,

much as the bilayer allows cells to have an aqueous interior and exterior.[19] Moreover, by using surfactants with "tails" that can be photopolymerized, it should be possible to encapsulate individual droplets into hard shells and actually make "sheets" or films of droplets, much like colonies on a Petri plate. Alternatively, organotypic matrices such as hydrogel lattices or nanoporous materials might be used for the encapsulation or immobilization of genetic circuits.

- Finally, efforts have led to methods for chemical, photoluminescent, and electronic signaling between small, independent biological or physical units. Transduction between biological recognition elements, signal transduction pathways, and signaling modalities sets the stage for the abiological networking of Xcells into Xcologies. Networked Xcells could evolve adaptive and emergent behaviors, similar in character to either molecular "predator-prey" systems [20] or to cellular automata.

The development of an Xcell will involve integration of all of these disparate advances: an Xcell must be an encapsulated compartment in which chemical or light energy is captured and used for biopolymer synthesis. The biopolymers must be capable of processing signals, sending signals, and performing programmed tasks based on communication with their environment. Communication between Xcells should lead to the competitive development of Xcologies that can perform a particular task or task set.

Beyond the technical achievements that will emerge from this program, there are fundamental scientific advantages to positing the development of Xcells and Xcologies, not the least of which is that it will better illuminate fundamental biological principles of living systems. Currently, the only example of living systems is cell-based, and there is no agreed upon or even self-consistent definition of "life" or its properties. By providing an example of a programmable, acellular ecology, this largely philosophical problem is deftly avoided and engineers can begin to understand what principles and practices are most suited to adapting living systems and their mimics to military tasks. Beyond this advantage, having one example of a programmable Xcell will allow engineering standards to finally be set in place for further development. While explorations of synthetic biology have not so far yielded anything equivalent to very large scale integration, it seems likely that the production of Xcell circuitry will engender a revolution similar to the one that occurred in electronics in the early 1980s.

Natural Proteins as Engineering Units

Proteins exhibit a wide diversity of physical and catalytic properties. The question of whether enzymes involved in known metabolic pathways can be isolated and reassembled as basic building blocks to generate new pathways has been pursued for several decades. Recombinant DNA technology has significantly aided the ability to pick and choose enzymes from various organisms and then direct their reassembly into a single organism, hence conferring the ability to synthesize an unnatural product in the host organism.[21] Here, bacterial strains are genetically engineered by inserting genes for various synthetic steps in the metabolic pathways from several different microorganisms into the host. A strain of *E. coli*, for example, has been developed that can synthesize an isoprenoid precursor for the antimalarial drug artemisinin using enzymes assembled from three different microorganisms.[22] Genes from two other organisms were introduced into the *R. eutropha* bacterium to produce a copolymer from fructose.[23]

Genetic Regulation and Signaling Pathways

A prime example of modularity in biological systems is evident in genetic regulatory and signaling pathways, where combinations of proteins can function as molecular switches affecting gene expression and ultimately the phenotype. Thus, the "wiring" of the many signal transduction pathways is based upon particular protein domains, present in transduction proteins that bring components of signal transduction pathways together. Here, ligand binding, chemical reactions, or the physical translocation of a component generates an event that can serve as an input for another switch.[24, 25] Using this approach, synthetic circuits have been synthesized that show AND and OR logic gating based on ligand input.[26, 27, 28] An artificial oscillatory network has also been constructed from three transcriptional repressor systems that are not naturally paired in any biological system,[29] and a molecular automaton that plays tic-tac-toe against a human has been designed based on 23 molecular scale logic gates using deoxyribozymes as the building module.[30]

Signal-transduction pathways have many common functional/architectural elements and hierarchical structure enabling the creation of "plug and play" signal transduction cassettes that can be incorporated into acellular systems (for example, ANTS). Due to their simplicity, microbes represent a source from which much can be borrowed for purposes of engineering acellular networks. Microbial quorum sensing is responsible for a variety of phenotypes and is rich in diversity and modes of action. As such, quorum sensing represents a "guide" for learning how signals can be translated to altered phenotype(s) as well as for designing of useful acellular prototype signal transduction networks. Quorum sensing may be the foundation upon which the more sophisticated intracellular communication found in higher order organisms has evolved. If this is the case, methods that incorporate native signaling architecture may be engineered for acellular network purposes, exerting greater control with less collateral damage to other resident, nontargeted processes.

Although there have been few technological or commercial applications derived directly from adapting or rewiring the quorum sensing signaling process, metabolically engineered signaling modules have been constructed to alter phenotype giving rise to recombinant gene products.[31, 32] For example, Bulter et al.[33] created an artificial genetic switch using acetate for modulating cell to cell signaling in *Escherichia coli*. Neddermann et al.[34] developed a hybrid expression system by incorporating the quorum circuitry of *Agrobaterium tumefaciens* into a eukaryotic transcriptional controller for HeLa cells. Weber et al.[35] implanted the *Streptomyces* bacterial QS system for tuning heterologous protein expression in mammalian cell culture and mice (human primary and mouse embryonic stem cells).

The molecular basis by which quorum sensing works varies among different species, both in the synthesis of the signaling compounds and in the modes of their perception. However, this diversity affords a vast array of potential acellular applications—the types of rewired systems for achieving specific outcomes can vary greatly and are limited by imagination and ability to abstract regulatory features and reconstruct them into modular, transplantable controllers. There are also many modes by which the regulatory function is translated into altered gene expression. Most of the well-studied systems modulate protein transcriptional regulators. Additionally, there is also evidence that small RNAs are involved as riboregulators, and posttranscriptional regulation has been linked to quorum sensing systems and quorum sensing signal transduction.

Devices

A device is construed from decomposition of a system into basic functional parts that perform physical processes. Devices process inputs to produce outputs by performing metabolic or biosynthetic functions and interface with other devices and the environment. Devices can represent one or more biochemical reactions involving transcription, translation, ligand/receptor binding, protein phosphorylation, or enzymatic reactions. Devices can be composed of few reactants (for example, a protein phosphorylation device consists of a kinase and substrate), or multiple reactants and products (for example, a regulated gene, transcription factors, promoter site, and RNA polymerase compose a transcriptional device). Devices can be extracted unaltered from nature, such as transcription factors and promoter of a regulated gene, or by modifying biochemical reactions. Transcription or translation control devices are controlled by the specificity and efficiency of nucleotide interactions and are relatively easy to build based on the specified order of the reaction. Transcriptional control systems have been employed for signal amplification, control of downstream devices, and multifactorial control by multiple transcription factors.[36, 37, 38] While transcription and translational devices are fairly easy to connect and can offer great logical complexity, changes in output are relatively slow (minutes) and consume a large amount of cellular resources to support the protein synthesis events that mediate the output.

Devices employing protein-ligand or protein-protein interactions have different input-output characteristics and usually require more complex modification of their natural substrates. Protein switches have been been created by inserting allosteric domains into existing enzymes, replacing protein domains randomly followed by directed evolution, or using computational chemistry to model receptor binding sites and selectively mutate key residues to confer altered activity. Multiple examples exist of modifying proteins/enzymes to significantly alter their ligand/substrate response and serve as the initial input device for extended metabolic cascades.[39, 40, 41, 42, 43] Designing and building protein interaction devices carry several caveats. The proteins must be well characterized to allow the designer to determine where deletions, additions, or replacements of key domains should or can occur so as not to negate the protein's 3–D structure that plays a critical role in protein interactions. Connecting protein devices also poses more difficulty than connecting translation/transcriptional devices. Protein devices must be well matched and one must validate that information transfer between devices is predictable and reliable. The benefits of protein interaction devices are that output is very fast (output in sub-second timescale), signals can be readily amplified by other cell reactions (that is, kinases and enzymes), and proteins can mediate repeated interactions while consuming nominal cellular resources.

Modules

A *module* is defined as a distinct set of individual devices with well-defined interconnections and functions that can perform a more complex task than a device. In the natural cell, modules are defined in the form of pathways, such as a metabolic or signal transduction pathway. Synthetic biology seeks to understand and establish the rules of composition whereby the function of a whole pathway can be derived from the additive function of each component part, which can be engineered as a device. Constructing modules from wild-type devices found in naturally occurring systems is rarely successful, since evolution has already selected wild-type devices to perform optimally in their natural context. Modeling and rational mutation of devices have proven effective in altering wild-type devices to properly interact with each other when combined into a module function, as long as the properties of the individual devices are well known. In the absence of well-known characteristics of a device, where rational mutation of selected sites would be expected to yield improved properties, directed evolution has proven useful in optimizing device and module functionality. The onus of directed evolution is having some a priori idea of which features of the system need to evolve and being able to apply selective pressure toward those criteria.

Synthetic transcriptional regulatory networks are the best characterized modules created to date and include cascade, feed-forward, and feedback motifs. Cascade modules generally generate a steady-state output that is monotonic to its input, though physical variations have shown their ability to control temporal sequencing of gene expression and the ability to attenuate gene expression noise. Two feed-forward transcription modules have been constructed and allow for a transient non-monotonic response to a step-like stimulus. Incoherent feed-forward modules accelerate response to the step-up in stimulus, but not the step-down. Coherent feed-forward modules delay response to the step-up in stimulus, but not the step-down. These constructs have yielded modules that can function as persistence detectors, delay elements, and pulse generators, which conceivably could be used to construct parallel circuits to carry information at different rates.[44, 45, 46] Applying regulatory feedback to a module can provide state and memory functions. Positive feedback produces bistability, while negative feedback can yield oscillation. Genetic "toggle" switches and various oscillators have been produced using regulatory feedback designs.[47, 48, 49]

Synthetic protein signaling modules have focused on modifying and assembling signal transduction devices. Synthetic metabolic networks use transcriptional and translational control elements to regulate the expression of enzymes that synthesize or break down metabolites. Here, metabolite concentration serves as an input for other devices. In the most publicized demonstration of this approach, Martin et al. transplanted the entire mevalonate isoprenoid pathway for synthesizing isopentyl pyrophosphate from *S. cerevisiae* into *E. coli*, which, when coupled with an inserted synthetic amorpha-4, 11-diene synthase produced the precursor to the antimalarial drug artemisinin. Artificial circuits have also been demonstrated where metabolic pathways have been connected to regulatory proteins and transcriptional control elements.[50]

Two major metrics of success in constructing functional devices and modules in synthetic biology are predictability and reliability, where the system performs the function it was designed to and it performs that function at the expected level of fidelity each time. It is prudent to consider that functional modules will typically reside in a cell environment that is subject to a range of biological uncertainties and cell "noise" stemming from other cell functions/

processes. Synthetic biology does not demand that cells behave perfectly, only that a significant number of the cell population performs the desired task. Design and fabrication methods that account for this uncertainty will greatly facilitate the on-demand engineering, versatility, and robustness of biological devices and modules.

Synthetic Organelles

Nature provides exquisitely refined and accurate capabilities that are thus far impossible to recreate in abiotic structures, two of which are ultra-high-resolution recognition and the assembly of multicomponent enzymatic complexes. Such processes can be mimicked using abiotic structures, creating a generic, bioinspired enzymatic technique that enables the template-directed synthesis of a biochemical pathway within a compartmentalized abiotic multicomponent macromolecule.

The targeted assembly of proteins is an essential aspect of life. Within a cell, protein assembly is often directed through the compartmentalization of individual proteins into specific organelles or membranes. Compartmentalization provides a method for concentrating and therefore enhancing the assembly of functional macromolecular structures, such as the multicomponent light harvesting structures found in chloroplasts or the transcriptional complexes located in the nucleus. In almost all cases, specific signal peptides direct this targeted compartmentalization within a cell.

Signal peptides are generally short, 3 to 60 amino acids in length. Each class of signal peptide interacts with an organelle-specific protein to facilitate its import and/or retention. Using this process as a model, engineered signal peptides could be used for the bioinspired compartmentalization of proteins within molecularly imprinted polymers. Molecular imprinting is a technique that creates synthetic materials containing highly specific receptor sites having an affinity for a target molecule. Three-dimensional cavities are created within a polymeric matrix complementary to the size, shape, and functional group orientation of the target molecule. The size and shape of the cavity allow the target molecule or similar molecules to occupy the cavity space, while the functional group orientation within the cavity will preferentially bind in specific locations to only the target molecule and not to similar molecules. Molecular imprinted polymers (MIPs) can mimic the recognition and binding capabilities of natural biomolecules such as antibodies and enzymes. MIPs have several advantages over natural biomolecules, such as modular synthesis that can provide functional hydrogels in high yields and selectivities.

The MIP hydrogel can be treated like a cellular organelle that could be engineered to bind specific signal peptide sequences. Thus, in order to compartmentalize any set of proteins into the MIPs, one would need to attach a specific signal sequence to them in order to mimic enzymatic chemistry inside the imprinted hydrogel. This method, if successful, would have the benefit of being universal—that is, any set of proteins could be assembled as long as they contain the right signal peptide. Thus, the potential to do multicomponent enzymatic chemistry directly in the hydrogel would be considerably higher. Additionally, MIPs could be imprinted with multiple tags, perhaps even patterned in some way to create a desired network. By arranging the spatial configuration of enzymes in an MIP cavity, one may be able to alter product specificity, much like reengineering the active site in directed evolution studies. To accomplish this, it is necessary to make signal peptides of specific geometries that, when imprinted into a polymer hydrogel, can later be recognized for assembly of a multiprotein complex.

Materials for Membranes and Support

One of the challenges will be to select and develop appropriate technologies for the environmental interface, support structure, and/or membranes within and through which the functional components work. Because the functional components are the heart of the system, the specifics of these supports need to be tailored to the structure and function of the components, yet the successful interactions of the components are likely to be at least partially dependent on the membranes and support materials. Thus, it is essential to take a holistic approach and address promising membranes and support technologies that could support any of a number of likely scenarios. This will provide the agility to construct an encapsulation and support system appropriate to the functional technologies as they develop and mature.

During the development of the functional components, there will be the need for testing and integrating well before technologies are fully downselected. In fact, these tests and integration experiments will be an integral part of the downselect process. Using an agile approach to materials technology will facilitate these evaluation steps and maximize the potential for each technology.

Biotic Sensors

Historically, one of the failing points of biotic sensors was the environmental needs associated with maintaining cell viability. In general, biochemical activity is carried out in aqueous solutions that make it incompatible with long-term unattended operation in real-life scenarios. Abiotic functional components of ANTS may share some of these issues, depending on the fidelity with which the abiotic synthesis mimics natural biology. The matrix materials will need to address the range of environmental dependence and consider the possibility of hybrid matrices to address different levels of dependence within one sensor. This poses several scenarios:

- The functional components are subject to the same or equivalent environmental requirements as their biological homologues and require the same or equivalent through-membrane and intracellular mobility.
- The functional components are not dependent on environment but do require through-membrane and intracellular mobility.
- The functional components neither depend on the environment nor require through-membrane mobility, but they do require intracellular mobility.
- The functional components depend on neither the environment nor intracellular mobility, but they do require through-membrane mobility.

Historically, biotic sensors were based on living cells or tissues, and the morphology of these cells and tissues was fixed by nature. In contrast, the ANTS morphologies will be driven by functional aspects of how the evolved system is to be deployed, operate (functional components), and communicate. The morphology issue will have some bearing on appropriate matrix material, while again the matrix material requirements of the functional components may partially dictate the available morphologies and deployment/communication options. This situation will also benefit from the holistic approach.

Recognition, transduction, and associated devices will be evolving concurrent with the evolution of an appropriate support matrix. The purpose of the support matrix research is two-fold; first, to provide appropriate materials in which to study evolving synthetic biology products and, second, to work toward hardened materials that include both functionalized and support aspects, tailored to the abiotic cell product.

Research will begin by supporting BRE research at the current state of the art. Initial goals will be to rapidly provide matrices in which early BREs and transduction devices can be studied. For these purposes, we will build on existing technologies associated with channel-based sensing, starting with lipid bilayers and suspended liposomes. Stabilization of these "conventional" membranes using nanoparticle technology will be investigated for three purposes: to extend the viability of these membranes to allow temporally extended experiments on device function; to elucidate fundamental facts concerning the distinction between lipid-suspended particles and lipid-wetted agglomerates (that is, at what particle loading does a liposome lose its properties vis-à-vis BRE functionalization and how do these limits affect device function); and to determine whether nanoparticles can function as rafts in a lipid bilayer system. The results of this stage of research are unlikely to result in a hardened matrix for a final product; they will facilitate synthetic biology work toward the required devices and may provide essential parts of an inhomogeneous or layered "cell."

Beyond the fragile lipid-based matrices are supported membranes, hydrogels, and chitosan products. All of these approaches have been used in conjunction with synthetic biology or related research and all provide some stability advantage over simple lipid membranes or liposomes. Hydrogels and supported membranes are likely to provide the "breadboards" for device development and integration, while the unique properties of chitosan make it a likely candidate both for use in evolving breadboard systems and as a component in hardened cells.

Moving toward Xcells hardened against the environment and suitable for incorporation into myriad military systems, membrane and matrix support directions will need to be tailored to evolving functional components. At this stage, the rapid advances in directed and self-assembly will be investigated as manufacturing tools, while materials including porous inorganics, polymerosomes shells, chitosan, cellulosic materials, and aerogels will be considered for their advantageous properties. Structure analysis is expected to be a central part of this research, along with functionalization and device interface. Hybridization of materials to provide an appropriate mix of properties is likely to be required where assembly technology alone cannot provide the necessary functions in homogeneous materials.

Chitosan

Individually, biotechnology and microelectronics revolutionized our ability to detect and communicate threats. Biotechnology provides the tailored biological components to recognize threats to specific physiological targets. Microelectronics provides autonomous devices that are cheap, robust, and small and that offer data and signal processing capabilities for near-instantaneous communication over vast distances. Coupling biotechnology and microelectronics will allow unprecedented capabilities for recognizing and responding to a plethora of threats, but this coupling will require the seamless integration of biological components into microelectronic devices.

Biology and microelectronics communicate through markedly different signaling modes. Electronic devices signal through the flow of electrons or the transmission of light, while biology signals with ions (for example, Na^+, K^+, and Ca^{2+}) and molecules (such as hormones and neurotransmitters). Bridging this communication gap is essential to fusing the individual strengths of biology and electronics.

Chitosan possesses unique properties that enable it to interconnect biology and devices.[51] Chitosan can be "connected" to device addresses because the pH-responsive film-forming properties of this aminopolysaccharide allow electrodeposition in response to localized electrical signals.[52,53] Chitosan's electrodeposition is spatially[54] and temporally[55] selective, while the electrodeposited films are stable in the absence of an applied voltage. Chitosan can be "connected" to bio-components (for example, nucleic acids,[56] proteins,[57,58,59] and virus particles[60]) in two different ways—chitosan's nucleophilicity facilitates covalent conjugation,[61] while the electrodeposited film can entrap nanoscale components (such as proteins).[62] Thus, chitosan allows device-imposed electrical signals to guide the interconnection of biological components to devices.[63]

In addition to enabling biodevice interconnection, chitosan possesses properties that allow biological signals to be transduced into signals readily processed by common devices. Specifically, chitosan films are permeable to ions and small molecules allowing signaling through electrical modes.[64,65] Further, chitosan films are transparent, which facilitates signaling through optical modes (for example, through integrated waveguides).[66] Finally, chitosan films are stimuli-responsive, enabling mechanotransduction.[67] Thus, chitosan can icrosyste environmental stimuli into multiple signaling modes.

Power

Biological systems operate on energy transductions in the mitochondria. That is, the energy powering the cells is derived biochemically. In the case of an abiotic system performing selected cell-like functions and modeled almost exclusively on biotic cell architecture, the biochemical approach will provide the most directly usable form of energy. It will be necessary to first develop complete understanding of the energy transfer to, and energy use by, the cell functions being duplicated in the abiotic system. In order to duplicate the energy-producing biochemical functions in the abiotic cell, one must determine energy requirements of the abiotic system, develop chemical reactions that are triggered by recognition events, and couple that energy to the signaling event. These small-scale energy requirements differ markedly from traditional power and energy issues faced by macroscopic and even mesoscopic equipment.

On the small scales associated with cell function, whether artificial or biotic, it is possible to consider energy on the basis of thermodynamics of individual molecular interactions. In the simplest of terms, some will require energy and some will release energy. In a living eukaryotic cell, ATP is synthesized from ADP and P_i and oxidative phosphorylation involving input energy derived from dissipation of a proton gradient or electrical potential. However, this is not the only source of ATP generation. Phosphorylation of ADP to ATP is also observed coupled to oxidation via thioester intermediate formation and dehydration reactions resulting in generation of enol phosphates that have high phosphoryl transfer potential as typified by the two ATP generating steps in Glycolysis. It is noteworthy that the above two reactions represent the sole source of de novo synthesis of ATP in the red blood cell. Thus, engineering of circuits involving the respective glycolytic enzymes which are cytosolic (soluble) within a compartment could serve as source of ATP generation for other ATP requiring circuits as long as a few key metabolites (3-Phosphoglyceraldehyde, NAD^+, P_i, and pyruvate) were provided the system. A lactate dehydrogenase "recycling" circuit can even be included to recycle reduced NAD^+ (that is, NADH) as well as pH control. Therefore, acellular "substrate level phosphorylation" circuits are well within reach due to the soluble nature of this system and its not requiring elaborate membrane bound oxidative phosphorylation functions (that is, electron transport). Thus, in an abiotic system adopting cell-like functions, the energy required to drive these functions can be engineered to include such functions and capabilities. Substrate phosphorylation alluded to above constitutes a recognition event resulting in a thermodynamically spontaneous reaction that can supply enough power for subsequent signal transduction events.

Additionally, an external energy source can be coupled to a catalytic wheel and provide the activation energy for ion pumping.[68] Understanding biological energy balances and duplicating them in abiotic systems is a promising approach to powering the cellular functions mimicked in an abiotic cell. The goal in the case of an abiotic cell is to use elegant thermodynamic design as well as partitioning of the reactions involved in recognition, signaling and transduction, taking advantage of catalytic enzyme activity, substrate binding, and conformational events.

Although ideally the cells would "find," convert, and use energy on an as-needed basis, in the practical use of power, storage requirements need to be considered. Nanocapacitors[69] have been demonstrated to be practical approaches to this problem and have in fact been used on the meso scale to power telemetry functions. In separate investigations, carbon nanotubes have been used to enhance the efficiency of ultracapacitors. The possibility exists through combined technologies to develop a nano-enhanced nano-ultracapacitor to store and supply energy converted from environmental solar or thermal energy.

In a living system, interactions between the organism and the environment typically involve cooperative functioning of many cells. In the special case of single-cell organisms, these interactions are either limited by or dependent on the multiplicity of organisms—that is, bacteria respond to their environment by multiplying and affect their surroundings by the cooperative effects of many cells.

In the case of abiotic unicellular systems, it is likely that the energy requirements for signal amplification and communication to macroscopic indicators or telemetry will not be served by simulated biochemical energy. It is also an undeniable fact that biological systems, efficient though they are, do require a certain amount of "input" fuel, and it may not be possible to load an abiotic cell with sufficient fuel to support functioning throughout its lifetime. For these situations, alternate energy sources must be considered.

With the expanding volume of research in nanotechnology and microelectromechanical systems, there have been significant strides in the areas of nanocapacitors, nanowires, nontraditional solar photovoltaics, thermal photovoltaics, and biomimetic solar energy conversion. These, combined with conductive polymer technology and recent successes in the area of molecular switching, provide ample technology for powering and controlling signaling and telemetry operations in an abiotic cell. For these requirements of tailored application of nontraditional but universally applicable power sources, we will rely on the research and technology advances of existing energy and power programs.

The level of amplification of the desired signal and the process of telemetry are not native functions of typical biotic cells, and it will be necessary to link the cell signal to a non-native cell function. This may be achieved by the use of nanowires and nanoswitches. A use of nanowires to detect events in single cells has been demonstrated for the case of live neuron activity and photon activity.[70, 71] Biological molecules have been identified that operate as molecular switches, and the switching mechanisms have been demonstrated up to and including the complexity of a 4-way switch.[72]

Harnessing of ubiquitous environmental energy to power amplification and telemetry will be essential to a self-maintaining system. The two most amenable sources of power for these applications are solar and thermal, and the energy conversion processes of photosynthesis can be mimicked abiotically.[73] Like the mitochondrial-based oxidative phosphorylation systems, photosynthetic systems also utilize a proton gradient in oxygenic photosynthesis. There are numerous photosynthetic bacterial systems from which abiotic circuit components can be modeled. Solar photovoltaics based on nanoparticles have been demonstrated at the macro scale and in fact are mature to the level of commercialization. Engineering these materials to a smaller scale has not been reported, although this is likely an artifact of the phenomenon that the challenge in solar energy is to scale up rather than down. Likewise, novel thermal photovoltaic systems are under investigation on the nano scale.

Signal Transduction

A transducer is a device that converts information from one domain to the other. Any analytical measurement is a series of conversions from nonelectrical and electrical domains. Understanding the data domains involved in a particular sensing operation can help in understanding its operation, applications, limitations, and adaptability. Information through a transduction event could be encoded as the magnitude of an electrical signal, continuous in both amplitude and time (analog); signal fluctuations around a threshold; or as a simple two-level system (ON or OFF). Biological recognition elements can translate information about target analyte concentration directly into a mechanical signal (for example, shape change[74, 75] or change in pore size),[76] optical signal, another molecule or ion more easily sensed, or directly into an electrical signal. Research is needed in these areas to determine the tradeoffs of each approach and to determine where technologies can be combined for greater effect.

Conclusion

It is clear that a biomimicry approach in which one designs an artificial organism that possesses all of the key receptors and metabolic machinery required to detect and respond to toxicants in the external environment would address the needs of biological and chemical detection for military, medical, environmental, and industrial process control applications. It is also clear that advances in receptor biochemistry, synthetic and systems biology, metabolic engineering, bioinformatics, protein-protein and protein-surface interactions, and micro-manufacturing technologies have reached a point where it would be possible to design a completely synthetic and intelligent sense and respond system.

NOTES

1 Philip Morrison, "The Silicon Gourmet," *Scientific American* (April 1997), 113–115.

2 W. Thomsen et al., "Functional Assays for Screening GPCR Targets," *Current Opinion in Biotechnology* 16, no. 6 (2005), 655–665.

3 The National Science Foundation has established a Center for Biosurfaces at the State University of New York–Buffalo.

4 W.J. Cooper and M.L. Waters, "Molecular Recognition with Designed Peptides and Proteins," *Current Opinion in Chemical Biology* 9, no. 6 (2005), 627–631.

5 P. Hollinger and P.J. Hudson, "Engineered Antibody Fragments and the Rise of Single Domains," *Nature Biotechnology* 23, no. 9 (2005), 1126–1136.

6 J.Z. Hilt and M.E. Byrne, "Configurational Biomimesis in Drug Delivery: Molecular Imprinting of Biologically Significant Molecules," *Advanced Drug Delivery Review* 56, no. 11 (2004), 1599–1620.

7 B.R. Conway and K.T. Demarest, "The Use of Biosensors to Study GPCR Function: Applications for High-Content Screening," *Receptors Channels* 8, no. 5–6 (2002), 331–341.

8 T.H. Park and M.L. Shuler, "Integration of Cell Culture and Microfabrication Technology," *Biotechnology Progress* 19, no. 2 (2003), 243–253.

9 S.A. Benner, ed., *Redesigning the Molecules of Life* (Heidelberg: Springer, 1988), 115–175.

10 W.W. Gibbs, "Synthetic Life," *Scientific American* 290 (2004), 74–81.

11 R. Breslow, "Biomimetic Chemistry," *Chemical Society Reviews* 1 (1972), 553–580.

12 W.F. Degrado et al., "Design, Synthesis, and Characterization of a Cyto-Toxic Peptide with Melittin-like Activity," *Journal of the American Chemical Society* 103 (1981), 679–681.

13 A.J. Kennan et al., "A De Novo Designed Peptide Ligase: A Mechanistic Investigation," *Journal of the American Chemical Society* 123 (2001), 1797–1803.

14 K. Johnsson et al., "Synthesis, Structure and Activity of Artificial, Rationally Designed Catalytic Polypeptides," *Nature* 365 (1993), 530–532.

15 S.H. Park, A. Zarrinpar, and W.A. Lim, "Rewiring MAP Kinase Pathways Using Alternative Scaffold Assembly Mechanisms," *Science* 299 (2003), 1061–1064.

16 A.D. Griffiths and D.S. Tawfik, "Miniaturising the Laboratory in Emulsion Droplets," *Trends in Biotechnology* 24 (2006), 395–402.

17 O.J. Miller et al., "Directed Evolution by In Vitro Compartmentalization," *Nature Methods* 3 (2006), 561–570.

18 J. Chelliserrykattil and A.D. Ellington, "Evolution of a T7 RNA Polymerase Variant that Transcribes 2'-O-methyl RNA," *Nature Biotechnology* (2004), 1155–1160.

19 E. Mastrobattista et al., "High-throughput Screening of Enzyme Libraries: In Vitro Evolution of a Beta-galactosidase by Fluorescence-activated Sorting of Double Emulsions," *Chemistry and Biology* 12, no. 12 (2005), 1291–1300.

20 T. Ellinger et al., "In Vitro Evolution of Molecular Cooperation in CATCH, a Cooperatively Coupled Amplification System," *Chemistry and Biology* 4 (1998), 729–741.

21 J.R. Jacobsen and C. Khosla, "New Directions in Metabolic Engineering," *Current Opinion in Chemical Biology* 2 (1998), 133–137.

22 V.J. Martin et al., "Engineering a Mevalonate Pathway in Escherichia coli for Production of Terpenoids," *Nature Biotechnology* 21 (2003), 796–802.

23 T. Fukui, H. Abe, and Y. Doi, "Engineering of *Ralstonia eutropha* for Production of Poly (3-hydroxybutyrate-co-3-hydroxyhexanoate) from Fructose and Solid-state Properties of the Copolymer," *Biomacromolecules* 3 (2002), 618–624.

24 W.A. Lim, "The Modular Logic of Signaling Proteins: Building Allosteric Switches from Simple Binding Domains," *Current Opinion in Structural Biology* 12 (2002), 61–68.

25 J. Hasty, D. McMillen, and J.J. Collins, "Engineered Gene Circuits," *Nature* 420 (2004), 224–230.

26 S.H. Park, A. Zarrinpar, and W.A. Lim, "Rewiring MAP Kinase Pathways Using Alternative Scaffold Assembly Mechanisms," *Science* 299 (2003), 1061–1064.

27 K.E. Prehoda et al., "Integration of Multiple Signals through Cooperative Regulation of the N-WASP-Arp2/3 Complex," *Science* 290 (2000), 801–806.

28 J.E. Dueber et al., "Rewiring Cell Signaling: The Logic and Plasticity of Eukaryotic Protein Circuitry," *Current Opinion in Structural Biology* 14 (2004), 690–699.

29 M.B. Elowitz and S. Leibler, "A Synthetic Oscillatory Network of Transcriptional Regulators," *Nature* 403 (2000), 335–338.

30 M.N. Stojanovic and D. Stefanovic, "A Deoxyribozymebased Molecular Automaton," *Nature Biotechnology* 21 (2003), 1069–1074.

31 J.C. March and W.E. Bentley, "Quorum Sensing and Bacterial Cross-talk in Biotechnology," *Current Opinion in Biotechnology* 15, no. 5 (2004), 495–502.

32 H. Tjalsma et al., "Engineering of Quorum-sensing Systems for Improved Production of Alkaline Protease by Bacillus Subtilis," *Journal of Applied Microbiology* 96, no. 3 (2004), 569–578.

33 T. Bulter et al., "Design of Artificial Cell-Cell Communication Using Gene and Metabolic Networks," *Proceedings of the National Academy of Sciences of the United States of America* 101, no. 8 (2004), 2299–3304.

34 P. Neddermann et al., "A Novel, Inducible, Eukaryotic Gene Expression System Based on the Quorum-sensing Transcription Factor TraR," *EMBO Reports* 4, no. 2 (2003), 159–165. Erratum in *EMBO Reports* 4, no. 4 (2003), 439.

35 W. Weber et al., "Streptomyces-derived Quorum-Sensing Systems Engineered for Adjustable Transgene Expression in Mammalian Cells and Mice," *Nucleic Acids Research* 31, no. 14 (2003), e71.

36 U. Baron et al., "Tetracycline-controlled Transcription in Eukaryotes: Novel Transactivators with Graded Transactivation Potential," *Nucleic Acids Research* 25 (1997), 2723–2729.

37 R. Lutz and H. Bujard, "Independent and Tight Regulation of Transcriptional Units in Escherichia Coli via the LacR/O, the TetR/ O and AraC/I1-I2 Regulatory Elements," *Nucleic Acids Research* 25 (1997), 1203–1210.

38 R. Lutz et al., "Dissecting the Functional Program of Escherichia Coli Promoters: The Combined Mode of Action of Lac Repressor and AraC Activator," *Nucleic Acids Research* 29 (2001), 3873–3881.

39 J.E. Dueber et al., "Reprogramming Control of an Allosteric Signaling Switch through Modular Recombination," *Science* 301 (2003), 1904–1908.

40 G. Guntas and M. Ostermeier, "Creation of an Allosteric Enzyme by Domain Insertion," *Journal of Molecular Biology* 336 (2004), 263–273.

41 G. Guntas et al., "Directed Evolution of Protein Switches and their Application to the Creation of Ligand-binding Proteins," *Proceedings of the National Academy of Sciences of the United States of America* 102 (2005), 11224–11229.

42 M.A. Dwyer and H.W. Hellinga, "Periplasmic Binding Proteins: A Versatile Superfamily for Protein Engineering," *Current Opinion in Structural Biology* 14 (2004), 495–504.

43 L.L. Looger et al., "Computational Design of Receptor and Sensor Proteins with Novel Functions," *Nature* 423 (2003), 185–190.

44 S. Mangan and U. Alon, "Structure and Function of the Feedforward Loop Network Motif," *Proceedings of the National Academy of Sciences of the United States of America* 100 (2003), 11980–11985.

45 S. Mangan, A. Zaslaver, and U. Alon, "The Coherent Feedforward Loop Serves as a Sign-sensitive Delay Element in Transcription Networks," *Journal of Molecular Biology* 334 (2003), 197–204.

46 S. Basu, "Spatiotemporal Control of Gene Expression with Pulse-generating Networks," *Proceedings of the National Academy of Sciences of the United States of America* 101 (2004), 6355–6360.

47 T.S. Gardner, C.R. Cantor, and J.J. Collins, "Construction of a Genetic Toggle Switch in Escherichia Coli," *Nature* 403 (2000), 339–342.

48 B.P. Kramer, C. Fischer, and M. Fussenegger, "BioLogic Gates Enable Logical Transcription Control in Mammalian Cell," *Biotechnology and Bioengineering* 87 (2004), 478–484.

49 M.B. Elowitz and S. Leibler, "A Synthetic Oscillatory Network of Transcriptional Regulators," *Nature* 403 (2000), 335–338.

50 Bulter et al., 2299–2304.

51 H. Yi et al., "Biofabrication with Chitosan," *Biomacromolecules* 6, no. 6 (2005), 2881–2894.

52 L.Q. Wu et al., "Voltage-Dependent Assembly of the Polysaccharide Chitosan onto an Electrode Surface," *Langmuir* 18, no. 22 (2002), 8620–8625.

53 R. Fernandes et al., "Electrochemically Induced Deposition of a Polysaccharide Hydrogel onto a Patterned Surface," *Langmuir* 19, no. 10 (2003), 4058–4062.

54 L.Q. Wu et al., "Spatially-selective Deposition of a Reactive Polysaccharide Layer onto a Patterned Template," *Langmuir* 19 (2003), 519–524.

55 H. Yi et al., "Signal-directed Sequential Assembly of Biomolecules on Patterned Surfaces," *Langmuir* 21, no. 6 (2005), 2104–2107.

56 H. Yi et al., "A Robust Technique for Assembly of Nucleic Acid Hybridization Chips Based on Electrochemically Templated Chitosan," *Analytical Chemistry* 76, no. 2 (2004), 365–372.

57 R. Vazquez-Duhalt et al., "Enzyme Conjugation to the Polysaccharide Chitosan: Smart Biocatalysts and Biocatalytic Hydrogels," *Bioconjugate Chemistry* 12, no. 2 (2001), 301–306.

58 T.H. Chen et al., "In Vitro Protein-polysaccharide Conjugation: Tyrosinase-catalyzed Conjugation of Gelatin and Chitosan," *Biopolymers* 64, no. 6 (2002), 292–302.

59 T. Chen et al., "Nature-Inspired Creation of Protein-Polysaccharide Conjugate and its Subsequent Assembly onto a Patterned Surface," *Langmuir* 19 (2003), 9382–9386.

60 H. Yi et al., "Patterned Assembly of Genetically Modified Viral Nanotemplates via Nucleic Acid Hybridization," *Nano Letters* 5, no. 10 (2005), 1931–1936.

61 A.T. Lewandowski et al., "Tyrosine-based 'Activatable Pro-Tag': Enzyme-catalyzed Protein Capture and Release," *Biotechnology and Bioengineering* 93, no. 6 (2006), 1207–1215.

62 L.Q. Wu et al., "Chitosan-mediated and Spatially Selective Electrodeposition of Nanoscale Particles," *Langmuir* 21, no. 8 (2005), 3641–3646.

63 J.J. Park et al., "Chitosan-mediated In Situ Biomolecule Assembly in Completely Packaged Microfluidic Devices," *Lab on a Chip* 6 (2006), 1315–1321.

64 L.Q. Wu et al., "Biomimetic Pattern Transfer," *Advanced Functional Materials* 15 (2005), 189–195.

65 L.Q.Wu et al., "Mimicking Biological Phenol Reaction Cascades to Confer Mechanical Function," *Advanced Functional Materials* 16 (2006), 1967–1974.

66 M.A. Powers et al., "A Fabrication Platform for Electrically Mediated Optically Active Biofunctionalized Sites in BioMEMS," *Lab on a Chip* 5 (2005), 583–586.

67 S.T. Koev et al., "Mechano-Transduction of DNA Hybridization and Dopamine Oxidation through Electrodeposited Chitosan Network," *Lab on a Chip* 7 (2007), 103–111.

68 T.Y Tsong and C.H. Chang, "Catalytic Wheel, Brownian Motor, and Biological Energy Transduction," *Association of Asian-Pacific Physical Sciences Bulletin* 13, no. 2 (2005), 12–18.

69 K.B. Shelimov, D.N. Davydov, and M. Moskovits, "Template Grown High-density Nanocapacitor Arrays," *Applied Physics Letters* 77, no. 11 (2000), 1722–1724.

70 J. Wang et al., "Highly Polarized Photoluminescence and Photodetection from Single Indium Phosphide Nanowires," *Science* 293 (2001), 1455–1457.

71 F. Patolsky et al., "Detection, Stimulation, and Inhibition of Neuronal Signals with High-density Nanowire Transistor Arrays," *Science* 313 (2006), 1100–1104.

72 V. Iancu and S.-W. Hla, "Realization of a Four-step Molecular Switch in Scanning Tunneling Microscope Manipulation of Single Chlorophyll-a Molecules," *Proceeding of the National Academy of Sciences of the United States of America* 103, no. 37 (2006), 13718–13721.

73 D.A. LaVan and J.N. Cha, "Approaches for Biological and Biomimetic Energy Conversion," *Proceedings of the National Academy of Sciences of the United States of America* 103, no. 14 (2006), 5251–5255.

74 M.L. Clarke and Z. Chen, "Polymer Surface Reorientation after Protein Adsorption," *Langmuir* 22, no. 21 (2006), 8627–8630.

75 H. Frauenfelder et al., "Protein Folding is Slaved to Solvent Motions," *Proceedings of the National Academy of Sciences of the United States of America* 103, no. 42 (2006), 15469–15472.

76 M. Carrillo-Tripp et al., "Ion Hydration in Nanopores and the Molecular Basis of Selectivity," *Biophysical Chemistry* 124, no. 3 (2006), 243–250.

chapter 6

BIOSENSORS and BIOELECTRONICS

Dimitra Stratis-Cullum and James Sumner

The focus of this chapter will be on biosensors and bioelectronics and the relevance to Department of Defense (DOD) applications. The overwhelming majority of the current biosensors market is made up of glucose monitors that are used by diabetics for monitoring blood sugar, but medical applications will not be covered here. Since understanding the biorecognition and transduction processes is critical to understanding biosensors, extensive discussion of these topics is included. Bioelectronics is an exciting new field that brings together multiple disciplines of science including biology, chemistry, physics, engineering, and material science. New and emerging bioelectronic research areas that have the potential to impact future DOD operations and applications are also introduced in this chapter.

Biosensors

A *biosensor* is a self-contained integrated device that is capable of providing specific quantitative or semi-quantitative analytical information using a biological recognition element (biochemical receptor) that is retained in direct spatial contact with a transduction element. However, a common working definition is that a biosensor is an analytical device that uses biological macromolecules to recognize an analyte and subsequently activate a signal that is detected with a transducer. Some authorities subdivide biosensors into affinity biosensors and catalytic biosensors based upon the activity of the biorecognition element. Thus, affinity biosensors have as their fundamental property the recognition (binding) of the analyte by the biorecognition element (for example, antibody-antigen), whereas catalytic biosensors have as their biorecognition element proteins (or microorganisms) that not only bind the analyte but also catalyze a reaction involving the analyte to produce a product (for example, glucose biosensors).

Ideal Biosensor Characteristics

In many cases, biosensors are single use, although for many applications regeneration is desirable. The ideal biosensor is frequently described as being robust, selective, reproducible, and sensitive with a large dynamic range, an ideal that is seldom attained. Bioanalytical systems such as immunoassay may well use the same elements (that is, bioreceptors, transduction devices) but require additional processing steps such as reagent addition and washings, and usually involve reading the completed assay on some separate piece of instrumentation. These are to be distinguished from biosensors. Thus, a very large volume of related literature exists on the use of immunoassays, nucleic acid–based assays, and so forth for detection of various analytes where the techniques described in such assays may well have applications to the development of actual biosensor devices.

Assay Formats

In general, a variety of assays can be employed with the many different biorecognition and transduction elements described below. Assay formats can be either direct or indirect. In the direct methods, analyte binding to the biorecognition element is detected directly by such techniques as mass changes, changes in refractive index or impedance, pH, and so forth. For indirect techniques, an additional reaction (other than the biorecognition binding event) has to occur in order to detect the binding of analyte and bioreceptor. Another type of assay format is a competitive but indirect strategy where the analyte and a competitor molecule (labeled in some fashion) compete for limited numbers of biorecognition binding sites. There is also a noncompetitive indirect assay format in which a second biorecognition element with a label is added to the analyte sample, so that signal is only generated when the analyte binds to the immobilized biorecognition element and the second biorecognition element also binds to the analyte. For the indirect type of assay, the label can be optical-, electrochemical-, or mass-related. In the case of direct assays and noncompetitive indirect assays, the signal is directly proportional to the analyte concentration, while in the case of competitive indirect assays, the signal is inversely proportional to analyte concentration. Frequently, biosensor assay formats involve some form of amplification, the rationale for this being increased sensitivity of the assay. Either the analyte molecules themselves may be amplified (for example, polymerase chain reaction [PCR] of DNA sequences) or the biorecognition event may be amplified (for example, ELISA immunoassays).

Applications

Bacterial Biosensors for Environmental Monitoring

Several types of biosensors can be used for environmental monitoring applications, but whole cell bioreporters that respond to certain pollutants/ toxicological conditions are beginning to find favor in environmental sensing. Applications range from detection of contaminants, measurement of pollutant toxicity, monitoring of genetically engineered bacteria released into the environment, and even uses for detection of stresses such as ionizing radiation and oxidative damage. In general, such whole cell sensors should not be called biosensors unless the cells are integrated somehow into a standalone sensing/transduction/readout system. Some whole cell sensors use genetically engineered bacteria into which a *lux* gene construct (coding for an active luciferase) are under the control of an inducible promoter. The general sensing strategy is to put the luciferase construct under control of a promoter that recognizes the analyte of interest, although in some cases, the bioreporter organism can be engineered to respond to many analytes (broad specificity) by using a heat-shock promoter.

For example, the seminal work in this arena described a naphthalene biosensor organism (genetically engineered *Pseudomonas fluorescens*) that detected the bioavailability of naphthalene (serving as a surrogate for polyaromatic hydrocarbons) and salicylate in contaminated soil by induction of the *lux* reporter, and subsequent bioluminescence as a result of expression of the reporter gene. The naphthalene biosensor described above is an example of a catabolic bioreporter. Catabolic bioreporters produce luminescence only in the presence of a specific contaminant, since the *lux* genes are linked to a degradation pathway promoter.

Metabolic bioreporters, on the other hand, have reporter genes downstream of a strong constitutive promoter, and thus luminescence is directly affected by toxic agents that interfere with metabolism. These concepts have been subsequently used to develop a variety of bioreporter organisms. Other investigators have used genetically engineered bacteria in which the green fluorescent protein, gfp (or derivatives thereof) from the jellyfish *Aequorea Victoria*, functions as a bioreporter in the same manner as the *lux* gene. Some arguments have been advanced that the expression of bioluminescence is a more direct bioreporter than gfp expression, since the expression of gfp is only detected upon excitation with the correct wavelength of light. However, in practical terms, bioluminescence must be quantified by some sort of photodetector/luminometer, so the difference between that and a fluorescence excitation/emission detection system seems minimal.

Beta-galactosidase enzymes have also been used as biomarkers. Others have used naturally luminescent bacteria (that is, *Vibrio fischeri*) as bioreporter organisms, since toxic compounds can disturb metabolic processes, thus causing a reduction in their bioluminescence.

Microbial biosensors have been developed to assay biological oxygen demand (BOD), a value related to total content of organic materials in wastewater. These sensors often measure BOD by evaluating the rate at which microbial cells deplete oxygen. It is not possible to provide a comprehensive list of whole cell sensor applications in this chapter. Very seldom are these bioreporter organisms integrated into a complete sensing package where a microculture environment, integrated luminometer or fluorometer, and light tight-enclosure can produce a true biosensor device.

Food Biosensing

Food biosensing applications range from monitoring the quality of the food to food safety and security applications. The assessment of food quality can be subjective, as factors such as appearance, taste, smell, and texture may enter into the overall evaluation. Sensor technologies that measure specific parameters such as sugar content and composition, total titratable acidity (for example, for fruit), and specific chemicals such as glucose, sucrose, lactate, alcohol, glutamate, and ascorbic acid provide more objective means of evaluating food quality and freshness or spoilage.

Also, quantitative detection of food contaminants, such as growth hormones fed to animals, antibiotics, and pesticides, is important. Testing for genetically modified organisms (GMOs) will probably become more frequent with regard to food, and some authorities expect this area to see the fastest growth of any testing market in the food industry. Biosensor research for GMO detection presently seems to be focused on DNA-based detection technologies. Finally, there is the issue of pathogen and/or pathogen toxin contamination of food. This last is of great concern because of the potential for outbreaks of foodborne illnesses, such as are seen periodically. Fatalities can result from such exposures, and recent such outbreaks have generated much media interest.

There is a clear role for biosensors in many areas related to food quality and safety including pathogen detection in food products. Recent reviews of statistics regarding pathogen testing (all types of tests, including biosensors) in the food industry prove interesting in regard to future biosensor development. For example, for 1999, approximately 16 percent of all microbial tests in the industry were for specific pathogens, ~16 percent were for yeast and mold, roughly 31 percent were for coliform and *E. coli*, and ~37 percent were for total viable organisms. The same statistical survey found that microbial testing in each of the food sectors was as follows: 36 percent for the processed food sector, 10 percent for fruits and vegetables, 22 percent for meat, and 32 percent for dairy. The U.S. food industry performed ~ 144.3 million microbiological tests in 1999.

Unfortunately, heavy reliance is still placed upon conventional culturing techniques, so there are significant time gaps between sampling for microbial contamination and detection of microbial contamination, and these techniques are often labor-intensive and require expertise in interpreting results. Other investigators also point out the importance of pathogen detection in the food industry—that is, that the food industry accounts for ~38 percent of total research in the field of pathogen detection.[1] The other major areas of research interest for biosensors are water and environment quality control (16 percent) and clinical applications (18 percent). In terms of research articles describing biosensors applied to pathogen detection, a review article summarizes a number of interesting points.[2]

Thus, of research articles concerning pathogen detection, techniques for detection of Salmonellae species are the most abundant (33 percent), followed by *E. coli* (27 percent), with Listeria (14 percent) and Campylobacter (11 percent) species the other major pathogens. Legionella account for 7 percent of research articles discussing detection techniques, with all other bacterial species accounting for 8 percent. This relative abundance of research articles published with regard to detection schemes clearly indicates the emphasis upon pathogen detection for food safety. This same review also discusses some relevant statistics regarding detection technologies for pathogens, gleaned from a review of the relevant literature over the past 20 years. The most popular methods by far relate to culture/colony counting with PCR following closely behind. ELISA assays are in third place in terms of abundance, with biosensors following in fourth place. In the biosensor category, the most used techniques are optical (35 percent), followed by electrochemical (32 percent), piezoelectric (16 percent), and all other categories (16 percent).

The tried and true methods of microbiology involving concentration of microbes from food samples; plating, culture, and growth in selective media; and colony counting are still the gold standard with regard to identification and quantification. Nevertheless, these techniques are time- and labor-intensive, and there is clearly a need for rapid, low-cost techniques that provide automated or semi-automated pathogen analysis. Even better would be systems that integrate all aspects of the analysis from sampling to final quantitative result. Biosensors seem well suited to fill at least some of this niche in the food safety testing market.

A major problem facing all such attempts to move pathogen testing from classical microbiological procedures to biosensor-based analysis is the issue of sensitivity. Recent literature indicates that the infectious dose of Salmonella or *E. coli* 0157:H7 is 10 organisms, while the existing coliform standard for *E. coli* in water is 4 cells per 100 milliliters. Culture methods can and do attain this sensitivity. From a review of recent biosensor literature that the authors carried out, this level of sensitivity is not achievable by most of the biosensor research devices currently proposed. It seems clear that biosensors will only see their full potential realized with regard to pathogen detection for food safety, when specificity and sensitivity can compare to established methods and such biosensors can also be cost-competitive (or cost-saving) with current techniques.

Biodefense Biosensing Applications

Events of recent years have indicated the need for sensors of pathogens/toxins that could be used by military enemies or for terrorist purposes. Much research effort is currently focused on analytical strategies to detect these agents. A major difficulty with biological attacks is actually determining in a timely manner whether the attack has occurred, to enable early response and minimize casualties. The ideal sensor for biological warfare agents (BWA) that could be used for attack would provide highly sensitive and selective identification of threat organisms in virtually real time. Also, since the potential release of BWA could be on the battlefield or in urban settings, the surrounding environment/atmospheric milieu may be highly complex (smoke, dust, particulate matter), and the analytical technique must be able to detect organisms of interest without interference from background material. To date, these requirements have provided a daunting analytical challenge.

Attempts to meet this challenge have employed molecular techniques that can identify chemical markers of BWA. In general, detection of BWAs can follow one of two paths. In one, positive identification of a BWA must be obtained in a few hours, the so-called detect-to-treat option. Such detection would give medical personnel the means to successfully treat individuals that have been exposed to BWA. The time frame of several hours makes the analytical task easier than in the case with the other detection scenario. In this latter case, a detect-to-warn sensor must be able to provide a warning within a few minutes that a BWA release has occurred. If such sensors could be developed, then perhaps therapeutic treatment of all members of an exposed population might be an option. The variety of BWA that potentially could be used, such as bacteria (vegetative or spores, gram-negative or gram-positive), viruses, and toxins, adds difficulty to the sensing task. Specificity and sensitivity are both critical for useful BWA sensors. False positives are unacceptable due to the consequences attendant upon an assumed BWA incident.

False negatives are also unacceptable for obvious reasons. It is true that nucleic acid–based biosensing technologies have been demonstrated to have the sensitivity to detect one organism or one spore. Also, in general, nucleic acid–based sensing systems are more sensitive than antibody-based detection systems. Such analyses, however, depend upon amplification of the target nucleic acid, a process that takes time (although intensive efforts have been and are being made to shorten this step). Further, nucleic acids from nonrelevant sources may interfere in PCR amplifications, and nucleic acid–based detection will not work for toxins. Detection of BWA with extreme sensitivity is important since some of the organisms manifest infectivity at extremely low levels. Nucleic acid–based detection also offers the best chance for detection of novel, engineered organisms or so-called stealth organisms, where virulence genes are inserted into an otherwise innocuous microbe. In these cases, the nucleic acid sequence of the virulence elements still provides a diagnostic signature.

Integration of all steps (lysing cells/spores to prepare nucleic acids, amplification of target sequences, and detection of the same) required for nucleic acid sensing is crucial for a true biosensor, and so far, reducing the entire assay time to a few minutes has been unreachable. Thus, there has been interest in nucleic acid–based techniques that provide gene-based specificity, but do not require amplification steps to attain detection sensitivity to the required levels. Such exquisitely sensitive detection almost precludes techniques like surface plasmon resonance (SPR) and recommends procedures like fluorescence detection.

This difficulty has led others to suggest that detectors based on surface features (for example, immunogenic molecular species) of the threat organisms may be a more tractable route for development of detect-to-protect devices. Such devices must have a rapid response, must not give false positives, and must have high sensitivity and the ability to detect the target in aerosol samples, when the natural environment for most detecting biomolecules is aqueous. The major factors for successful immunosensing of BWA are the efficiency of the antigen-antibody complex formation and the ability to sensitively detect these complexes. Nonspecific binding can be a problem for such techniques, particularly as sensitivity is pushed. There is also an urgent need for multiplexed detection in BWA sensors, since the list of potential agents, while not extensive, still numbers more than a handful.

Biorecognition Elements

The biorecognition element of biosensors has classically been some natural biomolecule, antibodies being the most common example. However, biorecognition elements also include bioreceptors such as nucleic acids, enzymes, cell membrane receptors, and even larger biological units, such as whole cells. The biorecognition element plays a crucial role to the overall biosensor performance, imparting selectivity for a particular analyte. If the definition of biorecognition elements only includes natural biomolecules (up to the collection of biomolecules comprising a whole cell), such artificial recognition elements as imprinted polymers or peptide nucleic acids would be excluded. This rigorous definition seems too restrictive, however, and in this chapter the authors include biosensors using such biorecognition elements as aptamers, peptide nucleic acids, liposomes, and molecularly imprinted polymers. A brief description of the various types of biorecognition elements follows.

Antibodies

Antibodies are proteins produced by the immune system that have as their unique feature antigen recognition sites that bind antigens by noncovalent interactions in an exquisitely specific fashion and often with relatively high affinity. The antigen-binding portions of the molecule are the V_H (heavy chain variable region) and the V_L (light chain variable region) regions, which both fold to provide a sort of "lock and key" fit for the specific antigen. Included in antibody categories would be polyclonal and monoclonal antibodies. Antibodies are probably the most commonly used of all biorecognition elements for sensor applications. Frequently, sensor design makes use of antibody "sandwiches," where, for example, captured antibodies immobilize the analyte, while other labeled detector antibodies bind to the analyte and thus provide a sensing signal. It is also common to use some sort of amplification technique (for example, enzymes-ELISA assays) to provide a more robust signal of the recognition process. Ingenious combinations of nucleic acid and immunological techniques (such as immuno-PCR, immuno-rolling circle amplification) can also be used for amplification.

Antibody Fragments/Engineered Antibody Fragments

For biosensing applications, the molecular recognition site(s) of the antibody (the antigen combining sites) are of overwhelming importance, while the function of the fixed chain region, of importance physiologically, may not be required for sensing applications. Also, antibodies are often immobilized on solid substrates for sensing applications, and here the orientation of the antibodies is critical, as the antigen combining site should ideally be oriented in a spatially accessible manner. Thus, smaller antibody fragments may have certain advantages, such as more defined points of immobilization through thiol groups liberated after cleavage of the whole molecule, and also these fragments could be more densely packed upon a sensor surface, thus increasing the epitope density over that found with intact immunoglobulin and so perhaps enhancing sensitivity. Enzymatic cleavage of immunoglobulin molecules to produce fragments such as an antigen-binding fragment is one way to produce these smaller recognition elements.

Using the techniques of phage display or ribosome display for antibody gene cloning and expression, antigen-binding fragments of antibody molecules called single chain variable fragments (scFv) can be obtained. These fragments feature the V_H and V_L domains joined by a flexible polypeptide linker that can be engineered to have desirable properties for immobilization (for example, a thiol group of cysteine). Some studies have found the scFv antibody fragments to be superior to either the antigen-binding fragment or the whole immunoglobulin molecule for sensor applications.

Enzymes

Enzymes are catalytic proteins that have an active site that shares some of the features of the antigen-binding site of antibodies—that is, exquisite specificity for certain molecular structures, referred to as their substrates. Again, genetically engineered enzymes with altered properties can be used. Enzymes are not frequently used as biorecognition elements per se, but are usually a component of a multiple molecular biorecognition package, where the enzyme is included to provide amplification (such as in ELISA immunoassays or coupling to an electrochemical detection transduction system [for example, horseradish peroxidase, urease for electrochemical detection schemes]). An example where an enzyme functions as a direct biorecognition element would be the use of cholinesterase enzymes to detect certain organic phosphate pesticides/chemical warfare agents that are cholinesterase inhibitors.

Proteins/Peptides with Specific Binding Properties (Phage Display)

Using techniques developed in recent years, it is now possible to produce and evolve peptides/proteins with strong binding affinity for specific proteins. These techniques include phage, bacterial, and ribosome display. It is outside the scope of this article to provide an in-depth review of these procedures. In each technique, iterative rounds of affinity purification enrich proteins/peptides with desired binding properties for a specific ligand. The unique feature of phage display is that the production of 10^7 to 10^8 short peptides on the surface of filamentous bacteriophages can be achieved through the fusion of an epitope library of DNA sequences with one of several coat proteins (the minor coat protein, phage gene III, was used initially). An epitope library of such cloning can display 10^7 or greater peptides that can be panned for peptides that bind specifically to an analyte of interest. Proteins made from the phage display technique have been used to detect bacteria, bacterial spores, toxins, and viruses.

Nucleic Acids

Included here are oligo- or polynucleotides such as DNA and RNA. DNA is used more frequently due to its inherent greater stability under a variety of conditions. The biorecognition process consists of noncovalent interactions between bases of complementary nucleic acid strands by Watson-Crick base pairing rules, and is manifested by hybridization of two single strands having such complementary base sequences. Exquisite specificity can be obtained due to the cumulative specific interaction of many bases to their complementary units on the opposite polynucleotide chain. Nucleic acid sensors are frequently coupled to schemes that utilize amplification of diagnostic nucleic acid sequences (for example, PCR, nucleic acid sequence-based amplification, rolling circle amplification, and so forth) with the potential for extreme sensitivity, since a large number of copies of an initially low copy number sequence can be obtained by amplification techniques.

Aptamers

Aptamers can also be included as nucleic acid biorecognition elements, since they are made up of single strands of nucleic acid (DNA or RNA), but here the biorecognition is not via base pairing but by folding to form a unique structure. In contrast to base pairing as the biorecognition feature, the aptamer folds in such a manner that it forms a highly specific 3-dimensional structure that recognizes a specific analyte molecule, somewhat similar to the nature of the antibody-antigen interaction.

Aptamer selectivity and binding affinity can be similar to the specificity and affinities of antibodies. An advantage of aptamers over antibodies is that the aptamer structure is inherently more robust than the antibody quaternary structure, so that aptamers can be subjected to numerous rounds of denaturation and renaturation, thus providing for easy regeneration of the sensor as well as an overall more robust biosensor. Antibodies are fundamentally products of animal immune systems, with all the limitations that implies.

Aptamers are produced by chemical synthesis after selection rather than via immune system cells, and so can be readily modified to enhance stability and affinity. Also, aptamers can be selected against certain targets that can cause the immune system to fail to respond, such as bacterial toxins or prions. Because of all these advantages, aptamers are being used increasingly as biosensor elements.

A recent development has been designing RNA sensors that combine aptamers with ribozymes (catalytic RNAs) to produce RNA molecules (these structures have been called aptazymes) whose catalytic activity depends in some manner upon binding of a specific ligand, the analyte. The binding of the analyte can either stabilize or destabilize the catalytic domain and so adjust the catalytic activity. Thus, detection of the analyte species is signaled by changes in the enzyme's activity.

Peptide Nucleic Acids

Peptide nucleic acid (PNA) structures are not strictly natural biorecognition molecules, since they are a hybrid of a series of N-(2-aminoethyl)-glycine units making up the backbone structure (instead of the sugar-phosphate backbone of natural nucleic acids) with the standard bases bound to the peptide backbone. All intramolecular distances and configurations of the bases are similar to those of natural DNA. Thus, hybridization to DNA or RNA molecules is readily achieved. Since the PNA backbone is uncharged, PNA–DNA duplexes are more thermally stable than DNA–DNA duplexes, and so single-base mismatches have a more destabilizing effect in PNA–DNA hybrids. The PNA structure is stable over a wide temperature and pH range and resistant to nuclease and protease activity and so has obvious applications in the biosensor area. Also, PNAs can be used for the detection of double-stranded DNA directly, eliminating the requirement for thermal denaturation, due to the ability of PNAs to form higher order complexes (3- and 4-stranded) with DNA.

Also to be mentioned are another type of nucleotide analog called locked nucleic acids (LNA). These nucleic acid analogs have modified RNA nucleosides where a ribonucleoside has a methylene linkage between the number 2-oxygen and the number 4-carbon. These LNA have the usual hybridization properties expected of nucleic acids with other desirable properties such as highly stable base pairing, high thermal stability, compatibility with most enzymes, and predictable melting behavior.

Molecular Beacons

This is a subset of nucleic acid biorecognition molecules, generally synthetic oligonucleotides that are designed to have a hairpin (stem-loop) structure. The loop contains a nucleotide sequence complementary to the analyte sequence to be detected, while the stem structure has only a small number of complementary bases, forming a short double-stranded portion. Fluorophore and quencher molecules are held in close proximity on the two ends of the double-stranded stem, when the molecular beacon is in the closed position. Upon binding of the analyte sequence to the loop portion, the stem opens, the quenching effect on the fluorophore is relieved, and the resulting fluorescence signals the presence of the analyte sequence. Aptamers can also serve as molecular beacon biosensors in that they can be used to detect nonnucleic acid analytes, such as proteins or small organic molecules. PNAs can also be used to form molecular beacon-type structures and these can be stemless, in contrast to the usual stem-loop structure found in natural DNA molecular beacons. This stemless structure has advantages (less sensitive to ionic strength, quenching not affected by DNA-binding proteins) as compared to DNA beacons.

Cell Surface Receptors/Glycoproteins, Glycolipids/Glycans

Many of the molecular recognition events having to do with cell-cell interaction, pathogen attack upon cells, and so forth take place in the glycocalyx layer of the cell membrane. The glycocalyx layer consists primarily of oligosaccharide headgroups of glycoproteins and glycolipids. The lipid or protein portion of these molecules is embedded in the membrane, while the hydrophilic oligosaccharide chain is extended into the outer environment. As the molecular biology of pathogen and toxin interaction with the cell surface is elucidated at the molecular level, these receptor molecules are receiving increasing attention as biorecognition elements.

One of the major difficulties heretofore with exploitation of these receptors is the complexity of the cell membrane, making mimicry of this structure a daunting task. Simpler models of cell membranes with defined chemical compositions include liposomes (vesicles) and Langmuir-Blodgett monolayers. Recent advances have formulated biorecognition molecules into these synthetic lipid structures.

In some cases, pathogens or toxins produced by pathogens target specific cell surface receptor sites. These same cell-based receptors can then become biosensor recognition elements. Thus, biorecognition elements seeing use in sensors include gangliosides, glycosphingolipids, which are localized upon certain cell types, as well as carbohydrate/oligosaccharide structures found on the cell surface. For example, ganglioside GT1b has been used as a recognition element for botulinum toxin, while N-acetylneuraminic acid and N-acetylgalactosamine bound cholera and tetanus toxin, respectively, in a semi-specific manner. Uropathogenic *E. coli* strains attach to uroepithelial cells by adhesion to the glycolipid globotetraosylceramide (globoside), and so sensors for these *E. coli* strains can be formulated using this globoside. Human gastric mucin has been used in a biosensor format to study the interaction of *Helicobacter pylori* with the extracellular matrix components. Lectins, a broad class of proteins/glycoproteins that bind with high affinity to specific carbohydrate moieties, have also been incorporated into biosensing devices.

Often, oligosaccharides form primary molecular components and markers on pathogen surfaces, and thus may form the basis for diagnostic sensors for specific bacteria or groups of bacteria. For example, the lipopolysaccharide molecules (commonly known as bacterial endotoxin) that are found on the cell surface of Gram-negative bacteria can serve as targets for biorecognition by lectins or other proteins.

Whole Cells

In this chapter, the term *whole cells* refers exclusively to bacterial cells. In some ways, whole cells do not fit the strict definition of a biosensor biorecognition element, since they are most often used to function as bioreporters. That is, the cells are genetically altered to synthesize a certain marker that produces a detectable signal of some sort (luminescence, fluorescence) when they encounter specific compounds/environmental conditions. Such whole cell sensors have been applied to environmental analysis and monitoring. A brief description of such applications is included later in this chapter. There are, however, some examples of whole cells being used as biosensors, particularly microbial cells. Often in these instances, the whole cellular machinery functions in a sense as a biorecognition/processing system for the analyte, as, for example, using bacterial cells induced to constitutively produce an amino-acid metabolizing enzyme to detect that particular amino acid. Another innovative approach using whole cells is to genetically engineer cells of the immune system to respond to specific pathogen binding events by producing a detectable signal from some genetically inserted reporter construct.

Bacteriophage

Bacteriophage, by virtue of the fact that they recognize in a specific manner their host bacteria, can become biorecognition elements. The bacteriophage themselves may be used, with appropriate labels (for example, fluorescent dyes) to detect the binding event, or the phage may be selected for binding to a specific analyte via fusion proteins on the phage coat (that is, phage display). This latter paradigm has even been extended to expression of the biorecognition peptide as an N-terminal add-on to the major coat protein of the phage. In this case, there is a display of thousands of copies of the recognition protein on the phage surface, thus potentially greatly increasing the sensitivity of the assay by increasing the density of the epitopes displayed. Thus, the phage themselves become both biorecognition (by virtue of peptides displayed on their surface) and amplification (multiple copies of fluorescent dye per phage) elements. This type of biosensor could be considered a whole cell biosensor (bacteriophage) that has a bioreceptor on its surface.

Biomimetic Materials

Liposomes. Liposomes are synthetic lipid vesicles, mimicking in some respects cell membranes, although they are not nearly as complex. Liposomes are frequently used as components of biorecognition systems, but usually the liposomal membrane structure is modified with a standard biorecognition element such as an antibody or oligonucleotide, and these molecules provide the specific recognition. In such cases, the liposome may have encapsulated fluorescent dyes, enzymes, electrochemically active species, and so forth, which, upon biorecognition, may be released, and thus the liposome itself functions as an amplification system. More direct involvement of liposomes as sensing elements per se are occasionally also encountered. For example, insertion of the ganglioside GM1 into the liposomal membrane has been used to develop a sensor for cholera toxin, since the GM1 has specific affinity for it. In a similar vein, liposomes containing cholesterol in the membrane could function as sensors for the toxin listeriolysin.

Molecularly imprinted polymers (MIPs). A recent development that may in the future provide artificial macromolecular bioreceptors is the technique of molecular imprinting. In this procedure, a target molecule (that is, the analyte), acting as a molecular template, is used as a pattern to direct the assembly of specific binding elements (that is, monomeric units), which are subsequently polymerized with a cross linker. The interaction of the monomeric units with the template molecule may be either through noncovalent (the more usual method) or covalent bonding. Following polymerization, the target molecule is removed from the polymeric matrix (by suitable solvent or chemical cleavage), leaving a cavity within the polymeric matrix, which is a molecular "negative" image of the target molecule. The monomeric subunits that interact with the target molecule in a noncovalent molecular imprinting scheme do so by interactions such as hydrogen bonding, van der Waal's forces, hydrophobic interactions, and so forth, just as antibody-antigen interactions take place. The subsequent interaction with the cross-linker freezes the binding groups in a specific orientation.

Ideally, once the target molecule is removed from the polymer, a binding site exists that is complementary in size and chemical functionality to the original target molecule. In a number of studies, MIPs have been shown to have binding characteristics similar to those of antibodies and other bioreceptors. A potentially important advantage of MIPs is enhanced stability compared to natural biomolecules. The use of MIPs for the analysis of small molecules is becoming established, but detection of larger biomolecules (for example, proteins and protein toxins) is more problematic. Here, the formation of the template cavity to conform to the 3–D image of the total protein may be difficult (for example, individual antibodies to proteins respond to one antigenic determinant or epitope). Similarly, use of MIPs for sensing whole microorganisms can be challenging.

Biomimetic chromic membranes. A specific biomimetic system that has been used for convenient colorimetric sensing of bacterial toxins, bacteria, or viruses is the employment of cell membrane–mimicking materials ("smart" materials) into which cell surface receptor residues are inserted. These systems use polydiacetylenic membranes (Langmuir-Blodgett or vesicle bilayers) containing either cell surface gangliosides, sialic acid, or carbohydrate residues to detect the presence of toxins such as cholera toxin or influenza virus, the binding event being signaled by a color change (bichromic shift). The specificity of the sensor resides in the ganglioside or sialic acid residues in the polymeric assembly. Gm1 ganglioside is the primary target of cholera toxin on the intestinal cell surface, while GT1b gangliosides are located at the neuromuscular junction and form the target for botulinum toxin. In the case of influenza virus, the sialic acid residues on glycolipids form targets for the viral surface hemagglutinin of the influenza virus, by which pathway the virus is endocytosed into the cell. When these receptor molecules are incorporated into the polydiacetylenic structure, binding of the virus or bacterial toxins apparently induces a distortion of the conjugation plane of the polymer, leading to the color change.

Transduction Mechanisms

As described above, the transducer is the portion of the biosensor responsible for converting the biorecognition event into a measurable signal. It is possible to exploit a change in a number of physical and chemical properties (mass, temperature, electrical properties, and optical properties) to allow for different transduction formats. The four basic transduction types—electrochemical, piezoelectric, calorimetric, and optical—are reviewed here, with emphasis on the advantages and disadvantages to each, and recent advances toward idealizing biosensor performance.

Electrochemical

Electrochemical transduction is one of the most popular transduction formats employed in biosensing applications. One of the main advantages of biosensors that employ electrochemical transduction is the ability to operate in turbid media and often in complex matrices. Another distinct advantage of electrochemical transduction is that the detection components are inexpensive and can be readily miniaturized into portable, low-cost devices.

In general, electrochemical-based sensing can be divided into three main categories: potentiometric, amperometric, and impedance. Potentiometric sensors typically rely on a change in potential caused by the production of an electroactive species that is measured by an ion selective electrode. For a biosensor system, this change in electroactive species concentration is usually brought about by an enzyme. In an amperometric sensor system, a change in current is directly measured. Electrochemical sensors based on impedance most commonly utilize impedance spectroscopy since controlled AC electrical stimulus over a range of frequencies is used to detect variations in the sensor surface properties (that is, charge transfer and capacitance at the interface layer). In this way, the resistance to flow of an alternating current is measured as voltage/current. For example, metabolic changes (for example, growth) have been shown to correspond to an increase or decrease in impedance. Some of the many variations of potentiometric, amperometric, and impedance biosensors that provide for improved biosensor performance include field effect transistors (FET) and electrochemiluminescence (ECL).

Field effect transistors. Many researchers have recently looked to field effect transistors as a means to miniaturize potentiometric sensors, while providing increased sensitivity due to minimal circuitry. A particularly promising FET advance includes the light addressable potentiometric sensor (LAPS), which consists of an n-type silicon semiconductor-based sensor and an insulating layer that maintains contact with the biological solution. Potential changes occurring at this silicon interface are monitored as differences in charge distribution between the insulator surface and the FET. To accomplish this, alternating photocurrents generated by a light emitting diode are utilized, so that changes in potential can be transduced into voltage per time differentials. Successful application of the LAPS has been demonstrated with commercially available systems for the detection of biological threats of interest to defense applications, as well as foodborne pathogen threats of concern to food safety applications.

Eletrochemiluminescence. ECL combines the advantages of chemiluminescence (high sensitivity and low background) with electrochemical transduction. ECL utilizes a controlled voltage potential at the surface of an electrode to power a luminescent redox reaction. The redox reaction most commonly employs a ruthenium (II) trisbipyridal chelate coupled with a tripropyl amine, although recent studies have demonstrated success with other fluorophore species.

Mass

Biosensors that detect the change in mass due to target and biorecognition element interactions predominantly rely on piezoelectric transduction. Piezoelectric transduction relies on an electrical charge produced by mechanical stress, which is correlated to a biorecognition binding event causing a change in the mass on the piezoelectric device. The main advantage to the piezoelectric transduction (that is, mass sensor) approach includes the ability to perform label-free measurements of the binding events, including real-time analysis of binding kinetics. Advantages and disadvantages of the various types of mass transduction are discussed below.

Bulk wave. The most commonly employed transducer is the quartz crystal microbalance (QCM), which relies on a bulk wave effect. A QCM device consists of a quartz disk that is plated with electrodes. Upon introduction of an oscillating electric field, an acoustic wave propagates across the device. The change in mass associated with bioreceptor-target interactions causes a decrease in the oscillation frequency that is directly proportional to the amount of target. This transduction format can be coupled to a wide variety of bioreceptors (for example, antibody, aptamer, imprinted polymer, and so forth), provided that the mass change is large enough to produce a measurable change in signal. Not surprisingly, QCM transduction is not capable of small molecule detection directly and usually requires some sort of signal amplification to be employed.

Surface acoustic wave. Changes in the overall mass of the biomolecular system due to association of the bioreceptor with the target analyte can be measured using alternative piezoelectric transducer devices, which offer some advantages over bulk wave sensing. For example, surface acoustic wave (SAW) devices exhibit increased sensitivity compared to bulk wave devices and transmit along a single crystal face, where the electrodes are located on the same side of the crystal and the transducer acts as both a transmitter and receiver. SAW devices can directly sense changes in mass due to binding interactions between the immobilized bioreceptor and target analytes and exhibit increased sensitivity compared to bulk wave devices. However, the acoustic wave is significantly dampened in biological solutions, limiting its utility for biosensing applications. Some improvements using dual channel devices, and special coated electrode systems allowing for non-contact SAW devices that can function in biological solution interfaces have been produced. However, reliable biosensor application incorporating these devices is still under pursuit, as improvements in sensitivity are still required for specific microbial analyses.

Micro- and nano-mechanical devices. Micro- and nano-mechanical cantilevers can be manufactured through silicon fabrication techniques developed in the electronics industry. Under ambient temperature conditions, the cantilever devices naturally oscillate, and this resonant frequency can be monitored. Through modifications with bioreceptors, target interactions can be monitored as changes in mass, and hence the resonant frequencies, upon binding. Obvious advantages of this form of transduction include low cost and mass production. However, practical applications to biosensing are limited due to oscillation dampening in liquid solutions, compared to air- and vacuum-packed cantilever systems. To address this limitation, there have been recent advances toward incorporating hollow channels inside the microcantilever to immobilize the bioreceptors internal to the device. In other words, high resonant efficiencies obtained through vacuum packaging can be maintained, while specific mass changes can be monitored by flowing the sample through the inside of the device. Further signal enhancements can be obtained through use of nanoparticle and magnetic bead amplification to cause larger frequency shifts and produce a more sensitive analysis.

Flexural-plate wave. A flexural-plate wave (FPW) transducer contains a thin membrane that can propagate a surface acoustic wave. The FPW is in contact with a thin film of liquid that vibrates with the membrane, and hence the change in mass of a vibrating element can be detected to indicate biological interactions. This technique is not very sensitive but, like SAW devices, can provide for real-time analyses.

Thermal/Calorimetric

Calorimetric sensors utilize thermistor probes to monitor changes in temperature due to exothermic chemical reactions. Many biological reactions are exothermic (for example, enzyme reactions), and hence calorimetric detection allows for a near-universal transduction format. One key disadvantage of this approach is that environmental temperature fluctuations must be shielded from the sensor system. Calorimetric biosensors traditionally have been large and bulky, although advances in silicon microfabrication technologies and microfluidics have allowed for miniaturization and improved performance.

Optical

Due to a number of advantages, optical transduction is one of the most widely used biosensor transduction formats. For example, optical transduction can be very rapid where the limiting factor for the speed of detection is often a diffusion-limited process of the biomolecular recognition event, rather than the optical transducer. Another advantage of optical transduction is that the interferences that can hinder electrochemical transduction measurements (such as voltage surges, harmonic induction, corrosion of electrode elements, and radio frequency interferences) are not present. Some of the disadvantages of using optical transduction formats include detection challenges when analyzing turbid samples and the cost associated with detection system components. A wide variety of optical transduction formats have been employed, where changes of the interaction of light with the biomolecular system are used to produce a measurable signal. These changes can be based on differences in refractive index, production of chemiluminescent reaction products, fluorescence emission, fluorescence quenching, radiative and nonradiative energy transfer, temporal changes in optical emission properties, and scattering techniques, well as other optical effects.

These effects can be monitored using a variety of optical platforms including total internal reflectance and evanescent wave technologies, interferometric, resonant cavities, and biochip devices. The following paragraphs review the most common as well as most popular emerging optical transduction formats: fluorescence, interferometry, chemiluminescence, surface plasmon resonance, and surfaced enhanced raman.

Fluorescence. Fluorescence is the most popular form of optical transduction due to the high sensitivity that is fundamental to this type of optical process. Another advantage of fluorescence-based methods is that they generally do not have the interference issues that SPR and other refractive-index based methods possess. However, in most cases, the intrinsic sample fluorescence is not sufficient for analysis, and a fluorogenic reporter is used to label an affinity reagent to create a bioreporter. By monitoring the intensity of the fluorogenic reporter, it is possible to determine the presence and concentration of the target analytes, as illustrated in the bioassay techniques already described.

It is also possible to monitor shifts in the wavelength of the fluorophore reporter, as well as energy transfer phenomena, and time dependence of the fluorescence emission, all of which can be related to binding interactions depending on the assay employed. The distinguishing features between biosensors, besides the above mentioned properties that can be monitored, include the optical detection format used. For example, it is possible to utilize fiber-optic probes to immobilize bioreceptors at the tip of the fiber and use total internal reflectance properties of the fiber to transmit excited and emitted light. Total internal reflectance can also be employed in an evanescent wave format, where a residual amount of (evanescent) light at the reflectance point that escapes is used to excite immobilized bioreceptors only in close proximity to the surface, rather than in bulk solution. This format allows for controlled excitation and can allow for minimal fluorescence background.

However, a key disadvantage is the lack of evanescent excitation power, and sometimes poor coupling of the emission when using similar collection geometries. Fluorescence detection can be used with a wide variety of detection formats. For example, it is routinely coupled with flow cytometry and microfluidic platforms or imaging array systems such as biochips that utilize spatial patterning of biological recognition elements to match fluorescence location to target species.

Interferometry. Interferometers can measure biomolecular interactions that take place at a surface within an evanescent field that causes a refractive index change and a corresponding change in the phase of the propagating light. To accomplish this, two channels are used, one of which serves as a continuous reference that does not experience refractive index changes due to the target analyte's presence. By combining light from the reference and sample beams, an interference pattern is created, which can be used to determine the presence of the associated target species. Although this approach has been in existence for quite some time, it is primarily a laboratory technique, since it is not a very robust biosensor technology and suffers from significant false positive results.

Chemiluminescence. Chemiluminescence is a type of optical sensor technology that relies on a series of chemical reactions, usually employing oxidation of a luminol reactant, to produce a reaction product that gives off characteristic light. The intensity of this light is correlated to a sample target of interest, by coupling some of the reaction products to biorecognition elements, to serve as chemiluminescent reporters. The advantage of chemiluminescent transduction is the very low optical background. The only background from the biosensor may be due to any cross-reactivity from the biorecognition element with nonanalyte species, leading to a very sensitive (high signal-to-background) measurement. The reaction usually employs signal amplification and bioassays with multiple washing and labeling steps, which leads to a somewhat time-consuming analysis.

Surface plasmon resonance. Surface icrosy resonance is a phenomenon that can exist at the interface between a metal and a dielectric material such as air or water. A surface plasmon (commonly excited by optical radiation) of the electrons at the surface of the metal results in an associated surface bond and evanescent electromagnetic wave of optical frequency. This evanescent wave has a maximal intensity at the interface and decays exponentially with distance from the interface. In SPR-based biosensing, changes at the interface (that is, biological recognition element and analyte binding) cause changes in the local refractive index, which in turn causes changes in the resonance excitation of the surface icrosy.

SPR is a form of reflectance spectroscopy, where change in a surface icrosy resonance that occurs during optical illumination of a metal surface is harnessed to detect biomolecular interactions. The SPR chip consists of a prism with a thin gold film upon which the bioreceptors are immobilized. Light is totally internally reflected from the metal-coated prism face, and the changes in reflectivity are measured. Surface plasmons are excited at a specific incident angle and result in a massive reduction in reflectivity at that angle. Changes in the refractive index at the interface result in a change of the optimal angle required for excitation.

Any change in the optical properties of the medium in contact with the immobilized layer will cause changes in the characteristics of the surface plasmon wave. Specifically, these changes can be measured as changes in wavelength, intensity, phase, or angle of incidence. SPR techniques are widely popular in contemporary biosensor development because it is a surface-sensitive technique that can measure real-time interactions between unlabeled species.

Although SPR is a simple technique with a number of advantages, it is not the most sensitive. However, one variation of SPR includes a resonant mirror format that utilizes a series of polarizing filters to block light that is internally reflected. At a particular incident resonant angle, the light is diverted through a spacer layer that has a low refractive index into a higher refractive index guide so that the signal peak appears on a dark background. Despite these advances, one of the primary limitations of SPR-based biosensors is that anything that alters the refractive index at the sensing surface will interfere with the analysis, including nonhomogenous (complex) sample matrices and nonspecific binding interactions.

Surface enhanced raman. Another type of plasmonic spectroscopy that is gaining popularity in biosensing applications is surfaced enhanced raman spectroscopy (SERS). SERS is an enhanced form of raman spectroscopy, where a nano-roughened metal surface is used to enhance the scattered raman signal. The SERS enhancement is thought to be the result of a combination of intense localized fields arising from surface icrosy resonance in metallic nanostructures and chemical effects.

Advantages of this approach include the ability to obtain a spectroscopic fingerprint, similar to infrared absorption spectroscopy, while being relatively interference-free from water. Also, since SERS is an enhanced technique, it can be used to see very low concentrations of even biological materials, as evidenced by recent interest in a variety of immuno-SERS methods under development. The difficulties often encountered with SERS biosensors stem primarily from signal reproducibility challenges, which are directly tied to the reproducibility of nanostructured SERS substrate fabrication. Of the transduction methods covered in this chapter, SERS is one of the least established for biosensor development but shows much promise for continued development.

Biosensors: Summary

This discussion of biosensors has shown that the field is an emerging one that has not by any means reached its full potential. Unfortunately, the gap between biosensor prototypes that work in the laboratory and commercial biosensors that can see routine use in real-world environments is often still large. The many varied biorecognition elements, transducers, and detection strategies lead to a wealth of potential applications.

The field of biosensor research/applications is still probably in the stage where it is not clear which biorecognition elements/transducers/detection schemes will be most productive, and indeed, the field may well continue to evolve in different directions as techniques such as MIPs become more established. Also, the large number of potential applications may demand a varied armamentarium of biosensor techniques. In general, robustness of the biorecognition element continues to be a concern for many applications. The use of transducers/detection mechanisms that involve sophisticated/expensive instrumentation also will preclude such devices from many routine applications. Simpler is not necessarily always better, but dipstick sensors for certain applications are attractive.

This section will end with a few comments regarding two other areas that the authors believe will have increasing importance in the field of biosensor applications. Array biosensors, biochips, labs-on-a-chip—that is, devices that permit multi-analyte detection (multiplex analysis) and ideally that can integrate all processing steps into a micro-analytical system—will certainly see increased interest in the future. Such integrated devices have a number of advantages, such as the ability to assay large numbers of samples in minutes, savings on expensive reagents, and reduction in time for individual assay steps due to decreased assay volumes.

Unfortunately, a number of obstacles still need to be overcome with regard to these systems. Among these are such things as further development of microfluidics systems, sample preparation modules, sensitive detection modules, and robust assay methodologies. Another area receiving much recent interest is in the area of nanotechnology—more specifically, the use of nanomaterials, such as nanoparticles, in biosensor development. Nanoparticles do not serve as biorecognition elements per se, but can be modified by attachment of appropriate molecular species (for example, antibodies or DNA fragments) to become nanosized carriers of biorecognition elements that can also serve valuable transduction functions (that is, magnetic properties, fluorescence). The unique properties of nanoparticles (colloidal gold and quantum dots) also recommend them for novel sensing applications. A number of investigators have mentioned the potential use of nanoparticles, particularly quantum dots, as molecular bar codes. Metallic nanoshells and semiconductor quantum dots are highly promising markers for use in high-throughput screening and bar coding by virtue of spectral multiplexing. However, much work remains to be done before the full promise of these materials is applied to practical and useful biosensors.

Bioelectronics

Overview

The journal *Biosensors and Bioelectronics* published an excellent description of bioelectronics: "The emerging field of Bioelectronics seeks to exploit biology in conjunction with electronics in a wider context encompassing, for example, biological fuel cells and biomaterials for information processing, information storage, electronic components and actuators. A key aspect is the interface between biological materials and electronics."

Bioelectronics is not as developed as the field of biosensors; however, DOD is making research investments to find out what possibilities lay ahead for future military applications. Bioelectronics is a multidisciplinary field involving researchers from biology, chemistry, physics, materials science, and engineering. This diversity is necessary in order to understand the nuances of biology as electrical systems. In this section, we will seek to describe several research areas including the phenomena being exploited, level of development, and future applications.

Biological Fuel Cells

Biological fuel cells come in two main varieties: enzymatic and microbial. The goal of using these technologies is to develop environmentally friendly power sources capable of operation on fuels that would be readily available in the environment, generated in the field, or already included in the logistics train. These fuels would also be nontoxic and yield harmless emissions.

Two electrodes are required for a fuel cell reaction: the anode and the cathode. An example of a reaction in either type of biological fuel cell would be reacting glucose, a sugar, with water at the anode, producing electrons and protons that can be used to do work. At the cathode, the protons react with oxygen and the electrons are passed through a load to produce water. The main difference between enzymatic and microbial fuel cells is which biological system is catalyzing the reaction.

Enzymes are proteins generated by all life in order to catalyze chemical reactions. These reactions can be either catabolic—breaking larger molecules into smaller ones—or anabolic—linking smaller molecules to form larger ones. By selecting the appropriate catabolic enzyme, a desired fuel, such as a sugar, alcohol, or organic acid, can be broken down to liberate electrons to be harvested by the anode. These selected enzymes are produced in microorganisms, extracted, and purified for use. Enzymatic fuel cells are much more developed than microbial fuel cells and have an energy density that is higher by orders of magnitude, and some U.S. companies are already starting to market the technology after having learned how to protect the enzymes during storage and operation.

Microbial fuel cells utilize whole microorganisms instead of purified enzymes as the biological catalyst. This means that they are capable of using a variety of fuels instead of just one specific fuel such as in enzymatic systems. Some of the fuels that have been demonstrated include but are not limited to sugars, organic acids, cellulose, waste water, ground water contaminated by fuel oil, and decomposing organic material in aquifers.

The Naval Research Laboratory has invested several years of work on sediment-based microbial fuel cells that utilize naturally occurring organisms and fuels by burying the anode in the anaerobic region of aquifer sediments and leaving the cathode above the sediment. These types of fuel cells have been used to operate meteorological sensors and transmit the results to the laboratory for years at a time.

Microbial fuel cells can also be designed in a similar fashion to conventional chemical fuel cells where the organisms are contained in an anode compartment that is kept anaerobic while the cathode is exposed to oxygen. This type of platform is commonly used in laboratory experimentation and for applications such as complex fuel utilization and waste remediation. These have been used, for example, to obtain power and reduce organic content in agricultural waste water and starch runoff from paper plants.

Microbial systems have a large disadvantage in terms of power production, but they hold some promise for other reasons. First, as enzymes, fail they cannot be replaced, but whole organisms are self-replicating. Microbial systems are also capable of utilizing a variety of complex fuels rather than just a single fuel. This opens the door to performing operations such as power reclamation from wastes, cleaning of waste water flows, and possible production of other simpler fuel molecules such as organic acids, sugars, and alcohols, which could then be used in enzymatic or conventional chemical fuel cell technologies.

Electronic Biomaterials/Components

There is great interest in biology's capability of assembling complex and well-ordered structures through "bottom-up" fabrication. As our knowledge of fundamental biotechnology expands, we are poised to harness this knowledge and these processes for future military capabilities. Three broad application areas through the inclusion of biotechnology are catalysis, sensing, and photonic structures. Specific research areas such as self-assembly of periodic 3–D structures using DNA templates, enzyme-assisted nanolithography, virus templating, directed protein assembly, and biomineralization could be harnessed to push the limits of nanofunctionality, develop sense and respond surfaces, and develop bottom-up assembly of electronic and photonic structures.

Electronic biomaterials and components can be described in two main classifications: using biologicals as templates and relying on the biological/ biologically derived material itself. Templates that have been demonstrated include DNA, viruses, proteins, and short synthetic polypeptides. Researchers at the University of Utah utilized DNA as a template to link microscopic electrodes with gold nanoparticles, while others at the Massachusetts Institute of Technology have used mutated viruses to attach multiple types of metals and semiconductors to the short proteins expressed on its surface.

Researchers at the University of Washington have taken this to the next step through an Army Multidisciplinary University Research Initiative where they have developed a polypeptide tool box for linking metallic and semiconductor materials. The short poly peptides operate free from the organism, are short enough to synthesize, and are capable of being combined so that dissimilar materials can be linked to each other. It has even been found that doping organic light-emitting diodes with DNA has also proven to increase brightness, possibly due to bringing order to the organic semiconductors.

The use of biomaterials as electronic materials has also been studied. While many of these experiments are far from a technological reality, much knowledge has been gained from these studies. Recently, it has been reported that some bacteria are capable of producing nanofilaments that appear to be conductive. These are reported to be biological nanowires ranging from 10 to 100 nanometers in diameter depending on the microorganism. Many researchers have also attempted to modify field effect transistors with a variety of proteins. Some have shown semiconductor characteristics that change depending on environmental conditions, but only one has been constructed into a demonstration device, and this will be discussed in more detail later in the chapter.

Another area where biologically directed assembly will capitalize on is the use of *biomineralization*, the accumulation and assembly of inorganic/ organic or purely inorganic materials by biological organisms or purified enzymes. Bacteria, archaea, and algae have been studied extensively in this area and a host of pure metals, pure metalloids, oxides, sulfides, and selenides have been produced from the whole organisms. Over 30 elements including transition metals, actinides, lanthanides, and main group elements have been studied and reported in the literature, many with interesting electronic, optical, or catalytic characteristics. These organisms or purified enzymes could be immobilized and provided nutrients containing the desired elemental reactants to produce nanoparticles of specific compositions and morphologies on solid supports and could be extended to flexible substrates in the future. This work could be performed in conjunction with nanostamping or dip-pen lithography to form polycrystalline infrared microlenses or nanostructured catalysts for rapid and efficient decontamination.

Information Processing/Storage

The ultimate goal in biotechnology for information processing and storage is to develop technologies that would eventually approach the abilities of the brain. It would be desirable to approach the brain's low power consumption and high level of processing. All of this is performed with simple on/off functionality but with a yet unmapped complex connectivity, logic, and redundancy. To achieve this goal, much research is needed in the realm of not only neurobiology but also biological materials to understand how this high level of processing is truly obtained.

In the meantime, biological materials have been selected in an attempt to start understanding the manipulation, utilization, and connections of biologicals to the inorganic world of electronics. The most studied class of these materials are proteins from the rhodopsin family. Rhodopsins are surface-bound proteins expressed in microorganisms that typically live in high salt or acidic environments and are utilized as proton pumps. These pumps are light activated and remove excess protons from the inside of the cell. Early application attempts were to design biological photovoltaics. While concentrating charge across a gradient was achieved, building practical photovoltaics failed, due partially due to the fact that the system was based on protons instead of electrons and no practical/efficient conversion was developed.

In the studies of how rhodopsin proteins operated, the discovery was made that light caused the protein to change to a conformation with different spectroscopic characteristics. Some of these conformations could be locked in placed for years if maintained in a controlled environment, meaning that it was possible to build a light-activated memory device that could be written to and read like a CD or DVD—except in this case, the memory material would be biological and could be reset to its original state instead of etched.

Actuators

A few examples of bioelectronic actuators have been described in the literature. Most have come from studies of animals such as pit vipers, beetles, and butterflies, which are extremely heat sensitive. Pit vipers have pores in the snout that seem to collect temperature information to allow discrimination of just fractions of a degree, while butterflies can collect sunlight on their wings to find warm locations in the environment, and certain beetles have demonstrated the ability to detect heat from a forest fire over 2 kilometers away. These creatures have been studied extensively for clues on how to design and construct infrared sensor technology that would be inexpensive, flexible, and very sensitive.

One of the few devices that have actually been constructed utilizing biological material is based on the semiconductor gel from a shark's Ampullae of Lorenzini, or sensing organs. This gel has been identified as a key component that allows a shark to track its prey based on the electromagnetic radiation released from swimming and injured fish. Upon exposure to electric, magnetic, and thermal gradients, this gel has been demonstrated to expand and contract. The device developed from this substance under an Air Force contract was an infrared sensor where the gel was layered between two very thin layers of gold. Upon irradiation, the gel would swell and the gold layers would be spread further apart, thus increasing the capacitance of the device. While it was not a particularly sensitive device or capable of constructing small enough pixels for a high-resolution camera, it was the first demonstration of using a biological material in an application previously only performed by inorganic materials.

Conclusion

While bioelectronics is a very young field, there are many lessons to be learned that will yield a variety of useful technologies for defense applications. The earliest applications will likely be bio-inspired. This is where we learn how a certain task is performed successfully in biology and design the physical solution from those lessons. Next, using the capability of life to self-organize, we will be able to direct the assembly of electronic structures. Later, we may actually see limited use of biological materials with that ultimate goal of capitalizing on biology's abilities of efficient operation, replication, and repair.

NOTES

1 See O. Lazcka, F.J. Del Campo, and F.X. Munoz, "Pathogen Detection: A Perspective of Traditional Methods and Biosensors," *Biosensors and Bioelectronics* 22 (2007), 1205–1217.

2 Ibid.

chapter 7

BIOENZYMES and DEFENSE

ADRIENNE HUSTON

The Age of Biology

Over the past several decades, scientific and technological advances have challenged long-held views about genetic diversity and biological processes. Advances in culturing techniques have allowed a greater proportion of the Earth's microscopic inhabitants to be grown and studied in laboratory environments, while advances in DNA sequencing capabilities have enabled the analysis of an organism's entire genome and large fragments of genomic material collected directly from the environment. Our knowledge of biological processes now extends to the molecular scale as we seek to understand the physiological limits of life on this planet and continue to search for novel tools and materials to address societal needs.

Numerous uses of biomolecules in the agricultural, defense, energy, food, medical, structural material, and textile manufacturing sectors that would have been written off as science fiction merely a decade ago have been realized. As humans investigate ways to move away from a hydrocarbon- toward a bio-based economy, and novel applications for the gene products found in the Earth's diverse ecosystems continue to be realized, access to genetic diversity as a resource will likely be of great societal and national security interest in the future.[1]

Humans are most familiar with organisms that thrive in moderate climes. Due to our physiological limitations, our visions of a hospitable environment tend to be limited to moderate conditions of temperature, salt, and pH. Microorganisms are much more adaptable, however, and the majority of the Earth's biosphere is located in environments exhibiting extreme conditions.

Such environments include the hydrothermal vents at the bottom of the ocean whose temperatures exceed 100°C, polar sea ice floes whose temperatures can dip below -20°C, sulfurous environments (pH values close to 0), soda lakes (pH values reaching 14), hypersaline environments such as the Dead Sea (with salinities reaching 6 M NaCl), and deep sea environments (where hydrostatic pressures approach 0.1 megapascals). In particular, 75 percent of the Earth's biosphere is estimated to be harbored in stably cold environments whose temperatures never exceed that of our kitchen refrigerators, including the little-explored deep sea, polar, and alpine regions, caves, and the upper atmosphere. The microorganisms that successfully inhabit such environments, aptly named *extremophiles*, possess numerous physiological adaptations that enable growth and activity under these conditions. Already, extremophiles and their biomolecules (that is, enzymes) occupy a significant niche in the global industrial enzyme market, whose current value is estimated at $2.1 billion, and conservative estimates forecast a growth curve of 10 to 15 percent annually.[2]

Enzymes are catalytic, globular proteins that are the product of specific genes contained within an organism's genome. Their activities are highly specific, accelerating the particular biochemical reaction to which they were designed. Most of us are familiar with degradative/digestive enzymes—such as the ones present in our saliva and digestive tracts that aid us in the breakdown of food products. However, enzymes are also necessary for virtually every biochemical reaction that takes place in all living organisms, including DNA replication, metabolic regulation, and transformations resulting in the synthesis of new chemicals.

Because enzymes are crucial for the maintenance of life, among the known adaptations to extreme environments are molecular changes within an enzyme's structure enabling the maintenance of vital catalytic powers across a broad range of conditions. An organism found in hydrothermal vent areas, for example, possesses thermostable enzymes capable of withstanding temperatures that would render most human-derived enzymes inactive. On the other end of the temperature scale, enzymes produced by bacteria found in polar environments such as sea ice are able to catalyze chemical reactions at near-freezing temperatures, but they denature (lose their appropriate structural configuration allowing activity) at temperatures nearing our body temperature. Other types of structural adaptations are found in relation to environmental variables such as pressure, salinity, water availability, and pH. Nature thus has provided us with a wealth of enzymes capable of catalyzing virtually every biochemical reaction necessary for life under every habitable environment present on Earth.

Perhaps the most famous and lucrative application of an extremophile-derived product to date is Taq DNA polymerase isolated from *Thermus aquaticus*, a thermophilic bacterium first isolated from a geothermal spring in Yellowstone National Park.[3] DNA polymerases catalyze the polymerization of nucleic acids into a DNA strand using an existing, complementary DNA template during the replication process. The thermostability of Taq DNA polymerase is the crucial feature that enabled the automation and commercial success of the polymerase chain reaction, the method currently used to exponentially amplify a specific gene or DNA fragment in laboratory environments without using living organisms.

This technique has revolutionized the field of molecular biology and is used in numerous processes such as detection of infectious and hereditary diseases, detection of disease-causing organisms in a given environment, screening and identification of specific genes within gene libraries, forensic identification (genetic "fingerprinting"), and generation of mutations in genes and transgenic organisms. The world market for this enzyme reaches approximately $500 million per year, and it plays a central role in the agricultural, biomedical, defense, food, and forensic industries.

Mining the Earth's Biosphere for Diverse Enzymes

Traditionally, the method of identifying novel enzymes, such as Taq polymerase, from the Earth's microbial biosphere relied upon the ability to cultivate and study live microorganisms in laboratory settings. However, most microbial communities are very complex, requiring currently unknown growth factors and consisting of different species within consortia whose growth is interdependent. These complexities have made it very difficult to isolate and cultivate microorganisms in laboratory conditions. Traditional cultivation techniques are estimated to recover less than 1 percent of microorganisms in a given environment,[4] which has resulted in culture collections that significantly underrepresent the genetic diversity and biotechnological potential of the microbial biosphere.

Fortunately, novel culturing methods are enabling the cultivation and isolation of a greater proportion of the microbial community. Additionally, the emerging field of metagenomics (also known as community or environmental genomics) has recently enabled scientists to clone, sequence, assemble, and express gene products and to screen large amounts of genomic DNA extracted directly from the environment,[5] thus bypassing the need to cultivate microorganisms. By unlocking the world's vast hidden microbial biodiversity, the field of metagenomics is revolutionizing the study and understanding of microorganisms and already has revealed novel genes whose products have found economical applications.

For example, in a recent study of microbial genomic DNA collected from the Sargasso Sea, a low-nutrient, open-ocean ecosystem, over 1 million previously unknown genes were discovered, many of which are believed to encode novel enzymes.[6] Sea ice environments in particular are hypothesized to be a rich source of novel gene products due to the convergence of numerous stressors such as fluctuations in temperature, water availability, ultraviolet irradiation, salinity, and pH.[7, 8] Studies are currently under way to investigate the microbial biodiversity contained within this fragile environment.

A further benefit of metagenomics approaches is the ability to perform high-throughput screening of gene libraries to identify enzymes with a specific desired activity. Using these methods, hundreds of variants of a similar enzyme can be identified that have evolved to catalyze reactions under environmental conditions ranging from the soil in your backyard to hydrothermal vents and polar regions. Having such diversity on hand enables enzyme-based technologies (for example, environmental biosensors that can quickly detect the presence of disease-causing organisms such as anthrax) to be easily adapted to different environmental or combat conditions by substituting appropriate enzyme variants.

When natural genetic diversity is not sufficient, protein engineering techniques enable the mutation and fine-tuning of natural enzymes to address specific industrial or defense needs. Using the natural enzyme as a backbone, a variety of protein engineering approaches can produce a designer enzyme with enhanced properties by introducing mutations in a controlled, measurable manner. These approaches include rational protein design, whereby mutations are intentionally made at targeted sites, or "directed evolution" techniques, whereby mutations are made randomly in a gene whose product is then subjected to a selection or screening step to identify enzyme variants exhibiting desired properties. Using these approaches, scientists have successfully engineered variants of natural enzymes to possess altered rates of activity, substrate selectivity, and environmental tolerance.

Industrial and Defense Enzyme Applications

Industries are facing increasing global competition in addition to political and consumer pressure to strive toward streamlined, sustainable, and innovative technologies. Traditional reliance on processes that require high temperatures, pressures, or harsh chemical additives to drive reactions has resulted in high energy consumption and harmful byproducts that require significant financial investments to address. As environmental concerns arise, industrialization in emerging regions continues, and energy costs continue to soar, biological tools are increasingly replacing harsh chemical and physical means of synthesizing chemicals and processing materials, and are harboring promise for creating cost-effective, renewable energy and chemical product alternatives to those derived from petrochemical sources.

Enzymes in particular offer numerous advantages over chemical processes in that they are highly specific in their catalytic activity (thus producing fewer unwanted byproducts), exhibit fast reaction rates, operate under specified conditions of temperature and pH, are derived from renewable resources, and are biodegradable. Due to their diverse properties and high specific activities, minute quantities can be used for applications ranging from large-scale organic synthesis of pharmaceutical building blocks, production of fuels, plastics, and fine chemicals from biological feedstocks, degradation of natural polymers as well as toxic compounds, and biosensor technology, to name a few.

In the same way that industries are looking to streamline their processes and gain a competitive edge, the U.S. military can also optimize its warfighting practices and shorten the logistics tail by taking advantage of the vast powers of enzymes. What follows are examples of potential applications of diverse enzymes to benefit industrial practices, while highlighting some of their uses on the battlefield.

Enzymes to Fight Infectious Agents

Very few new antibiotics have been developed over the last several decades. While medicine's arsenal of antibiotics has remained relatively stagnant, new infectious diseases are constantly emerging and antibiotic resistance is becoming increasingly widespread. As antibiotics lose their ability to control diseases, scientists are looking to nature for new and effective drugs. One approach includes taking advantage of the millions of years of experience the microbial world has had while engaging in its own version of biological warfare.

Bacteriophages—viruses that specifically infect bacteria—represent an enormous resource from which to obtain agents to control pathogenic bacteria. Following infection and replication inside a bacterial host, bacteriophages must exit the bacterial cell in order to disseminate their progeny phage and continue subsequent infection cycles. To enable this step, they have evolved enzymes called lysins that degrade the host bacterial cell wall from within, thus resulting in cell lysis and death.

Because bacteriophages are highly specific in the bacterial hosts they are able to infect, their lysins are also highly specific in their ability to kill certain bacterial strains. Using such specific mechanisms to control the growth of pathogenic bacteria would be a great advantage over the current use of broad-range antibiotics in that the unwanted side effects of killing "friendly" bacteria in the human body and enabling further development of antibiotic resistance in the bacterial community would be reduced.

Due to the fear of increasing antibiotic resistance, there are no direct preventative measures to control the carrier state of pathogens, and the current practice instead is to provide treatment after infection has occurred. This practice leads to increased health risks for immune-compromised patients, insufficient control of epidemics throughout populations, and reduced battlefield performance as warfighters recover from infection. The use of antimicrobial agents such as lysins that are specific for certain strains of pathogens promises a new, preventative approach to containing and fighting disease.

Indeed, several examples of the potential benefits of lysins have already been demonstrated. An icros has been isolated from a bacteriophage specific for *Bacillus anthracis* that could be used to specifically detect vegetative or spore forms of the bacterium in certain environments as well as prevent and treat anthrax infections.[9] Animal tests have also shown that lysins specific for the pathogens *Streptococcus pneumonia, Streptococcus pyogenes,* and group B streptococci can be used nasally or orally to prevent and reduce serious infections via a "lysis from without" mechanism.[10] The use of such preventative measures could significantly reduce the occurrence of disease by targeting the reservoir of pathogens in a specific environment (such as hospitals, nursing homes, or disease-infested communities) and could also protect civilian populations and warfighters against the ravages of epidemics and biological weapons.

Enzymes to Neutralize Toxic Chemicals

The biocatalytic destruction of organophosphates has become an important research area as technologies are sought for decommissioning chemical weapons, remediating contaminated sites, counteracting nerve agent attacks, and protecting against organophosphate pesticide poisoning. The chemical warfare (CW) agents that have received the most attention since World War II are the organophosphorus (OP) nerve agents such as the type "G" agents sarin, soman, and tabun, and the type "V" agents. The toxicity of these compounds is attributed to their inhibition of the enzyme acetylcholinesterase, which plays a crucial role in regulating concentrations of the neurotransmitter acetylcholine within the human nervous system. OP nerve agents are believed to be the most likely CW agent to be used during asymmetric warfare or terrorist attacks because they are relatively easy to produce and their possession can be difficult to detect.

Organophosphate-based CW agents have traditionally been decontaminated using harsh chemicals such as oxidizing compounds or bleaches. These treatment chemicals are often toxic themselves or can generate toxic breakdown products, thus posing a threat to response teams as well as the surrounding environment. Enzymes that catalyze the hydrolysis of OP compounds, on the other hand, can provide a safe, effective, and sustainable approach to CW agent decontamination. Two enzymes in particular, organophosphorus hydrolase and organophosphorous acid anhydrolase, have received considerable attention for their ability to hydrolyze nerve agents to stable, benign products.

Scientists at the U.S. Army Edgewood Chemical Biological Center have joined forces with Genencor to facilitate large-scale production and testing of advanced catalytic enzyme systems, resulting in a commercially available product line of decon enzymes. They are nontoxic, noncorrosive, easy to use, and environmentally friendly and have been shown to be stable and active in a variety of forms including mixed with both fresh and salt water, and in firefighting sprays, foams, and degreasers. They are also highly active, and small doses can achieve desired decontamination levels, thereby reducing the logistics tail in military and emergency response situations. For example, it has been shown that just 1 gram of enzyme can decompose 1.7 kilograms of sarin agent within 15 minutes of contact at room temperature.[11]

OP compounds are the most widely used insecticides, accounting for an estimated 34 percent of worldwide sales. Soil contaminated by pesticides as a result of bulk handling, application in the field, or accidental release may occasionally lead to contamination of surface and ground water. In addition to accidental release of OP compounds through agricultural or industrial use, drinking water distribution systems supplying soldiers at the front lines or large population centers must be seriously considered as potential CW targets for terrorists.

Because of their nontoxic, noncorrosive, and environmentally benign properties, enzymes may also provide an ideal method for cleanup or bioremediation of toxic agents, pesticides, or other chemicals in water systems. Due to the large volumes of water contained in water distribution and treatment systems, a decon enzyme would need to be active for a much longer time than would be required during direct military operations. Scientists are exploring ways to stabilize enzymes to extend their lifetime once released into a body of water, and are also investigating ways to immobilize and entrap enzymes within filters to treat flowing water systems. Using advances in biotechnology such as protein engineering and metagenomics, companies such as Genencor not only are altering the stability and lifetime of existing enzymes, but also are searching for a wider variety of enzymes capable of neutralizing chemical and biological weapons for different applications. For example, the destruction of biological agents in pipes could be performed by a combination of enzymes that bind to, attack, and degrade the polysaccharide matrix of biofilms.

Enzymes may also harbor promise for safely counteracting the activity of OP nerve agents in humans with minimal side effects. A number of multistep pretreatment and postexposure antidotes exist for preventing the lethal effects of OP nerve agents, but they do not prevent pharmacological side effects that can lead to performance decrements and in some cases permanent brain damage. The development of a single inhalable or injectable antidote that would remain active in the bloodstream could work to protect the warfighter against OP nerve agents and optimize performance by decreasing reliance on burdensome protective suits and preventing postexposure incapacitation.

Currently under investigation are several detoxifying enzymes derived from outdated human plasma that bind to or catalyze the breakdown of OP nerve agents, such as human butyrylcolinesterase and paraoxoniase-1. Because these bioscavenger enzymes are of human origin, they are unlikely to initiate an immune response and are stable for long periods in the bloodstream. However, there is not enough human plasma available to meet the demands of affordably producing bioscavenger enzymes for military or emergency response situations. Recombinant technology, such as producing human enzymes in the milk of other animals, will likely be needed to produce sufficient quantities.[12] Enzymes with catalytic anti-OP activity have also been identified from numerous bacterial sources. Ongoing protein engineering studies may yield variants of these enzymes, with structural alterations enabling improved binding and catalytic properties, greater stability during storage and in the bloodstream, and greater immunological compatibility for human use.

Enzymes to Process Pharmaceuticals and Food

Enzyme catalysis has proven to be an important tool in the industrial synthesis of bulk chemicals, chemical building blocks for pharmaceuticals, active pharmaceuticals, and processing of food ingredients. Current examples of large-scale industrial applications of enzyme catalysis include the thermolysin-catalyzed synthesis of aspartame (the low-calorie sweetener)[13] and antibiotic synthesis.[14] For any industrial process, numerous performance parameters must be met, and enzymes offer advantages over traditional chemical catalysis due to their specificity and ability to catalyze reactions that are challenging for classical organic chemistry practices. For example, the synthesis of compounds obtained from substrates that are not soluble in aqueous media can be achieved using enzymes that operate under low water conditions, such as enzymes adapted to low-temperature or high-salt environments.[15]

The increasing demand for enantiomerically pure drugs and asymmetric pharmaceutical intermediates has also led to a rapid expansion of the use of biocatalysis in organic synthesis. As chemical and pharmaceutical industries continually look for new biocatalysts to apply to biotransformation and biosynthetic processes, it is predicted that enzymes will provide a superior solution over classical chemistry in many existing processes.[16] Enzymes hold particular promise for the production of biobased platform chemicals for a variety of purposes,[17] including bioplastics, solvents, antifreeze compounds, and many more products that may find industrial and defense applications.

Enzymes are particularly useful for the processing and preservation of foods. Proteases can be used for tenderization and taste improvement of meat products, and pectin-depolymerizing enzymes can be used for the degradation of pectin compounds in fruits and vegetables. Cold-adapted galactosidases can improve the digestibility of dairy products for lactose-intolerant consumers and enhance sweetness at temperatures that minimize contamination. In baking, the combined activities of enzymes such as xylanases, proteases, amylases, lipases, and glucose oxidases can result in improved elasticity and machinability of dough, resulting in a larger volume and improved crumb structure.

Applications specific to the defense industry include food ration development, where nutrition, personnel acceptance, cost, shelf life, and performance enhancement are primary developmental parameters. For example, during the processing of ready-to-eat rations, meat items must first be enzyme-inactivated to become shelf-stable. Inactivation is in part accomplished by heating the product, which results in a partially cooked product that has a "warmed over" flavor upon reheating prior to serving. Addition of thermostable proteases or peptide hydrolases that are inactive at storage temperatures but reactivate solely upon reheating can produce protein hydrolysates that enhance the flavor of the meat product.

Enzymes can furthermore act as antistaling compounds for carbohydrate-based food items in ready-to-eat rations. Different types of amylases with different functionalities can be used in combination to convert starch into forms that resist firming.[18] Use of thermostable amylases that survive the baking step and remain active during storage can prolong the product's shelf life. Enzymes can thus effectively work to increase product quality and menu variety of ready-to-eat rations, boosting morale of the troops in the front lines and enhancing the acceptance of carbohydrates and meat products.

Food items can also be amended with enzymes of intermediate stability that remain active in the human digestive tract to enhance the digestibility and uptake of important vitamins and minerals. This practice could lead to enhanced warfighter performance through nutritional initiatives while decreasing the amount of food that needs to be ingested to provide sufficient energy. Enzymes can furthermore be used to convert native biomass such as plants and trees to hydrolyzed, digestible food products capable of supporting soldiers in survival situations.

Growing Toward a Bio-based Economy: Biofuels and Bioenergy

Over the next half century, the human population is predicted to grow to nearly 9 billion people. Given the current projections of fossil fuel shortfalls, there is increasing worldwide interest in developing alternative sources of energy. Biofuels, including ethanol made from the fermentation of carbohydrates produced in plant matter, represent a renewable energy source that can also provide increased energy security, a reduction in greenhouse gas emissions, economic benefits for rural communities, and decreased burden of agro-industrial residue waste disposal.[19]

However, to be economically competitive with traditional energy sources, the current cost of biomass processing for ethanol must be reduced. In the conventional ethanol production process, a starchy feedstock such as corn is first milled and slurried with water containing the heat-stable amylase. This slurry is then cooked at high temperatures (105° to 150°C) to gelatinize and liquefy the starch, and the resulting product is then cooled to ~60°C, and a glucoamylase is added to convert the liquefied starch to fermentable sugars. In the final step of the process, yeast is added to the mixture and incubated at temperatures below 35°C for 48 to 55 hours to ferment the sugars to ethanol.

The conventional ethanol production process is not energetically or economically efficient as it requires high heat levels and specialized equipment, and limits the production capacity of biorefineries.

Industrial enzyme companies are working to develop a low-energy ethanol production process involving raw starch hydrolysis, also known as cold hydrolysis. This process essentially eliminates the energy- and equipment-intensive liquefaction cooking step, resulting in energy savings, higher ethanol yields, fewer unwanted side products, and savings on capital expenses by reducing and miniaturizing the necessary equipment. Due to their high activities and specificities at the low temperatures compatible with fermentation (28° to 35°C),[20] cold-active enzymes derived from bacteria adapted to colder climates may afford a reduction in biocatalyst loading and prove to be an economically attractive alternative. Industrial enzyme companies are also pursuing methods for inexpensive ethanol production from low-cost lignocellulosic biomass, including agricultural waste, forestry waste, energy crops, and municipal solid waste. For military purposes, one might imagine the development and use of mobile biorefineries that use local biomass and waste products, possibly generated by the troops themselves, to produce ethanol, thus reducing the cost and logistical burden of transporting fossil fuels.[21]

Enzymes also play a key role in the biological production of hydrogen, recognized as another promising energy source for the future. Using available sunlight, water, and biomass, unicellular green algae, cyanobacteria, photosynthetic, and dark fermentative bacteria can generate hydrogen via several processes including biophotolysis and dark fermentation. Involved in each of these processes are the hydrogen-producing enzymes hydrogenase and nitrogenase. The primary technical challenge at present is to lower costs of production in order to make biologically produced hydrogen a commercially viable primary energy carrier. Several Federal agencies are involved in funding basic research to bio-prospect for hydrogen-generating microorganisms, characterize the hydrogen-generating apparatus, perform directed evolution experiments aimed at optimizing hydrogen-generating enzymes, and furthermore engineer metabolic pathways to eliminate unnecessary draining of energy equivalents away from the hydrogen-generating apparatus.

Scientists are also investigating the feasibility of integrating different photobiological processes and fermentative production of hydrogen to maximize solar spectral use, efficiently recycle biomass and waste products, and improve process economics. Sometimes the enzymes relevant to optimizing these processes are isolated from bacteria living in the unlikeliest of places, such as methane producers growing in sewage sludge, or cellulose-degrading bacteria living in a termite's hindgut.[22] The most important application of hydrogen to date has been in space programs, where the light fuel presents a significant payload advantage. As their power densities increase with technological advances, one can imagine hydrogen-powered fuel cells replacing batteries as a lighter electrochemical power source for a number of defense applications.

In addition to large-scale energy generation to power vehicles, living quarters, and factories, enzymes can also be used for small-scale energy generation to power tiny devices such as biosensors and other implantable electronic systems. As outlined earlier in this book, biosensors can be used for a variety of applications, including detection of dangerous agents in the environment, and as implantable devices for detection of key metabolites to determine the physiological state of soldiers.

To fuel such electronic systems, scientists are exploring the feasibility of producing miniature, membraneless implantable glucose-O_2 biofuel cells whereby "wired" enzymes capture energy from living tissues by simultaneous glucose oxidation and oxygen reduction.[23] The continuous power output observed from such cells is thought to be sufficient for operating implanted sensors and transmission of data to an external source for periods up to several weeks. When linked to a sensor-transmitter system, automated assessments of soldier readiness can be sent to a command network (such as proposed by Future Combat Systems) to enable highly informed decisionmaking processes. One can even imagine the eventual development of self-activating, implantable devices capable of delivering therapeutic agents upon sensing the presence of biological or chemical weapons or a change in physiological status.

Enzymes to Build Functional Materials

The discussion in this review has primarily been focused on enzymes produced by microorganisms, as their genomes are the most easily targeted by the functional screening tools available in metagenomics, and they are thought to represent the largest reservoir of biodiversity on this planet. However, eukaryotes may also hold the key to elucidating several important processes, particularly when investigating ways in which nature builds efficient structures.

In the field of materials science, chemical synthesis of mineral/organic composites, and silica-based materials such as nanostructured resins, molecular sieves and electronic materials have traditionally required harsh chemicals and extremes of temperature, pressure, and pH. In nature, however, large quantities of structural materials are synthesized biologically from molecular precursors at low temperatures, pressures, and neutral pH into complex nanostructures such as the skeletons of diatoms, radiolarians, and sponges, often with a precision that exceeds current engineering capabilities.[24] A study of the mechanisms enabling the biological synthesis of silica spicules in a marine sponge revealed that structure-directing enzymes were responsible for the polymerization of silica.[25] Since this discovery, numerous enzymatic and enzyme-inspired (biomimetic) methods for the synthesis of nanostructured materials at low temperatures and mild chemical conditions have been illustrated.[26]

Due to the mild synthetic conditions as well as the high selectivity and precision when compared with traditional chemical techniques, these methods can be used to synthesize a new generation of polymeric biomaterials containing functional organic molecules such as enzymes or antibodies. Such capabilities will enable the design of novel sensors with environmental and biomedical applications, the synthesis of functional materials such as "active" protective wear, and perhaps even the synthesis of self-replicating materials capable of wound healing or self-repair. Further investigations will surely reveal additional biological and bio-inspired methods of synthesizing nanostructured materials under mild conditions, resulting in inexpensive, environmentally friendly generation of novel functional materials.

Conclusion

Microorganisms are the most abundant and adaptable life form on Earth. Over the past several billion years, they have had the opportunity to evolve to survive in almost any environment. Now, metagenomic and genetic engineering tools exist to exploit this natural biodiversity and to adapt gene products and metabolic pathways for numerous applications. Because extreme environments contain such a large portion of the Earth's diverse biosphere, further investigations of these regions yield great promise for revealing biological molecules and processes that can find numerous applications.

Ever-changing economics of sequencing will enable rapid sequencing of warfighters' genomes to identity susceptibilities or physiological strengths in certain conditions, in addition to sequencing the transcriptome of soldiers in the field to quickly determine their current physiological state before deployment.

As technology advances, our economy will move toward a bio-based economy and away from our dependence on petroleum hydrocarbons for fuel and chemical feedstocks.

Research topics should be focused on innovation of biotechnologies to not only enhance the protection of personnel, but also expand warfighting capabilities by enhancing the physiological performance of soldiers and decreasing logistical burdens.

NOTES

1 R.E. Armstrong, *From Petro to Agro: Seeds of a New Economy*, Defense Horizons 20 (Washington, DC: National Defense University Press, 2002).

2 Presentation from the Ninth Chemical Industry Finance and Investments Conference, 2005, available at **www.novozymes.com/NR/rdonlyres/64054ABC-F525-4B62-A1EB-5BBE6A55468E/0/Chemicals_Conference_ML_2005.pdf**.

3 T.D. Brock and H. Freeze, "*Thermus aquaticus* gen. n. and sp. N., a Nonsporulating Extreme Thermophile," *Journal of Bacteriology* 98, no. 1 (1969), 289–297.

4 S.J. Giovannoni et al., "Genetic Diversity in Sargasso Sea Bacterioplankton," *Nature* 345, no. 6270 (1990), 60–63.

5 C.S. Riesenfeld, P.D. Schloss, and J. Handelsman, "Metagenomics: Genomic Analysis of Microbial Communities," *Annual Review of Genetics* 38 (2004), 525–552.

6 J.C. Venter et al., "Environmental Genome Shotgun Sequencing of the Sargasso Sea," *Science* 304, no. 5667 (2004), 66–74.

7 R. Cavicchioli et al., "Low-temperature Extremophiles and their Applications," *Current Opinion in Biotechnology* 13, no. 3 (2002), 253–261.

8 J.W. Deming, "Psychrophiles and Polar Regions," *Current Opinion in Microbiology* 5, no. 3 (2002), 301–309.

9 R. Schuch, D. Nelson, and V.A. Fischetti, "A Bacteriolytic Agent that Detects and Kills Bacillus Anthracis," *Nature* 418, no. 6900 (2002), 884–889.

10 V.A. Fischetti, "Bacteriophage Lytic Enzymes: Novel Anti-Infectives," *Trends in Microbiology* 13, no. 10 (2005), 491–496.

11 J.K. Rastogi et al., "Biological Detoxification of Organophosphorus Neurotoxins," in *Biodegradation Technology Developments*, vol. 2, ed. S.K. Sikdar and R.L. Irvine (Lancaster, PA: Technomic Publishing Company, 1998).

12 See D.E. Lenz et al., "Stoichiometric and Catalytic Scavengers as Protection against Nerve Agent Toxicity: A Mini Review," *Toxicology* 233, no. 1–3 (2007), 31–39.

13 K. Oyama, "The Industrial Production of Aspartame," in *Chirality in Industry*, ed. A.N. Collins, G.N. Sheldrake, and J. Crosby (Chichester: Wiley, 1997).

14 A. Bruggink, E.C. Roos, and E. De Vroom, *Organic Process Research and Development* 2 (1998), 128.

15 A.L. Huston, "Biotechnological Aspects of Cold-adapted Enzymes," in *Psychrophiles: From Biodiversity to Biotechnology*, ed. R. Margesin et al. (Berlin: Springer-Verlag, 2008).

16 P. Lorenz and J. Eck, "Metagenomics and Industrial Applications," *Nature Reviews Microbiology* 3, no. 6 (2005), 510–516.

17 P. Turner, G. Mamo, and E. Nordberg Karlsson, "Potential and Utilization of Thermophiles and Thermostable Enzymes in Biorefining," *Microbial Cell Factories* 6 (2007), 9.

18 T. Olesen, "Antistaling Process and Agent," patent no. US6197352.

19 C.E. Wyman, "Potential Synergies and Challenges in Refining Cellulosic Biomass to Fuels, Chemicals, and Power," *Biotechnology Progress* 19, no. 2 (2003), 254–262.

20 Y. Lin and S. Tanaka, "Ethanol Fermentation from Biomass Resources; Current State and Prospects," *Applied Microbiology and Biotechnology* 69, no. 6 (2006), 627–642.

21 R.E. Armstrong and J.B. Warner, *Biology and the Battlefield*, Defense Horizons 25 (Washington, DC: National Defense University Press, 2003).

22 F. Warnecke et al., "Metagenomic and Functional Analysis of Hindgut Microbiota of a Wood-feeding Higher Termite," *Nature* 450, no. 7169 (2007), 560–565.

23 A. Heller, "Miniature Biofuel Cells," *Physical Chemistry Chemical Physics* 6 (2004), 209–216.

24 D. Kisailus et al., "Enzymatic Synthesis and Nanostructural Control of Gallium Oxide at Low Temperature," *Advanced Materials* 17, no. 3 (2005), 314–318.

25 K. Shimizu et al., "Silicatein a: Cathepsin L-like protein in sponge biosilica," *Proceedings of the National Academy of Sciences of the United States of America* 95, no. 11 (1998), 6234–6238.

26 For review, see Huston.

chapter 8

BIOENERGY: RENEWABLE LIQUID FUELS

Michael Ladisch

Becoming independent of crude oil imports, mainly from the Middle East, is an urgent concern for many countries all over the world. In order to secure a sustainable energy supply, especially in the transportation sector, governments need to apply policies that promote the use of renewable energy technologies.

The dependence on crude oil imports decreases as the production of total energy from renewable sources (renewable liquid fuels) increases. There are two major factors that will influence the amount of final energy production from renewable sources.

First, sufficient quantities of biomass need to be acquired for conversion into liquid fuels. If a country cannot provide enough biomass, it has to compensate by importing it, which does not lead to energy independence. Biomass is a limited resource. Even though countries like the United States and Canada encompass large areas of rural land, agriculture and forestry have distinct impacts for the economy as well as for the environment. The use of biomass for conversion to energy competes with other purposes such as food supply and forestry products. Land availability and crop yields are key criteria in determining the quantities of biomass that can be obtained for energy use within a country. What is more, the production of biomass for energy use must be sustainable, which requires a thorough understanding of the environmental impacts and the likely demand scenarios for renewable fuels.

The amount of final energy obtained from domestic renewable energy also depends on the applied conversion technologies. The efficiency of processing biomass into liquid fuels affects the energy input-output ratio. The less energy needed to run the processes, the more that is available to enter the market. Moreover, a change from the current conservative conversion methods to a new generation of technologies will highly influence the total amount of liquid fuels derived from biomass in the long term. A significant characteristic of the new generation technologies is the fact that a larger portion of the available biomass is convertible into biofuels.

Current Role of Biomass for Liquid Transportation Fuels

National governments, particularly in North America and Europe, are promoting the use of liquid fuels from renewable resources. Since the consumption of biofuel is mainly driven by its market price, various means of promoting biofuels intend to make them economically feasible for the customer. These means include tax reductions on gasoline and diesel blended with bioethanol and biodiesel, respectively, as well as on vehicles that utilize biofuels. Some countries, such as Brazil and the United Kingdom, enforce legislation that mandate a minimum biofuel blending level. In the long term, financial support for research and development in renewable derived fuels is important to establish a sustainable biofuel industry. The U.S. Department of Energy (DOE), for example, has been funding research centers and the development of commercial cellulosic ethanol plants since 2007.[1]

Bioethanol and Biodiesel in the United States, Brazil, and Europe

An overview of the impact of renewable fuels for the transportation sector in the United States in comparison to Brazil and Europe (see table 8–1) shows U.S. bioethanol production to be comparable to Brazil, with Europe producing much less than either country. In comparison, biodiesel consumption in Europe exceeds that of the United States. Sources of feedstocks that are converted to biofuels vary by region.

table 8–1. BIOFUEL CONSUMPTION IN THE UNITED STATES, BRAZIL, AND EUROPE

	United States	Brazil	Europe
Bioethanol			
Output, 2006 (billion gallons)	4.9	4.7	0.42
Main sources	Corn	Sugar cane	Wheat, sugar beet
Market share	2.9 percent	40 percent	< 0.1 percent
Biodiesel			
Consumption, 2006 (million gallons)	263	n/a	1,200
Main sources	Soybean	n/a	Rapeseed

While the United States consists of a total of 2 billion acres of land, 33 percent are used as forestlands and 46 percent as agricultural lands, of which 26 percent is grassland or pasture and 20 percent is croplands. In 2006, 190 million dry tons of biomass per year were used for bioenergy and bioproducts. Most of the biomass is burned for energy; only 18 million dry tons are used to produce biofuels (mainly corn grain ethanol).[2]

In 2004, 1.26 billion bushels of corn (11 percent of all harvested corn) were converted to 3.41 billion gallons of starch ethanol (one-third of world ethanol production).[3] The ethanol was produced in 81 plants in 20 states. In 2006, 4.9 billion gallons of ethanol were produced,[4] and a 5.5-billion-gallon capacity was planned until 2007.[5] Although demand for fuel ethanol more than doubled between 2000 and 2004, ethanol satisfied only 2.9 percent of U.S. transportation energy demand in 2005.[6, 7] Over 95 percent of ethanol production in the United States comes from corn, with the rest made from wheat, barley, milo, cheese whey, and beverage residues.[8]

U.S. biodiesel consumption was 75 million gallons in 2005[9] and 263 million gallons in 2006.[10] Total diesel fuel consumption in 2006 was 63 billion gallons. While blended in conventional diesel fuel, the most important biodiesel source is soybean.[11]

Brazil reached a bioethanol output of 4.7 billion gallons in 2006, contributing to one-third of world ethanol production.[12, 13, 14] Ethanol currently comprises about 40 percent of the total vehicle fuel used in the country.[15] Brazil accounted for almost 90 percent of the ethanol imported into the United States in 2005.[16]

European countries produced 13 percent of the world's ethanol. In 2006, approximately 416 million gallons were consumed (= 557,000 tons/year).[17] Germany, Spain, and France were the main producers, using wheat and sugar beet as the predominant crops.[18, 19] Biodiesel consumption was approximately 1.2 billion gallons in 2006 (= 3,852,000 tons/year).[20] Rapeseed (84 percent) and sunflower (13 percent) were the main feedstock.[21] In 2005, the overall transportation energy use of liquid fuels was 290,000,000 tons/year.[22]

Figure 8–1 displays the high growth rate for biofuels in Europe and the dominant role of biodiesel. Policies in Europe vary among countries, but leading biofuel consuming countries have set targets to reach a total market share of about 6 percent within the next 3 years. Current market share is about 2 percent. There are significant differences in composition of market share within Europe. However, biodiesel has the biggest share in most European countries. Only a few countries, including Poland, Sweden, and Spain, consume more bioethanol than biodiesel. Sweden and Germany consume notable amounts of biogas (methane generated by anaerobic digestion) and pure vegetable oil for transportation use, respectively.[23]

Prospects of Biomass Use for Liquid Fuels within the United States

In 2006, 142 million dry tons of biomass were already used by the forest products industry for bioenergy and bioproducts. Residues from the industry include tree bark, woodchips, shavings, sawdust, miscellaneous scrap wood, and black liquor, a byproduct of pulp and paper processing. Other potential forestry biomass resources include logging and site-clearing residues (such as unmerchantable tree tops), forest thinning, fuel wood (roundwood or logs for space heating or other energy uses), and urban wood residues such as municipal solid waste (discarded furniture and packing material).[24]

Agricultural biomass resources include annual crop residues, perennial crops, miscellaneous process residues, and grain (primarily corn). Annual crop residues are mainly stems and leaves (such as corn stover and wheat straw) from corn, wheat, soybeans, and other crops grown for food and fiber. Perennial crops comprise grasses or fast-growing trees grown specifically for bioenergy.[25]

figure 8–1. PRODUCTION OF BIODIESEL AND BIOETHANOL IN THE EUROPEAN UNION 15*

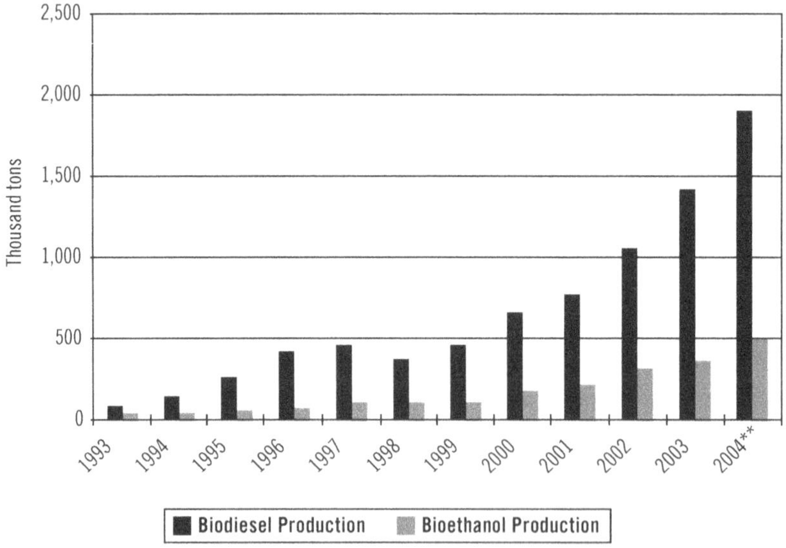

*Austria, Belgium, Denmark, Finland, France, Germany, Greece, Ireland, Italy, Luxembourg, The Netherlands, Portugal, Spain, Sweden, United Kingdom.

**Data for 2004 relates to EU25.
Source: Biofuels Barometer 2007, *Le Journal des Energies Renouvables*, no. 179, available at **http://ec.europa.eu/energy/res/sectors/bioenergy_publications_en.htm.**

Potential of Cellulose

A DOE and U.S. Department of Agriculture joint study examined whether land resources in the United States are sufficient to harvest over 1 billion dry tons of biomass annually, which would be enough to displace 30 percent or more of the Nation's liquid transportation fuels. This study projected that 1.366 billion dry tons of biomass could be available for large-scale bioenergy and biorefinery industries by the mid-21st century while still meeting demand for forestry products, food, and fiber. Of the 1.366 billion dry tons of biomass, 368 million dry tons are obtained from forests, and 998 million dry tons are sourced from agriculture.

The potential of the "billion-ton" study is based upon conservative assumptions. The scenario displayed in this study assumes "business as usual" and therefore does not mark an upper limit. The study assumes continuing improvements, such as in yields per acre and harvesting technologies. It does not take future technologies, particularly cellulose-to-ethanol, into account. Cellulosic ethanol has the potential to meet most, if not all, transportation fuel need since it is potentially possible to harvest the 1.366 billion tons of biomass required on an annual basis. Cellulosic material, available in large quantities, will provide the feedstocks needed for biomass processing (see figure 8–2).[26]

However, the second-generation conversion technologies that are implemented in demonstration facilities are not yet economically feasible.

figure 8–2. COMPARISON OF FOSSIL ENERGY RATIO FROM FOUR DIFFERENT ENERGY SOURCES

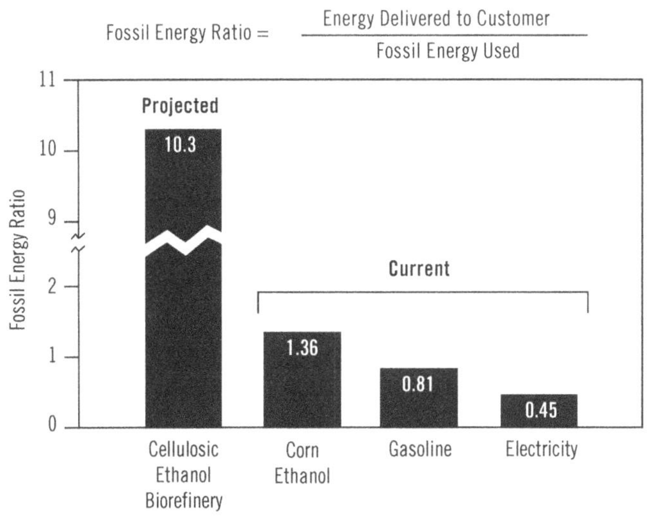

Today's cellulosic biorefineries convert the biomass in numerous complex, costly, and energy-intensive steps.

Production of Cellulosic Ethanol

Research on processing cellulosic material to ethanol has been conducted for many years. The following section gives an overview of its state of the art and will serve as an introduction to display some barriers in commercialization.

Characteristics of Lignocellulose

The strength and the complexity of lignocellulosic material, which makes hydrolysis difficult, are a result of its structure and composition. The strength of the cell wall is created by the network formed of hemicellulose and cellulose and the presence of lignin. The variability of these characteristics accounts for the varying digestibility between different sources of lignocellulosic biomass, which mainly consists of the polymers cellulose and hemicellulose, as well as lignin, which is nonfermentable.

Since cellulose is a heterogeneous substrate, it makes cellulolytic enzyme hydrolysis difficult to model. Cellulose is composed of chains of glucose (arranged in layered sheets) connected by beta 1–4 glycosidic bonds. The two most dominating characteristics of cellulose are as follows. The *specific surface area* (SSA) is defined as the amount of surface area per mass of cellulose. SSA is significant because it determines the number of enzymes that are adsorbed by the substrate. An increase in SSA leads to an increasing rate of hydrolysis. The crystallinity index states the relative amount of crystalline cellulose as opposed to the amount of amorphous cellulose. When endoglucanases (Egs) attack the amorphous regions, the crystallinity index increases.[27]

By using enzymes to break down the polymers, nearly theoretical yields of sugars can be obtained.[28] However, a pretreatment step is essential to achieve these high yields. In this step, cellulose is liberated from the lignin seal and from its crystalline structure, rendering it accessible for the enzymes during the hydrolysis (see figure 8–3).

Cellulose-to-Ethanol Process Overview

The first step of the process is biomass handling, where the size of the lignocellulose is reduced to make handling easier and ethanol production more efficient. During pretreatment, the lignin seal is broken, the hemicellulose is partially removed, and SSA is increased. After pretreatment, the cellulose and hemicellulose fractions are more accessible to enzymes in the forms of polymers and oligomers. Hydrolysis then breaks the chains into monomers.

The monomers can be fermented by natural yeast (glucose fermentation) or by genetically engineered bacteria (pentose fermentation). The result of fermentation is a mixture of water, ethanol, and residues, with carbon dioxide being formed and removed as a gas from the fermentation (see figure 8–4).

Finally, ethanol is purified in a distillation column while the residues can be either burned to power the process or converted to co-products.

figure 8–3. SCHEME OF LIGNOCELLULOSIC MATERIAL

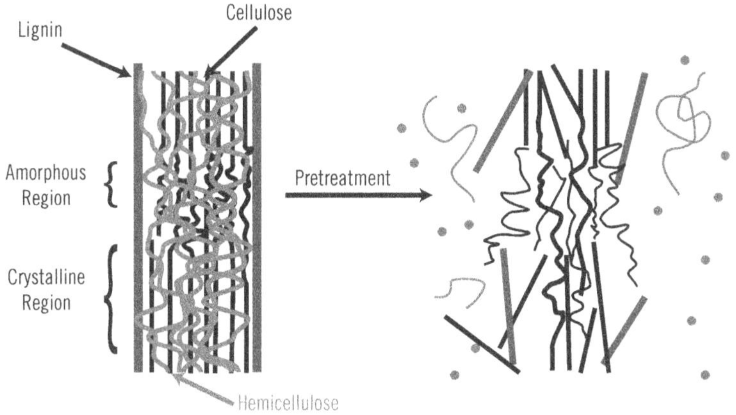

Source: T.A. Hsu, M.R. Ladisch, and G.T. Tsao, "Alcohol From Cellulose," *Chemical Technology* 10, no. 5 (1980).

figure 8–4. STAGE PRODUCTS OF CELLULOSE-TO-ETHANOL PROCESS

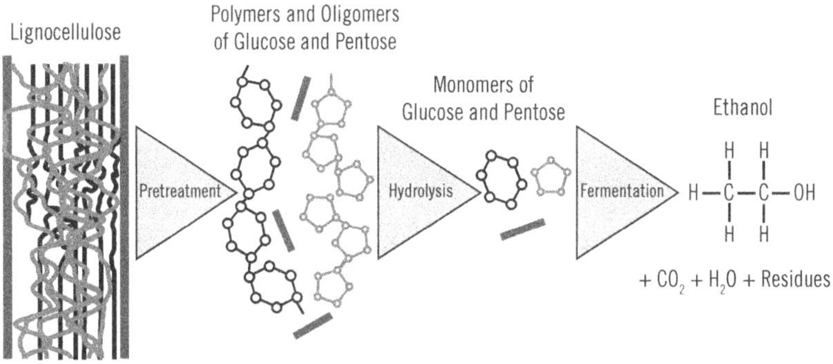

Hydrolysis and fermentation can be done separately or in the same tank, which reduces costs. In a biorefinery, nonfermentable material, such as lignin, is used to power the process. Furthermore, expensive catalysts are recovered for reuse. Carbon dioxide is recycled into plant matter through production agriculture. Figure 8–5 shows the process sequence of a biorefinery.

High concentrations of sugar and ethanol should be aimed at, so that the amount of energy needed for distillation can be minimized.

Feedstock Preparation and Pretreatment

Handling of the feedstock is an important step because it facilitates the biochemical conversion of lignocellulosic biomass. Size reduction makes the feedstock more accessible to process media and therefore reduces enzyme loadings as well as the amounts of water and chemicals needed. After the mechanical process of milling, the biomass particles are pretreated. The two main goals of the pretreatment step are improving the rate (amount of converted monomers per time unit) of enzyme hydrolysis, and increasing the yield of fermentable sugars from cellulose or hemicellulose.

Pretreatment is one of many steps in the cellulose-to-ethanol process, but it currently represents a critical and expensive step that must take place for hydrolysis to occur. Therefore, it cannot be engineered without regard to the other steps. An effective pretreatment has several attributes, including reducing particle size and conditions that avoid degradation of pentose from hemicellulose or glucose from cellulose, and limiting formation of degradation products that inhibit growth of fermentative microorganisms (which are needed for the fermentation process). Pretreatments should also limit energy, chemical, and/or enzyme usage in order to limit the cost of the pretreatment process itself.[29]

figure 8–5. UNIT OPERATIONS OF BIOREFINERY

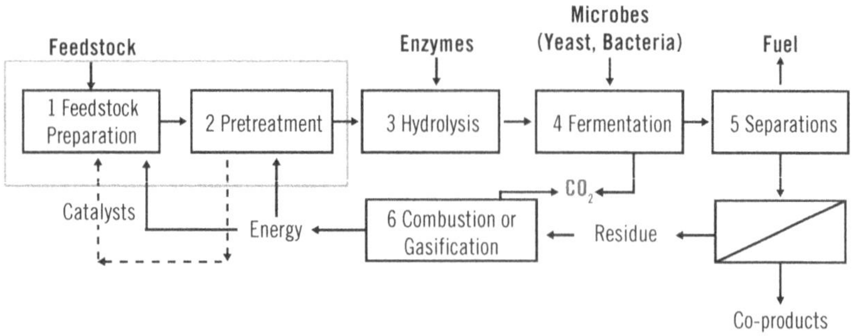

The question arises of how much economic effort should be devoted to pretreatment. To answer this question, the process parameters must be balanced against their impact on the cost of all other process steps before and after pretreatment. In addition, a tradeoff between operating costs, capital costs, and biomass costs have to be taken into account.

There are two different pathways for pretreatment: physical and chemical. Comminution, or mechanically reducing the particle size, and steam explosion (see below) are physical means. Chemical pathways are adding solvents to dissolve the cellulose (an effective but expensive method) or using other chemicals.[30]

Each pretreatment pathway has distinct effects on the chemical composition and chemical/physical structure of lignocellulosic biomass. Table 8–2 gives an overview of the effects caused by different pretreatment methods. One effect is the increase of SSA by creating pores in the cellulose structure, thus improving the micro-accessibility of the cellulases. Decrystalization of cellulose refers to disrupting the hydrogen bonds that tie up the cellulose chains. Decrystalization increases the macro-accessibility of the cellulases. The removal of both hemicellulose and lignin weakens the structure of lignocellulose, because it eliminates part of the network that accounts for the strength of the cell wall. The enzymes' accessibility to the cellulose fibrils is

table 8–2. EFFECT OF VARIOUS PRETREATMENT METHODS ON CHEMICAL COMPOSITION AND CHEMICAL/PHYSICAL STRUCTURE OF LIGNOCELLULOSIC BIOMASS

	Increases specific surface area	Decrystalizes cellulose	Removes hemicellulose	Removes lignin	Alters lignin structure
Uncatalyzed steam explosion	X		X		*
Liquid hot water	X	ND	X		*
pH controlled hot water	X	ND	X		ND
Flowthrough liquid hot water	X	ND	X	*	*
Dilute acid	X		X		X
Flowthrough acid	X		X	*	X
Ammonia fiber explosion	X	X	*	X	X
Ammonia recycled percolation	X	X	*	X	X
Lime	X	ND	*	X	X

X = Major effect   * = Minor effect   ND = Not determined

Source: M.S. Mosier et al., "Features of Promising Technologies for Pretreatment of Lignocellulosic Biomass," *Bioresource Technology* 96 (2005), 677.

improved when hemicellulose is solubilized and lignin is partially solubilized. Altering the lignin structure means partially cracking it and breaking the bonds between cellulose and its lignin sheathing. Different pretreatment methods emphasize certain effects, and the combinations of these effects increase the overall efficiency of the pretreatment.[31]

Pretreatment increases the yields of sugars and ethanol during hydrolysis and fermentation, respectively, when operated at optimal conditions. However, biomass pretreated at overly harsh conditions can lead to formation of degradation products, which decreases sugar and ethanol yields. The pretreatment model in figure 8–6 illustrates the process of cellulose saccharification and degradation.

Pretreatment renders the recalcitrant cellulose fraction of the biomass more accessible to enzymes by altering its structure. During this step, crystalline (C) and amorphous (C*) cellulose is converted into glucose oligomers (G_n). If the process conditions are too severe, the bonds in the cellulose chain will break further into sugar monomers (G), which will eventually degrade to toxic substances. The constant *k* represents the rate of reaction at each stage. The rates and the exposure time of the cellulose to the pretreatment environment determine the level of degradation.[32]

Pretreatment Technologies

Uncatalyzed steam explosion. High-pressure steam is applied to the biomass without addition of chemicals.[33] The rapid thermal expansion following the compression opens up the particle structure. Then, the water itself or SO_2 acts as an acid at high temperatures. During steam explosion, acids are released. This leads to removal of hemicellulose (from the complex lignocellulosic structure), improving their accessibility to the enzymes.

figure 8–6. PRETREATMENT MODEL

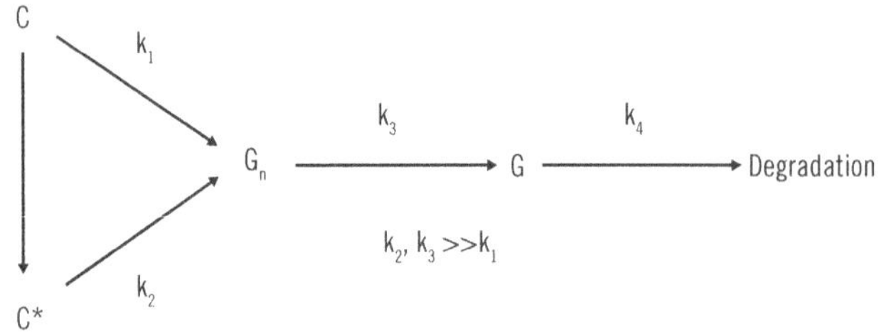

Source: N.S. Mosier et al., "Features of Promising Technologies for Pretreatment of Lignocellulosic Biomass," *Bioresource Technology* 96 (2005), 675.

Liquid hot water pretreatments. Cooking the biomass in hot water breaks lignocellulosic particles without adding acid. Pressure is needed for this process in order to maintain the water in a liquid state. However, water pretreatment requires no neutralization or conditioning of chemicals because of the absence of acid. A highly digestible cellulose results when enzyme is added, high yields of sugars from hemicellulose occur, and more than half of the lignin is dissolved during pretreatment. By maintaining the pH constant between 5 and 7, the hemicellulose fraction is converted into oligosaccharides while minimizing the formation of monomeric sugars. Thus, degradation reactions, which would lower the yield of fermentable sugars, are hindered.

There are three methods of transporting the hot water through the biomass: co-current, countercurrent, and flowthrough. This leads to diverse appliance constructions with distinct yields and concentrations of sugar.

Experiments with corn stover have shown that about 40 percent of the lignocellulose can be dissolved during pretreatment releasing sugar oligomers. In the following hydrolysis of the pretreated biomass, nearly 80 percent of the cellulose and 70 percent of the hemicellulose have been converted into glucose and pentose, respectively. This is due to the change of the chemical composition of corn stover. In addition, the physical appearance of corn stover is affected in such a way that the surface becomes perforated. The pores increase the enzyme-accessible surface area, which increases the enzyme digestibility. The result of pretreatment, in terms of yield and avoiding degradation, is mainly a function of temperature, pressure, pH, and time. While high temperature tends to reduce the required time for pretreatment, lower temperature may be preferred since the operational pressure, and therefore the energy demand, is lower.[34] The relevance of pH and the effect of degradation will be discussed in detail.

Acid pretreatment. Lignin is disrupted and hemicellulose is removed when dilute sulfuric acid is added to the heated biomass, enhancing digestibility of cellulose in the residual solids. The yields are very high, but so are the costs. The latter is due to:

- expensive construction materials (because of corrosion)
- the acid neutralization step (acid degrades the sugars)
- the release of natural biomass fermentation inhibitors
- the disposal of neutralization salts
- long reaction time
- high enzyme demand
- high energy demand for the cellulose grinding.[35, 36]

Flow-through acid pretreatment. Most of the disadvantages of acid pretreatment can be avoided when configuring the acid level at lower than 0.1

percent. The sugar yields are very high, and only few acids are released during the reaction. However, the system requires significant energy for pretreatment and product recovery as a result of the great amount of water processed.[37]

Lime pretreatment. Lime, calcium oxide, is an alkali and can be dissolved in water, where it becomes a medium strength base. Lignocellulose can be treated with lime water at ambient conditions. Low temperature and pressure will result in long reaction times, while high temperature and pressure will accelerate the reaction. The major effect of this technology is the removal of lignin from the biomass. A limitation occurs because some of the alkali is converted to irrecoverable salts or incorporated as salts into the biomass.[38]

Ammonia pretreatment. Liquid ammonia causes swelling of the lignocellulose, which increases SSA. In ammonia recycled percolation, ammonia reacts primarily with lignin. The ammonia depolymerizes the lignin and breaks the lignin-cellulose linkages. Ammonia fiber/freeze explosion simultaneously reduces lignin content and removes some hemicellulose while decrystallizing cellulose.

While the cost of ammonia and especially ammonia recovery drives the cost of this pretreatment, the sugar yields are very high and the enzyme loadings as well as the degradation are low.[39] Table 8–3 shows the main characteristics of the technologies described above.

table 8–3. REPRESENTATIVE PRETREATMENT CONDITIONS OF CONSIDERED TECHNOLOGIES

Pretreatment technology	Chemicals used	Temperature, °C	Standard atmospheric pressure, absolute	Reaction times (minutes)
Uncatalyzed steam explosion	Water	240	40	<15
Liquid hot water	Water	180–220	24–27	15–20
Dilute sulfuric acid	0.5%–3% sulfuric acid	130–200	3–15	2–30
Flowthrough acid	0.0%–0.1% sulfuric acid	190–200	20–24	12–24
Lime	0.05 gram–0.15 gram Ca(OH)2/gram biomass	70–130	1–6	60–360
Ammonia recycled percolation	10–15 weight percent ammonia	150–170	9–17	10–20
Ammonia fiber explosion	100% (1:1) anhydrous ammonia	70–90	15–20	<5

pH Monitoring during Pretreatment

Harsh pretreatment environments, like liquid water temperatures above 200°C and pH below 5, decrease the yield of fermentable sugars and thus the yield of ethanol. The pH decreases when the temperature decreases during pretreatment. At low pH, organic acids can form, thus leading to dissociation of hydrogen ions. The existence of hydrogen ions during pretreatment promotes autohydrolysis, or hydrolysis of cellulose to oligosaccharides and glucose in absence of enzymes. Autohydrolysis has a negative impact on the entire process for two reasons:

- Glucose tends to degrade to organic acids in the harsh pretreatment conditions. The organic acids such as furfur then form more hydrogen ions, which further accelerates utohydrolysis.
- Degradation decreases the extent of enzymatic conversion of polysaccharides and oligosaccharides to monosaccharides, which are needed for fermentation.

The reduction of pH can be avoided by means of adding KOH, which is a strong base. Controlling the pH at about 7 (neutral) results in a higher percentage of solid cellulose and a lower concentration of hydrogen ions after the pretreatment. A higher pH leads to bacterial contamination and must be avoided as well.

The formation of organic acids is relevant only to the cellulose fraction of biomass. The solubilization of hemicellulose during pretreatment, which occurs to almost 100 percent, does not result in the formation of significant amounts of acids.

The choice of chemicals used, pH, temperature, and pressure applied:

- affects the construction design of the pretreatment unit (cost of pretreatment itself)
- influences the yields from cellulose and hemicellulose
- releases cellulose and hemicellulose in distinct states of polymerization.

Due to the varying structure and composition of different lignocellulosic resources, the pretreatment methods applied should depend on the raw material used.[40]

Hydrolysis

The rigid crystalline structure of cellulose needs to be decrystallized in order to produce ethanol. There are two pathways of hydrolyzing: the use of acids and of enzymes (the latter being focused on here). Enzymatic hydrolysis has the potential to make ethanol, derived from cellulose biomass, competitive when compared to other liquid fuels on a large scale.[41] Cellulolytic enzymes, evolved from fungi and bacteria, are used in these processes as well as in other industries (such as the textile industry) to liberate the individual glucose monomers that will then be fermented to ethanol. Hydrolysis refers to a mechanism of breaking the bonds of the glucose chain by adding water.[42]

For improving the efficiency of the hydrolysis, it is important to gain knowledge in the substrate parameters (different crystal forms, varying degrees of crystallinity) and the enzymatic system applied (number of enzymes, synergistic effects, and product inhibition). Due to the high cost of cellulolytic enzymes, the enzymatic system needs to be optimized. The enzyme loading also depends on the results of pretreatment. The more accessible the cellulose is for the enzymes, the lower is the required enzyme loading.[43]

Cellulases are enzymes that catalyze cellulolysis (hydrolysis of cellulose). Cellulases are classified into three main groups: cellobiohydrolases (CBHs), endoglucanases, and beta-glucosidases. These groups play distinct roles in the hydrolysis. While CBHs attack the chain ends of the cellulose polymers, Egs attack the amorphous regions of the cellulose chains. After breaking the cellulose, beta-glucosidase hydrolyzes the generated cellobiose to glucose.[44] The three types of enzymes operate synergistically. The effect of the most obvious synergism, called endo-exo, is that Egs generate more chain ends by random scission for the CBHs to attack.[45]

The majority of the cellulases (CBHs and Egs) are modular proteins with two distinct, independent domains. The catalytic core is responsible for the hydrolysis of cellobiose from the cellulose chain. The cellulose binding domain (CBD) has a dual activity. The CBD is in charge of binding the enzyme to the cellulose layer, and it also affects the cellulose structure. After the enzyme has bound to the cellulose, it reduces the cellulose's particle size, therefore increasing SSA. Since the enzyme is connected to the cellulose layer via the CBD, the catalytic core can work independently from the CBD. This molecular architecture allows the enzyme to release the product while remaining bound to the cellulose chain. A linker region joins the independent catalytic core to the CBD (see figure 8–7).[46]

Fermentation and Separation

The glucose monomers that are released during hydrolysis can be fermented to ethanol using yeast. Saccharomoyces are natural yeasts that feed on the glucose to produce ethanol and are currently applied in the large-scale corn-to-ethanol and sugar cane-to-ethanol industries. However, ethanol production from lignocellulose requires fermentation of not only glucose, but also pentose sugars. Saccharomoyces are not able to ferment pentose. One way to manage fermentation of pentose is the utilization of genetically modified yeasts, specifically engineered for this purpose.[47]

Hydrolysis performed separately from fermentation is known as *separate hydrolysis and fermentation*. Cellulose hydrolysis carried out in the presence figure 8–7. MODULAR NATURE OF CELLULOLYTIC PROTEINS

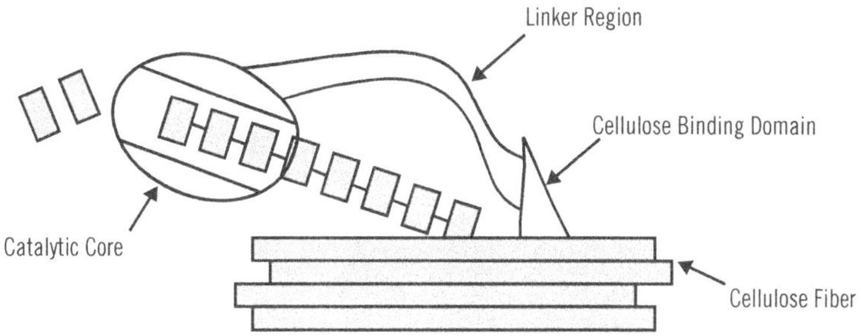

Source: N.S. Mosier et al., "Reaction Kinetics, Molecular Action, and Mechanisms of Cellulytic Proteins," in *Advances in Biochemical Engineering/Biotechnology*, vol. 65, ed. T. Scheper (Berlin: Springer, 1999), 33.

of the fermentative organism is referred to as *simultaneous saccharification and fermentation*. Simultaneous saccharification of both cellulose (to glucose) and hemicellulose (to pentose) and co-fermentation of both glucose and pentose are realized by genetically engineered microbes.[48]

Ethanol is recovered from the fermentation broth by distillation.[49] The residual lignin, unreacted cellulose and hemicellulose, ash, enzyme, organisms, and other components end up in the bottom of the distillation column. These materials may be concentrated and burned as fuel to power the process, or converted to various coproducts.[50]

Challenges

Fundamental research in biochemistry has been conducted in order to achieve the state-of-the-art technologies. However, feedstocks, pretreatment, and hydrolysis as well as the use of enzymes and yeast are not yet fully understood individually and as a system. This knowledge is critical in order to make current technologies more efficient so that cellulosic ethanol can be produced on a commercial scale.

Guideline and Time Frame for Establishing a Biofuel Industry

The usage of cellulose biomass has the potential to become the key source for meeting America's future transportation fuel needs. As mentioned above, a sustainable biofuel industry is based on supply of sufficient biomass and on cost- and energy-efficient conversion technologies. In 2006, DOE set up a technical strategy to develop a viable cellulosic biomass to biofuel industry over the next

5 to 15 years.[51] This strategy includes concurrent development of innovative energy crops with new biorefinery treatment and conversion processes.

The development is built on three successive phases. The research phase will provide the basis of knowledge in the fields of agriculture and biochemistry. This knowledge will be transferred into technology modules in the technology deployment phase. In the systems integration phase, all modules will be consolidated as a system to ensure productivity and sustainability. The roadmap is summarized in figure 8–8.

Research Phase

Fundamental knowledge in three critical areas—plant design, deconstruction of lignocellulose, and fermentation and recovery—has to be acquired.

The traits of current energy crops have to be optimized, since the crops play an important role in the bioenergy system. The biofuel industry strives for crops with high yields and improved robustness that are also uncomplicated to process. However, growing such crops in large quantities will require a sustainable soil ecosystem. Research in plant design will enhance understanding of cell-wall polymer organization and interactions.

Further research in pretreatment of cellulosic biomass will focus on improving hydrothermal and mechanical pathways. Also, a deeper understanding of enzyme function is required. This will lead to higher hydrolysis rates and increased icrosyst robustness. A goal is to be able to redesign cellulosome (containing the full repertoire of degrading various polysaccharides) for a diversity of feedstock.

An organism needs to be found that allows simultaneous saccharification and co-fermentation—that is, not only co-fermentation of pentose and hexose, but also incorporating hydrolysis and co-fermentation in one single step, thus reducing the number of processing steps. Due to the nature of the process, such an organism must have high tolerance to stress, inhibitors, and high alcohol concentrations.

Technology Deployment Phase

In this phase, plant domestication technologies will reach a mature state. This includes availability of multiple crops for distinct regional and global agroecosystems. Furthermore, improved pretreatment procedures, enzymes, and fermentation organisms will be deployed in existing biorefineries. First results of the concurrent development of feedstock and processing technologies will be recognizable. Deconstruction of lignocellulose should become easier to handle due to the domestication of the energy crops.

figure 8–8. STAGES AND CONTENTS OF DEPARTMENT OF ENERGY BIOFUELS ROADMAP

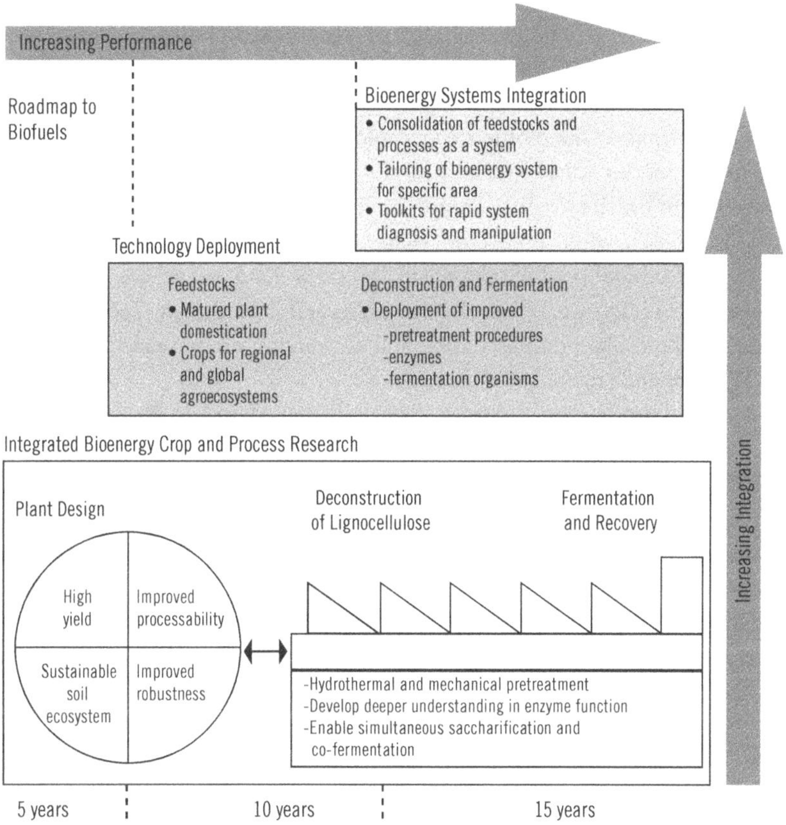

Source: U.S. Department of Energy, *Breaking the Biological Barriers to Cellulosic Ethanol: A Joint Research Agenda*, DOE/SC–0095 (Washington, DC: U.S. Department of Energy Office of Science and Office of Energy Efficiency and Renewable Energy, 2006), available at **www.doegenomestolife.org/biofuels**.

Systems Integration Phase

Feedstocks and processes will be fully consolidated as a bioenergy system and will accelerate and simplify the end-to-end production of ethanol. Bioenergy systems do not only function in a single area. They also are tailored for specific regional climate and soil characteristics. Toolkits will be made available that enable rapid diagnosis and manipulation of all critical aspects of the biosystem (field and facilities).

Conclusion

Successfully establishing a cellulose-to-biofuel industry requires understanding of a complex system of parameters. Biomass availability, technological progress, international fuel markets, and fuel utilization are parameters that will play major roles in the biofuel area.

In order to make biofuels economically feasible, several requirements have to be met. First, large quantities of biomass must be made available for biofuel use in order to reduce competition with other agricultural industries (food and feed), while high crop yields will reduce agricultural land use. Sufficient production of biomass and efficient land usage are important factors for avoiding the effect of price increase for feedstock. Technological progress in both plant domestication and conversion processes will drive this development and ensure sustainability.

Biofuel production and trade have to be considered on an international stage since production externalities affect the market price. Production costs vary among countries due to different fuel sources (such as sugar cane, bagasse, corn stover, sugar beet), land availability, labor cost, and so forth. Producing biofuels in an environmentally friendly and sustainable manner is important but also cost effective. However, there are not yet international legislations or incentive measures to ensure equal market opportunities for all producing countries. Low-cost biofuel imports to the United States will therefore have an impact on the feasibility of domestic production.

Action within the automotive industry can accelerate the biofuel market share. This includes providing the vehicle market with engines that are able to drive with a variety of fuels. Due to the nature of the conversion processes, biorefineries are economically feasible when they can generate a great diversity of products, including a variety of transportation fuels. However, this variety of liquid fuels is only marketable if there is a demand for a range of biofuels. Continuing the development of flexible fuel vehicles and promoting their acceptance in society will therefore be a key in establishing a biofuel industry.

NOTES

1 U.S. Department of Energy, *Breaking the Biological Barriers to Cellulosic Ethanol: A Joint Research Agenda*, DOE/SC-0095 (Washington, DC: U.S. Department of Energy Office of Science and Office of Energy Efficiency and Renewable Energy, 2006), available at **www.doegenomestolife.org/biofuels**.

2 Ibid., 9.

3 European Biomass Industry Association, "Bioethanol and the World," available at **www.eubia.org/212.0.html**.

4 U.S. Department of Energy, Energy Information Administration, "Petroleum Navigator," August 28, 2009, available at **http://tonto.eia.doe.gov/dnav/pet/hist/m_epooxe_yop_nus_1M.htm**.

5 *Breaking the Biological Barriers to Cellulosic Ethanol*, 12.

6 Ibid.

7 "Petroleum Navigator."

8 Barry D. Solomon, Justin R. Barnes, and Kathleen E. Halvorsen, "Grain and Cellulosic Ethanol: History, Economics, and Energy Policy," *Biomass and Bioenergy* 31, no. 6 (June 2007), available at **www.sciencedirect.com/science/article/B6V22-4N7YFWT-3/2/26b377d735b2a30e206fa35ed64903cf**, 417.

9 Doris de Guzman, "Biodiesel: Boom or Bust," *ICIS Chemical Business Americas*, February 5, 2007.

10 F.O. Lichts, *World Ethanol and Biofuels Report* 6, no. 1 (September 6, 2007).

11 de Guzman.

12 Solomon, Barnes, and Halvorsen.

13 "Bioethanol and the World."

14 U.S. Department of Energy, Energy Information Administration, "Country Analysis Briefs: Brazil," October 2008, available at **www.eia.doe.gov/emeu/cabs/Brazil/pdf.pdf**.

15 Solomon, Barnes, and Halvorsen.

16 Ibid.

17 877,000 tons/year = 36,831,000 gigajoules (GJ)/year (1 ton = 42 GJ); energy content of ethanol (LHV) = 0.0234 GJ/liter; 3.785 liters = 1 gallon.

18 "Biofuels Barometer 2007," *Le Journal des Energies Renouvables*, no. 179, available at **http://ec.europa.eu/energy/res/sectors/ bioenergy_publications_en.htm**.

19 "Bioethanol and the World."

20 3,852,000 tons/year = 161,787,000 gigajoules (GJ)/year (1 ton = 42 GJ); energy content of biodiesel (LHV) = 0.034 GJ/liter; 3.785 liters = 1 gallon.

21 "Biofuels Barometer 2007."

22 Available at **http://epp.eurostat.ec.europa.eu/portal/page?_pageid=1996,39140985&_dad=portal&_schema=PORTAL&screen=detailref&language=en&product=sdi_tr&root=sdi_tr/sdi_tr/sdi_tr_gro/sdi_tr1230**.

23 Ibid.

24 *Breaking the Biological Barriers to Cellulosic Ethanol*.

25 Ibid.

26 Ibid.

27 N.S. Mosier et al., "Reaction Kinetics, Molecular Action, and Mechanisms of Cellulytic Proteins," in *Advances in Biochemical Engineering/Biotechnology* vol. 65, ed. T. Scheper (Berlin: Springer, 1999), 26.

28 C.E. Wyman et al., "Coordinated Development of Leading Biomass Pretreatment Technologies," *Bioresource Technology* 96 (2005), 1961.

29 N.S. Mosier et al., "Features of Promising Technologies for Pretreatment of Lignocellulosic Biomass," *Bioresource Technology* 96 (2005), 675.

30 Ibid., 676.

31 Ibid.

32 J.R. Weil, "Unified Model for Hydrolytic Effect during Cellulose Pretreatment," master's thesis, Purdue University, 1993, 10.

33 Cristobal Cara et al., "Influence of Solid Loading on Enzymatic Hydrolysis of Steam Exploded or Liquid Hot Water Pretreated Olive Tree Biomass," *Process Biochemistry* 42, no. 6 (June 2007), available at **www.sciencedirect.com/science/article/B6THB-4NCSGST 2/2/66e5b58a04cca8acf97941b13dd5c21f**.

34 N.S. Mosier et al., "Optimization of pH Controlled Liquid Hot Water Pretreatment of Corn Stover," *Bioresource Technology* 96 (2005).

35 Wyman et al.

36 Mosier et al., "Features of Promising Technologies."

37 Ibid.

38 Ibid.

39 Ibid.

40 J. Weil et al., "Continuous pH Monitoring During Pretreatment of Yellow Poplar Wood Sawdust by Pressure Cooking in Water," *Applied Biochemistry and Biotechnology* 70–72 (1998).

41 C.E. Wyman, "Biomass Ethanol: Technical Progress, Opportunities, and Commercial Challenges," *Annual Review of Energy and the Environment* 24 (1990), 190.

42 Mosier et al., "Reaction Kinetics, Molecular Action, and Mechanisms of Cellulytic Proteins," 36.

43 *Breaking the Biological Barriers to Cellulosic Ethanol*, 31–32.

44 Mosier et al., "Reaction Kinetics, Molecular Action, and Mechanisms of Cellulytic Proteins," 24–25.

45 Ibid., 31.

46 Ibid., 33–36.

47 N.W.Y. Ho et al., "Genetically Engineered Saccharomyces Yeasts for Conversion of Cellulosic Biomass to Environmentally Friendly Transportation Fuel Ethanol," American Chemical Society Symposium Series, 2000, 144.

48 Mosier et al., "Features of Promising Technologies," 674.

49 M.R. Ladisch et al., (1984) "Cornmeal Absorber for Dehydrating Ethanol Vapors," *I&EC Process Design and Development* (1984), 23.

50 C.E. Wyman, "Ethanol from Lignocellulosic Biomass: Technology, Economics, and Opportunities," *Bioresource Technology* (1995), 50.

51 *Breaking the Biological Barriers to Cellulosic Ethanol.*

chapter 9

BIO-INSPIRED MATERIALS and OPERATIONS

Erica R. Valdes

We all feel humble in the face of the ingenuity and complexity of nature. Yet it need not prevent us from learning, exploring, and even attempting to construct clinically useful artificial systems having a few of the simpler properties of their natural counterparts.

—Thomas Ming Swi Chang[1]

In 1956, Dr. Chang, then an undergraduate, had the audacity to believe he could build artificial blood cells. Working in his own residence room, he proved himself correct.[2] Today, Dr. Chang has been joined by a legion of scientists and engineers working to duplicate materials and functions formerly restricted to the realm of nature.

It is impossible to address the full breadth of this field in a single chapter. The goals here are rather to present some of the main factors driving the recent explosion of bio-inspiration, to provide some specific examples of successes in the field, to address some of the more relevant topics in current research, and to provide some future possibilities and military applications.

Why Bio-inspired Materials?

The living world can now be viewed as a vast organic Lego kit inviting combination, hybridization, and continual rebuilding.

—Edward Yoxen[3]

This quotation can be considered the catch phrase of modern biotechnology; it applies to all bio-inspired materials and functions as well.

Nature has perfected countless astounding capabilities, and our role is to recognize them, learn from them, and use them in all the myriad combinations they present.

Strictly speaking, *biotechnology* is the use of biological science to create chemicals, materials, and organisms through the control of biological processes. There is no presupposition that the target products be naturally occurring biological or biochemical entities; in reality, it is more typical that the targets result from some degree of engineering to provide products not identical to those of nature. *Bio-inspired* is the current terminology used to describe the class of materials based either directly or indirectly on material structures and functions observed in the biosphere. Clearly, there is overlap where a biotechnological manufacture approach is adopted in the production of chemicals, materials, and organisms to be used for nonbiological applications and where a product is a hybrid of products of strict biotechnology and abiotic analogs of biological forms or functions.

Products of early biotechnology were often classified as *biomaterials*. This term is ambiguous. To a material scientist working in the area of materials intended for implant into the body, biomaterials are ones designed to be compatible with living tissue. The term was well established in that community before the current explosion in bio-inspired materials began. In response to this confusion, two other terms were adopted: *biotics* and *biomimetics*. Biotics refers to direct use of materials of biological origin, while biomimetics refers to the use of abiotic methods to mimic the structure and/or properties of biological materials. Currently, the term *bio-inspired* is used to encompass both biotic and biomimetic materials.

Biological systems excel over traditionally manmade systems in a number of arenas. In some cases, the advantages are primarily structural; nature has developed methods of achieving complex structures that are ideally suited to their applications. These structures provide inspiration both in direct emulation for similar applications and in paradigmatic studies allowing extension to distantly related applications. Other cases provide primarily functional advantages. These provide inspiration in specific solutions to functionality requirements in manmade systems.

Lastly, some cases provide processing and fate advantages. Natural synthesis processes are generally environmentally benign, and natural products are typically biodegradable. A range of areas in which biological systems excel and bio-inspired materials can provide advantages is identified according to the advantage category in table 9–1. Clearly, there is significant advantage to be gained by studying and emulating materials found in natural biological systems.

table 9–1. AREAS IN WHICH NATURAL BIOLOGICAL MATERIALS EXCEL

Structural Benefits	Functional Benefits	Processing/Fate Advantages
Broad range of polymer properties accessible	Adaptation to environment	Environmentally benign processing
	Signal amplification	
Atomic level control of structure	Biocompatibility	Biodegradation
	Range and control of color change	Self-assembly
Conformational changes	Catalysis	Ambient conditions
Hierarchical structure	Computation	
Lightweight materials	Energy conversion/conservation	
Micro- and nano-devices	Functional adaptation/evolution	
Self-repair	Lubricants	
Structure/function relationships	Adhesives	
	Membranes/transport control	
	Multifunctional materials	
	Smart materials/sensor and response	

Simple Examples

At the dawn of civilization, the only materials available to man were either biological or mineral, and prior to the development of tools there were limits on what could be done with rocks. Biomaterials were clearly the answer. Historically, people have been using and mimicking biological materials for ages. From the first fig leaf, people have been using biological material for clothing and shelter: plant and animal fibers for threads, yarns, and rope; animal hides and fur for warmth and shelter; reeds and grasses for woven containers; and wood for supports and structures. As civilization advanced and the written word evolved, stone tablets were replaced as a writing surface by papyrus, bark, animal skins, and eventually paper, bioderived materials that provided significant advantages over stone.

Nor is the use of materials to build artificial parts to replace biological parts new. Dentures, peg legs, glass eyes, and hand hooks illustrate the point that biocompatibility and "medical implants" have in fact been around for centuries.

From the standpoint of biomimetic bio-inspiration—that is, emulating a biological function in either an organism not naturally provided with that function or a nonbiological item—there are also historical examples. Man's obsession with the power of flight has been in evidence throughout folklore, mythology, and history. Likewise, our fascination with the aquasphere has led us to submarines and scuba gear.

The Beginning of Modern Bio-inspiration

In the 1980s, a revolution in bio-inspired materials began. Several aspects of natural biological materials came to the forefront for various reasons.

Bone

Biocompatible material and implant researchers were interested in developing a more natural artificial bone. Bone is unique in its structure and properties. In structure, it is a composite of organics (collagen fibers) and minerals (crystalline hydroxyapatite). The structural properties vary according to the location in the body. Overall, the material is a strong but generally brittle composite of varying ductility. In a living system, bone is also self-healing and sensing. Stress-generated electrical signals allow living bone to respond to demands placed on it by preferential growth, in essence customizing itself to its needs. Since the beginning of the interest in bone, scientists have succeeded in developing structural mimics using hydroxyatpatite crystals in composite with polyethylene. The more complex goals of making a sensing and self-healing bone replacement have yet to be achieved.

Soft Tissue

Soft tissue was recognized in the 1980s to have properties unheard of in the abiotic world. Collagen systems have the unusual characteristic of having mechanical properties that change with the hydration state of the material. Collagen shows the additional advantage of the ability for the individual fibrils to self-assemble. Thus, collagen works on several levels: first, the individual fibers are themselves connected to form larger fibers; second, the larger fibers have unique elastic properties; third, these fibers are variously incorporated into composite materials throughout the body. As the main structural fiber of the animal kingdom, it is found in applications ranging from tendons to skin and bone. Variations between the myriad applications come from the ways in which the fibers are incorporated into the composites.

A specific example of collagen in the body is the intervertebral disc. From a mechanical standpoint, they are particularly interesting in that they are designed to convert structural compressive stresses into tensional stresses in the disc material. This conversion is achieved by using what is termed *hierarchical structure*, in which the interaction between multiple levels of structure results in unique or favorable properties. Again, this is the reason behind broadly divergent applications of the same material in organisms.

In concluding the discussion of soft tissue concepts, consider the aortic valve. As an example of the astounding capabilities of biological tissue, this valve opens and closes at a rate of 60 times per minute nonstop. Over the course of a typical lifetime, this amounts to something on the order of three billion flex cycles, an enviable number to anyone concerned with material fatigue.

Exoskeletons/Invertebrates

During the same time frame, looking to invertebrates provided even more insight and inspiration. The insect exoskeleton is of particular interest to the field of structural composites. A major difficulty in the manufacture of composites is control of final product shape and the alignment of reinforcing fibers. Ideally, a composite part is produced in "net shape" rather than in a bulk form that is later cut. A typical process might include initial "lay-up" of the reinforcing fibers (in the case of fiber-reinforced materials), followed by impregnation processes driving matrix materials around and between the fibers, and a curing process to polymerize (carbon-carbon composites), flame (carbon-ceramic composites), or sinter and anneal (metal matrix composites) the matrix. The resulting material is likely to undergo shape changes that are difficult to control, and the desirable properties of the product make it difficult to machine to final state.

Insect exoskeletons are in fact fiber-reinforced composites with precise reproducibility. Characterization of the shell of the bessbug showed that it is a model composite with single-fiber thickness per layer with helical orientation. Copying the fiber orientation array of the bessbug shell resulted in polymer matrix composites that exit the curing process in a shape nearly identical to that of the shell, moving us a step closer to the goal of net shape control.

Another interesting material from the invertebrate world is nacre, also known as mother of pearl, which is also known for its impact strength. Microscopically, nacre is composed of layers of brittle platelets of the mineral aragonite (a calcium carbonate) separated and held together by sheets of elastic biopolymers including acidic macromolecules, silk-like proteins, and b-chitin. This classical composite approach combines the strength of the mineral with the ductility of the polymers to produce a tough and resilient material.

However, the story does not end there. In addition to layer strength with ductility, the mineral platelets in nacre are typically hexagonal platelets arranged in single layers that are staggered like brickwork with biolpolymer acting as the mortar. This arrangement is particularly useful in the inhibition of crack formation. Biologically, this structure is secreted by the epithelial cells of certain mollusks, either as a protective layer inside a shell (mother of pearl) or to encapsulate a foreign object (eventually forming a pearl). Abiotically, scientists attempt to mimic the composite structure in order to achieve similar control over mechanical properties.

Invertebrates also excel in the area of adhesives. The most familiar example of this is the barnacle, which spends all of its adult life cemented to one spot—so indelibly cemented, in fact, that it can damage naval vessels. Significant research has gone into understanding the barnacle adhesive in attempts to protect against it. A related effort has arisen to understand it with the intent of recreating it as a superior adhesive for marine applications. These studies have migrated to other biological adhesives in aquatic environments including that of the mussel, which has yielded the first example of a proteinaceous adhesive that is "cured" by cross-linkages that form when the organism extracts iron from the surrounding seawater. Again, from the biological side these studies are directed at understanding and preventing adhesion to naval vessels while the materials applications may yield unprecedented adhesives, particularly for use under water.

A final invertebrate worth mentioning for adhesive applications is the banana slug. Its slime is a versatile material that it uses for everything from lubrication to personal protection. This substance has also been claimed as the strongest adhesive found in nature. Attempts at understanding and recreating this slime have been unsuccessful to date.

Biomineralization/Ceramics

As with composites, minerals and ceramics can be very dependent on the microstructure of the material. In the case of ceramics and composites, the final material comprises small particles of component materials. Typical fabrication processes for ceramics use sol-gel processing, which begins with the generation of a liquid suspension of very fine particles called a colloidal suspension. This suspension then forms a gel in which the particles link together. The gel is then treated by a combination of compression, drying, and baking to produce the final ceramic. The properties of the product depend on the success in controlling or eliminating the void space between the particles. Thus, control of colloidal particles in the production of ceramics is an important challenge to ceramists.

Looking to biology, the biominerals found in, for example, bone and shells are grown rather than processed. They are built biochemically in a controlled and methodical way, leaving little to chance. This approach directly addresses several issues that regularly haunt ceramic processing ventures: room temperature sintering, limitations on packing density, undesirable interactions of particles in solution, and effects of additives on products. When incorporating ceramics into composites, another significant concern is the alignment of the particles. Depending on the application, it may be desirable for the particles to be either randomly oriented or, as in the case of near–net shape processing mentioned earlier, aligned very precisely in specific conformations.

This is another area in which biological systems excel. In many cases, the biochemical control not only produces near-perfect ceramic particles, but also produces them in the desired conformation and simultaneously produces the surrounding matrix material. A typical case is the abalone shell, in which a soft organic component is generated as a foundation that guides the growth of calcium carbonate plates. The level of control is on the size scale of 10 nanometers (nm) for the scaffolding and 200 nm for the mineral plates, or approximately 0.001 and 0.02 times the width of a human hair, respectively. In another case, that of diatoms, a single-cell algae produces a self-assembled organic scaffold that allows directed synthesis of a silica shell. The familiar microscopic shells remain long after the organism has died and decomposed. Studying and reproducing both the mechanical and biochemical aspects of these biological production processes can provide unprecedented control over ceramic and ceramic composite structure and function.

Current Research and Future Materials

Silk/Advanced Fibers and Spider Silk

For thousands of years, people have used the cocoon fiber of *Bombyx mori* as a textile fiber. The silkworm cocoon is in fact one continuous fiber approximately 10 microns in diameter, an order of magnitude smaller than a human hair, and close to 1,000 meters long, providing the staggering aspect ratio of 100 million. It is also an impressively strong fiber, silk fabrics having been observed to provide ballistic protection as early as the 1800s. The silkworm remained the main source of silk primarily because of the relative ease of farming the worms and harvesting the fibers.

In recent years, the incredible strength of spider silk has led researchers to investigate the possibility of reproducing it without the spider. Considering the strength of fibers, there are three properties typically considered. These are the stiffness, which is a measure of how much an object resists deflection when a force is applied; the strength, which is a measure of the force required to break it; and the toughness, which is its ability to withstand impact without breaking.

There are in fact several types of silk produced by spiders for specific purposes. Two of the more familiar are the dragline, which is recognizable as the support frame of the web, and the capture silk, which is the fine sticky silk that ensnares the prey. Although based on the same components, each type of silk has different properties, though all are at or above the strength of Kevlar, tough, processed in ambient conditions, and made of renewable resources. The military application most often mentioned with regard to spider silk is bulletproof clothing. Both dragline and capture fibers can absorb more energy than Kevlar. Dragline specifically can absorb as much as 10 times the energy, most of which is converted to heat.

Unfortunately, spiders are difficult to farm and harvesting of natural spider silk is impractical. Taking a genetic approach to this problem, scientists have identified the spider genes responsible for the fibroin proteins that comprise spider silk. Cloning those genes and placing them in goats has resulted in goats that produce spider silk fibroins in their milk. Media reports might lead one to think that this development has put spider silk armor right around the corner. In reality, this solves just a small part of the puzzle. As with so many biological materials, the morphology—the size and shape of the fiber—and the microstructure—the specific arrangement of the protein molecules within the fibers—of spider silk are complex and relevant to function. Spiders deal with these issues through a combination of mechanics and biochemistry in their spinnerets. Current challenges to the scientists in this field involve manipulating the goat-produced proteins to reproduce or exceed the physical properties of natural spider silk.

Resilin

Another important protein in arthropods is resilin, an elastomeric protein that confers on fleas the ability to jump more than 100 times their height. The protein is capable of many millions of cycles of extension and contraction, and as such is one of the most elastic proteins known. As much as 97 percent of the energy stored in compression of resilin can be recouped in the following expansion. This process allows cicadas to chirp, flies to flap their wings almost a million times an hour, and otherwise brittle insect shells to bend and then revert to their original shape.

In addition to this compression and expansion, resilin can stretch up to three times its length without breaking. These combined properties arise from the structure of resilin on the molecular level and make it more resilient than any current manmade material. Recent research has resulted in the production of pro-resilin by *E. coli*. The current challenge is to address the fact that in nature, resilin requires water to function properly. In nonbiological applications this is seldom practical, so the task remains to recreate the properties of resilin in a dry system.

Multifunctional Coatings

A common requirement in science and engineering is the modification of surfaces to allow, encourage, or enhance surface properties and interactions. Although numerous approaches are used to modify surfaces, most are limited by expense, complexity, or chemical specificity. Taking a bio-inspired approach, researchers have focused on the mussel *Mytilums edulis*. This organism forms adhesive bonds with almost any surface, including Teflon. Scientists are currently deciphering the mechanism that allows this adhesion. Using lessons learned, they have successfully demonstrated that a structural mimic of the relevant biochemicals allows deposition of thin polymer films on practically any surface.[4] Moreover, these films are amenable to fictionalization. This and similar approaches will significantly reduce the complexity, expense, and reliance on highly specific chemistries in the building of functional coatings.

Artificial Cell

The value of the artificial cell has been recognized for decades. Scientists, engineers, and industry continue to investigate this area for the biomedical community. In general, the goal is not to duplicate all cellular activity but rather to duplicate the desired activity in a cell or synthetic cell designed for use in the body. Extending this thought beyond the biomedical arena, scientists are beginning to investigate the possibility of reproducing specific cell functions in the environment that exists outside the body.

One of the challenges in this area will be to devise a support material that allows full operation of the cell-inspired functions in an environment radically different from a living body. In the cell, the support is the cell membrane, and in research directed at aqueous applications, this is typically mimicked by lipid bilayers and hydrogels. For the case of dry applications these systems need to be either stabilized in air,[5,6] enclosed by environmentally hardened materials that allow the encased material to interact freely with its surroundings, or replaced entirely by materials that allow or enhance the desired function. Some of the areas that can be expected to contribute to this process include chitosan, alginate hydrogels, molecular imprints, porous inorganic membranes, phosphorylcholine-polymer surface technology, and polymerosomes.

Octopus

The octopus, just one example of a familiar organism, has an impressive array of talents that are enviable from a military perspective. It is a structurally complex invertebrate, with its beak being its only one small hard feature. The structure and movements of the animal arise entirely from the muscles. As a result, its structure is quite changeable. Octopuses have been known to squeeze their bodies through small cracks—for example, between an aquarium container and its lid—or curl themselves up into a slightly ajar clamshell. Some species have been found to use two arms to "walk" while using the remaining arms for camouflage.[7] While walking is generally believed to require both muscles and skeleton, the octopus uses orthogonal bands of muscle to provide flexibility and motion while the constant internal volume lends support.

This atypical locomotion provides an unusual type of camouflage, allowing the kinematics of the animal to mimic other organisms. Unlike just shape and color change camouflage, this is effective even when the animal moves relatively rapidly.

The more conventional method of camouflage, hiding by matching the surroundings with shape and color changes, is another area in which the octopus excels. Using a complex chromatophore system that includes reflective iodophores as well as hue-changing cells, the octopus can change not only the color but also the pattern and even the texture of its skin, and it can make these changes in as little as 3 seconds.

In addition to impressive mechanical capabilities and camouflage strategies, the octopus has a sense of taste 1,000 times that of humans, vision that allows detection of polarization, and a far better sense of touch than humans. The suckers on the arms of the octopus have chemical receptors as well as conventional touch receptors, allowing them to "taste" what they touch. Even the well-known ink cloud used by the octopus to visually elude its enemies has lesser known qualities of interfering with the senses of the enemy, enhancing the effect of the obscuring ink.

Dry Adhesion and the Gecko

Often, biology relies on probability as a means to an end. This is evident in two examples of dry adhesion. The first of these is the hook and loop fastener, most recognized by the trade name Velcro. In basic terms, these fasteners involve two pieces: a soft array of loops and a wiry array of hooks. Contact between the two pieces causes some of the hooks to attach to some of the loops. The large numbers of hooks and loops on each piece ensure that enough will engage to provide the required adhesion. This technology precedes today's interest in biological inspiration by several decades. Nevertheless, it was inspired by a familiar biological system: the burr. Burrs and similar plant structures are covered in stiff hooks that attach to almost any passing fiber, from dog fur to clothing, an observation that ultimately led to the invention of hook and loop-type fasteners.

There has been much recent interest in the feet of geckos. Many species of this common lizard can readily climb walls and even walk across ceilings. The interesting facet of this is that unlike most adhesive organisms, geckos do not require any fluid adhesive to successfully defy gravity. Microscopic studies of gecko feet reveal that each footpad is covered with hundreds of thousands of setae, each of which is comprised of up to 1,000 spatulae that are held to surfaces by van der Waals attractions, generally considered to be fairly weak intermolecular forces. Scientists have been attempting to recreate this phenomenon using nanofibers. By ensuring that all the tips of nanofibers in an array can contact a surface simultaneously, they have succeeded in matching and even exceeding the adhesive powers of the gecko foot.

Although the gecko is the current hot topic in biomimetic access to extra-human locomotion, it is by no means the only organism from which we can learn. Mosquitoes and water gliders are well known for their ability to walk on water. This ability arises from hierarchical structures on their legs, fine hairs that are grooved on the nanoscale to trap air, effectively producing air cushions between the insect's feet and the surface of the water.

This mechanism is akin to that of gecko climbing in that the ability to accomplish astounding feats at the macroscopic scale arises from massive arrays of structures found at the microscopic or nanoscopic level. This effectively transfers the unusual properties associated with nanotechnology into significant abilities on more familiar size scales. This is a recurring theme in bio-inspiration and one that has only recently been investigated for practical applications.

Plants

Although much of the interest in bio-inspiration arises from the animal kingdom, there are striking examples of bio-inspired success from the plant kingdom, and much remains to be explored. A number of plants, among them the lotus and jewelweed, are recognized for their ability to repel water. The leaves have microscopic level roughness that reduces both the contact area of foreign particles on the leaf and the contact angle of water with the surface. The result is that particles do not adhere well to the leaves and water droplets roll around on the leaves, effectively picking up the particles and cleaning the leaves. This effect has been used commercially in the development of an exterior paint and textiles that are dubbed "self-cleaning."

Bio-inspired Computing and Decentralization

> *If you cut off a spider's head, it dies; but if you cut off a starfish's leg, it grows a new one, and that leg can grow into an entirely new starfish.*
>
> —Ori Brafman and Rod A. Beckstrom[8]

The comparison between a spider and a starfish illustrates decentralization at work. While the spider is controlled by a discrete, identifiable head as the ultimate authority, the starfish is not dependent on a single control center; rather, it is a decentralized organism. Within the biosphere, this phenomenon of decentralization is not limited to simple organisms lacking centralized nervous systems. It is equally evident on a community level in the functioning of ant colonies, bee hives, and bird flocks.

Until recently, the tendency even among scientists was to assume some centralized control node among these creatures: the flock of birds must be following a leader; the ants and bees must be getting orders from the queen. The observed behaviors in these and similar groups are now recognized to be the result of autonomous collectives. Each animal is operating independently according to a set of simple sense and respond rules, and the result is behavior that appears to be driven by an elegant master plan.

A simple example is a group of ants looking for food. The relevant sensing functions are the ability to sense food and the ability to sense pheromones. The relevant responses are to pick up the food and to release pheromones. The rules are simple: if you find food, pick it up and release pheromones, and if you sense pheromones, follow the trail emitting the highest level of them. With these simple rules, you have a situation where the first ant encountering food picks some up, releases some pheromones, and goes on its way. Other ants sense the pheromones and head toward the food, where they pick some up and release more pheromones, strengthening the signal to even more ants. A path to the food is quickly established as more and more ants are attracted. Soon the familiar army of ants marching back and forth between the food source and home is established, all with no centralized authority. After the food source has been depleted, there is no further enforcing of the pheromone trail, it erodes with time, and the trail of ants to that particular location dwindles. There is no "leader ant" that can be eliminated to put an end to the invasion. If the goal is to end the invasion, the only recourse short of reprogramming the sense and response functions is to eliminate enough individual ants that the pheromone trail erodes more quickly than it is generated.

Another element that often accompanies discussions and systems of decentralization and as such deserves brief mention is stochastics. In simplest terms, this refers to randomness or noise. In the case of the ants, this would involve a certain amount of aimless wandering and the collective discovery of richer food sources or new food sources when old ones are depleted.

Bio-inspiration in the world of computation predates bio-inspired materials or systems as a discipline by several decades. The two most recognized examples of this are artificial intelligence and neural networks. The first of these is best described as the field addressing synthetically intelligent machines. It generally uses conditional programming of the "if *X*, then *Y*" variety, in which *Y* can be either a classification/identification decision or an action to be carried out on the condition of *X*. In terms of biological inspiration, this is the programming approach used in task-oriented robotics designed to carry out specific human functions. Simple sensing elements are used to address "if *X*" and are linked to mechanical or electronic output elements to execute *Y*. This has become so ubiquitous that young children can build and program simple robots.

The second familiar example of bio-inspiration in the computational realm is neural networks. Unlike the linear single-task model of typical artificial intelligence applications, neural networks address massive interconnections between large numbers of information nodes. Neural nets started as a truly bio-inspired attempt to model and in part duplicate the activity of the human brain, with the goal of providing the ability for an artificial system to not only execute linear if-then processes but also to adapt in response to information flowing through the network. More recently, the field has moved away from the biological model to focus on statistical and signal processing aspects of specific practical applications. Arguably the major advance afforded by neural network efforts is the ability to design computing systems capable of making observations and inferring functions based on these observations—that is, computing systems with the ability to learn and evolve.

In both artificial intelligence and neural network science, the general paradigm is still one of centralized and, in most cases, linear processing. In this context, the term *centralized* refers to systems that feed into or out of a single node, such as a spider's head. That is, a centralized system is an organization, network, or organism in which a single "leader" receives all available input and is the source of all actionable output. For most of us, centralized systems are intuitively comfortable; they tend to agree with our instinctive feel for "the way things work" as it is developed by childhood familial and institutional environments and reinforced by our routine interaction with myriad examples of one-to-one cause-and-effect correspondence (often with manmade devices specifically built with this direct linear approach).

Presently, there is a pervasive movement away from centralization and linear function in almost all walks of our culture. The most familiar example of cultural decentralization is the Internet, the expanding influence of which throughout the world has encouraged decentralization in many other areas. Although beyond the scope of this chapter, the decentralization theme is often used to address organizations, cultures, business models, and even military strategies. For further investigation in these areas, a number of currently available books address decentralization for the layperson.[9, 10, 11, 12]

So what does this have to do specifically with bio-inspired materials? From the standpoint of computing, it is the concept of massively parallel processing, while from the standpoint of biological systems, it ranges from the concept of "swarm intelligence" to the difference between a spider and a starfish. The starfish, on an organism level, or the ant colony, on an organizational level, both reap great benefit from the decentralized paradigm. In all cases the recurrent themes are the absence of a functionally necessary control node and the presence of massive redundancy and massively parallel activity, each of which provide significant advantages over centralized systems.

Extending this inspiration from general philosophical analogies applied to computation and networking systems, bio-inspired structural and functional materials can be combined to build networks based on decentralized processing approaches. The result would be a purely abiotic system comprised of multiple interconnected subsystems and operating in a very organic, responsive, and adaptive way.

Applying the principles of decentralization to a traditional hierarchy such as the military would be folly. However, applications to specific military systems can have huge advantages. One very current example is outlined in chapter 5 of this book, "Abiotic Sensing." The heart of this system is a collection of bio-inspired materials conjoined to execute specific sense and respond rules in a platform of massively redundant and operationally parallel entities. Although the specific sense and response system described is best envisioned as a chemical or biological (CB) "detector," the model can be applied equally to other aspects of the CB defense arena. For example, a network of massively parallel entities can be deployed with the abilities of detecting agents, sending signals in response to positive detection, and detecting. Because these entities are identical, there is no individual entity that is crucial to the function of the network as a whole. A similar decentralized network can be, for example, built into personal protective equipment.

Multidisciplinary Process

Historically, much science and engineering has been conducted through specialization. For example, many years ago it was possible to be a chemist and as such to be relatively well versed in most areas of chemistry and to make significant advances using only chemistry. As science and technology expanded, specialization became increasingly important and the simple chemist became a biochemist, an organic chemist, a physical chemist, and so forth. Further specialization turned the organic chemist into a polymer chemist, a petroleum chemist, or any of an endless list of subspecialties, which can in turn be further subdivided. This was very useful for detailed understanding of narrow but significant topics. However, it resulted in an ever-widening gap between the major disciplines as the gaps widened between the subdisciplines.

As research and development workers increasingly recognize the distinct advantages of bio-inspiration in materials, it has become clear that this field straddles several scientific disciplines that are not traditionally linked. Extensive collaboration and interdisciplinary programs are required to "make it work." Typically, a capability is recognized in the biological realm and identified as desirable. Even at this very early stage, there must be cross-disciplinary activity. A naturalist might recognize the value of a biological structure and mention it to a structural engineer, or an adhesives chemist fishing at the dock may take note of the barnacles on the pilings and take it upon himself to locate a marine biologist.

In these and most cases, it takes more than one discipline simply to connect the biological phenomenon with the bio-inspired application. A research and characterization phase looks to fully characterize the materials and functions that lead to the capability. Again, the effort probably will require a number of different disciplines. A biologist is likely to provide the materials; a chemist, materials scientist, or biochemist to analyze the materials; a materials scientist, physicist, mechanical engineer, or other specialist to identify the relevant structures and mechanisms. Investigations are likely to lead to either a harvesting or a synthetic approach and again the success is likely to hinge on effective multidisciplinary activities. In harvesting, the biological materials are simply used as found in nature or farmed and extracted for use, while in synthesis, a production process is developed. In the case of synthesis, a decision is then required between a biosynthesis approach, using biotechnology to yield biological materials for abiotic applications, or a synthetic approach, applying the lessons learned from nature to traditional chemical or materials processing methods.

The challenge for scientists in the area of bio-inspired materials is to bridge the gaps not only between the subdisciplines in, for example, chemistry but also to build bridges between seemingly disparate subdisciplines in chemistry with subdisciplines buried deep in biology, physics, mechanical engineering, materials science, and computer and network science.

Conclusion

With advances in biotechnology, nanotechnology, and material characterization technologies, bio-inspired materials are rapidly growing in availability and application. Careful scientific pursuit of desired biological properties has been shown to yield successful results. However, the original inspiration is often unplanned. Forethought and planning can predict neither the unusual sources from which bio-inspiration might spring nor the elegant applications to which it is applied. The key to success in this area is multidisciplinary research and engineering, along with a willingness to consider some fantastic possibilities. A 2001 Board of Army Science and Technology report[13] identified a broad range of military applications for biotechnology, many of which address bio-inspired materials. Table 9–2 summarizes a number of these and other potentials for military applications.

table 9–2. AREAS FOR POTENTIAL MILITARY APPLICATION OF BIO-INSPIRED MATERIALS

APPLICATION	DESCRIPTION
Camouflage and concealment	Bio-inspired materials with stealth characteristics, nonilluminating coatings
Combat identification	Biomarkers
Computing	DNA computers, biological models, amorphous computing, decentralization
Lightweight armor	Bio-derived and -inspired materials for weight reduction and self-healing armor
Mobility enhancement	Bio-derived and -inspired materials for unprecedented mobility (people, vehicles, robotics)
Novel materials	Biodegradable consumables, bio-derived and -inspired materials
Performance enhancement	Prosthetics, biointegration
Power	Biological photovoltaics, synthetic photosynthesis
Radiation resistant electronics	Based on nanobiotechnology, biomolecular devices
Sensors	Biological processes, biomimetic processes in abiotic cell
Size/weight reduction	Molecular electronics, self-assembly, bio-inspired armor

NOTES

1 T.M.S. Chang, *Artificial Cells* (Springfield, IL: Charles C. Thomas, 1972), available at **www.medicine.mcgill.ca/artcell/1972bookCovercr.pdf**.

2 A. Picard, "The Red Blood Cell Man," *McGill News* 76, no. 4 (1996), 22–25.

3 E. Yoxen, *The Gene Business: Who Should Control Biotechnology?* (London: Pan Books, 1983).

4 H. Lee et al., "Mussel-Inspired Surface Chemistry for Multifunctional Coatings," *Science* 318 (2007), 426.

5 Helen K. Baca et al., "Cell Directed Assembly of Lipid-Silica Nanostructures Providing Extended Cell Viability," *Science* 313 (2006), 337.

6 Liangfang Zhang and S. Granick, "How to Stabilize Phospholipid Liposomes (Using Nanoparticles)," *Nanoletters* 6, no. 4 (2006), 694–698.

7 Christine L. Huffard, F. Boneka, and R.J. Full, "Underwater Bipedal Locomotion by Octopuses in Disguise," *Science* 25, no. 307 (2005), 1927.

8 Ori Brafman and Rod A. Beckstrom, *The Starfish and the Spider* (London: Penguin Books Ltd., 2006).

9 Ibid.

10 Mitchel Resnick, *Turtles, Termites, and Traffic Jams; Explorations in Massively Parallel Microworlds* (Cambridge: MIT Press, 1994).

11 David Freedman, *A Perfect Mess* (New York: Little, Brown, and Co., 2007).

12 R. Frenay, *Pulse* (New York: Farrar, Straus and Giroux, 2006).

13 Board on Army Science and Technology, *Opportunities in Biotechnology for Future Army Applications* (Washington, DC: National Academy of Sciences, 2001), available at **www.nap.edu/books/0309075556/html/index.html**.

part three

BIO-INSPIRED MACHINES

chapter 10

LEARNING UNMANNED VEHICLE CONTROL from ECHOLOCATING BATS

Eric W. Justh

With nightfall, the aerial hunters camouflaged in black emerge by the thousands from their base carved into a hillside. The first challenge they face is evading opposing forces attacking from the air as they emerge. Using advanced ultra-wide band transceivers for both communication and navigation, they traverse the darkness, dodging obstacles and flying through foliage with ease. They are on a search-and-destroy mission, and their quarry are flying distributors of a biological agent responsible for the deaths of half of the human population of the planet. After a night of carnage, they return to their base and regroup, preparing for more combat the following night.

Does this passage describe a futuristic battle between armies of unmanned aerial vehicles (UAVs)? Perhaps one day. But in fact, this scenario has been replayed nightly for several months out of the year for tens of thousands of years. The aerial hunters are Mexican free-tailed bats, their bases are caves spread from the southern United States (such as Bracken Cave near San Antonio) through Central America and northern South America. The opposing forces waiting to attack the bats as they emerge from the caves are hawks. The bats' advanced ultra-wide band transceivers are based on ultrasound, and their quarry are mosquitoes and other flying insects (of which they consume 200 tons per night near Bracken Cave alone). Malaria is the biological agent, which some believe is responsible for killing half of all the humans who have ever lived.[1] Bats have the ability to navigate in cluttered environments and can distinguish the wingbeats of prey insects within foliage.[2]

The above vignette is but one instance of how the natural world is replete with inspiration applicable to the engineering challenges of tomorrow's warfare. A confluence of trends in technology and university research makes the connection between biology and defense research a particularly timely topic. Of course, biology is a wide-ranging discipline and defense research encompasses many diverse activities, so the nexus is quite broad. Here, however, we focus specifically on one aspect of the interface between biology and the field of systems and control: sensing and feedback used by echolocating bats. The overall message is the power of biological analogy in framing certain problems in modeling and feedback control.

There are a variety of levels at which to study biology, and it is useful to distinguish among them because they take quite different points of view. A brief comparison among genomics, systems biology, and behavioral biology (the focus of this chapter) conveys an inkling of the richness of the synergy between biology and technology.

Genomics is a major field of research in which DNA sequences for a variety of judiciously selected plant and animal species (in addition to humans) are recorded (with varying degrees of statistical confidence, depending on the research objective), catalogued, and processed. Genomics has connections to micro- and nanofabrication (gene chips), as well as information technology for managing the vast amounts of information and associated degree of statistical uncertainty. An example of the depth and richness of bioinformatics based on this technology is the identification, cataloging, and tracking of the different species of *Vibrio* bacteria, which are responsible for the waterborne disease cholera.[3] These bacteria, which are endemic to bodies of water worldwide, are able to incorporate various genetic elements associated with toxicity and antibiotic resistance. Tracking different strains both geographically and in real time is key to managing outbreaks and pandemics but is quite a challenging task, since it is necessary to incorporate geography (for example, river flows), infrastructure (such as sanitation systems), medical data (for example, patients with digestive symptoms, their response to different antibiotics, and mortality rates), historical records (such as pandemic strains dating back to the early 20th century), as well as genetic microarray measurements of *Vibrio* samples from selected patients.[4]

Although quite complex and enormously useful, in the context of our discussion, genomics is essentially a descriptive exercise: the goal is primarily to identify particular genes, associate them with observable traits, and develop techniques for manipulating them in order to alter observable traits. Since the biomechanical systems by which genes lead to observable traits are only partially (if at all) understood, a trial and error process is required. While this approach has led to some stellar successes for the pharmaceutical industry (for example, human insulin of recombinant DNA origin in the 1970s) and therefore has undisputable economic and social importance, it is a field whose success does not depend on understanding detailed models of system dynamics.

Systems biology is a newly emerging discipline that seeks to explain the system dynamics of the cellular and biomolecular building blocks of life.[5] Loosely speaking, systems biology can be considered an attempt to understand the engineering principles underlying the "design" of living systems. A broad range of topics within systems biology has emerged. For example, biological systems are complex systems that inherently represent a juxtaposition of great robustness to threats for which they were designed (or evolved) but extreme vulnerability to threats that were not anticipated (or seen during the evolutionary process).[6] To the extent chemical reaction chains within cells can be teased out, there is often a seeming paradox that the standard equations for the chemical kinetics yield a narrow range of parameters over which an observed behavior (for example, a chemical timer) is stable, and yet the observed behavior may appear experimentally to be highly robust.[7] This suggests a flaw in using the standard chemical kinetics equations in this setting and has led to alternative modeling approaches that capture the robustness, but are not as clearly connected to the underlying chemical kinetics.[8] How quantization effects should be included in models at the cellular and subcellular level is also an open question. An example of a "success" of systems biology could be coopting structures within a cell in an artificial feedback system to perform a new task that "nature" has not envisioned.

Behavioral biology is a longstanding discipline, but one that is poised for transformation based on recent advances in off-the-shelf technologies. (Here, the term *behavioral biology* is used in the neuroscience and biophysiology sense, not in the deliberative or social behavior sense.) While systems biology is cutting-edge in terms of the advanced (and expensive) technology required for experimentation, behavioral biology can capitalize on the dramatic cost reductions of maturing technologies. Behavioral biology (for purposes of this chapter) deals with how sensorimotor tasks are performed at the organism level. Sensing modalities, neural processing, biomechanics of motion, communication with others, and environmental interactions are all encompassed. Behavioral biology has a descriptive side, analogous to genomics: animals can be categorized by their behavioral characteristics. But there is also a systems-modeling side, where the objective (in analogy to systems biology) is to understand the role of feedback in observed behaviors. For specialists in the field of systems and control, it is self-evident that modeling feedback is the essence of behavioral biology. For a broader audience, it is a point worth clarifying.

The echolocating bat's interaction with its environment, communication with other bats, evasion of hawks, and use of echolocation to find, track, and intercept insects all fall under the purview of behavioral biology. Genomics could provide insight into how particular echolocation strategies evolved over time,[9] and systems biology could eventually yield insight into how particular signaling processes within the bat's neural system operate. It is currently much simpler, and still fruitful, to observe, record, and measure from the bat itself, and then build models that capture the phenomenology. These measurements range from passive collection of position and echolocation data from free-flying bats to experiments in which constrained bats are subjected to stimuli while various structures in their brains are probed using electrode arrays.

Analogous studies have been conducted in other types of animals as well. The Mauthner cells in fish, a pair of large neurons that take vestibular and sensory inputs and coordinate the "tail flip" escape reflex, have been studied extensively.[10] Experiments in which insect visual systems are subjected to artificial scenes in order to tweeze out the design of their sensorimotor systems date back 20 years.[11] Historically, studies of the squid giant axon[12] and of the lamprey spinal cord[13] date back even further.

Current technology now allows "payloads" attached to the head of a flying bat to connect to probes in its brain, record vocalizations as they are made, and transmit the data wirelessly to a ground station in real time. Furthermore, high-frame-rate, high-resolution infrared cameras can be used to record the detailed motion of the bat's wings and body while in flight. In addition to dramatically expanding the ability to collect data, technology also affords the possibility of generating data in novel ways. Through the use of speakers, it already is possible to alter a tethered bat's perception of its surroundings based on echolocation. Taking the concept a step further, interacting with free-flying bats through automation (either an array of networked sensors or a flying vehicle) would allow an even richer class of experiments to be performed.

Modeling and Feedback

Feedback is intrinsic to all biological processes and systems, although different terms (like *regulation* or *compensation*) are often used. The motion of the echolocating bat, for example, is clearly "sensor driven" in the sense that the bat is continuously sampling its environment and responding accordingly—that is, using feedback. Thus, understanding behavior really means understanding a dynamical system model involving feedback.

While the history of feedback in engineered systems goes back at least to James Watt's flyball governor (for steam engines) in 1788, or even the Polynesian outrigger canoe more than 3,000 years ago,[14] the negative feedback amplifier for communications, credited to Harold Black at Bell Labs in 1927, is closer to the birth of systems and control as a distinct discipline. During the post–World War II period and the advent of the Cold War, both the United States and the Soviet Union supported basic research in the systems and control field for military purposes. From the early gun-director systems of World War II[15] through the Cold War Aegis shipboard air defense system (still in use today), experts in the systems and control field have provided the necessary mathematical techniques to close the feedback loop between sensors (radar) and actuators (munitions and launchers). During this period, feedback systems started to permeate the commercial sector as well. Motor vehicles are a prime example: computer-controlled fuel injection, antilock brakes, roll-stability control, cruise control, and, most recently, adaptive cruise control are all examples of rigorously engineered feedback systems that have entered society in a largely transparent way. In addition to transportation, models incorporating feedback have penetrated manufacturing, operations research (that is, supply-chain management and logistics), finance, medical equipment, and other areas.

In some sense, the success of the systems and control field is reflected by the extent to which its contribution to technology is ubiquitous but completely transparent to the vast majority of the population. However, there is still a pressing need for the Department of Defense (DOD) to maintain a knowledge base and research infrastructure in the systems and control area in order to be able to tackle its next-generation feedback control challenges. These include cooperative control of multiple unmanned vehicles, as well as systems to recognize, track, and counter hostile teams of weaponized manned and unmanned vehicles.

The landscape of scientific research evolves over time, changes in response to political imperatives, and is interconnected to some extent with society at large. Although change (such as declining DOD research funding) can result in immediate pain for individual researchers and programs, the effects on a discipline as a whole tend to only reveal themselves with the passage of time and are not easily predicted. Because of their mathematical grounding, researchers in the systems and control field generally are able to move into new application areas and carve out niches. Recent years have seen a number of researchers who have long been associated with engineering applications in the systems and control field start to publish on problems in systems biology. University bioengineering departments are also being established and are growing rapidly, partially offsetting declining enrollment in traditional engineering disciplines. On a lesser scale, collaborations involving behavioral biologists and researchers in systems and control have emerged, and this is where it is important for DOD not to overlook opportunities.

Bio-inspired Research as a DOD Resource

In the 21st century, biology is replacing electronics as the main frontier for research and innovation. However, *replacing* is not quite the right word, because it is the commoditization of 70 years of intense activity and technological advancement in electronics that is enabling the ongoing revolution in biology. Medical imaging systems, gene chips, automated DNA sequencing, and databases for storing and cataloging intracellular chemical reaction networks are examples of technologies that have had enormous impact on biological research.

However, these new technologies actually create new research challenges. Voluminous amounts of new data can be collected, but simply having a lot of data does not imply greater understanding. The real challenge ahead, for both systems biology and behavioral biology, is using the newly emerging technologies for acquiring information to actually improve our understanding of how various biological systems operate. What *understanding* really means in this context is having a model, generally involving feedback, that is consistent with experimental measurements of a phenomenon as well as with descriptive explanations of the phenomenon. This is no different from "understanding" in the context of engineered systems, and of course there are different levels of modeling at which different types of models apply. For example, a lawnmower can be considered a well-understood system: faithful, well-explained, and universally agreed-upon models for internal combustion engines (ICE) exist, and the coupling of the engine to the blade and the interaction of the blade with the outside world (grass, rocks) can be modeled as well. Temperatures and pressures from the ICE model yield mechanical requirements on the materials used, helping to further explain the morphology of the lawnmower. Different ICE models can be used, ranging from a torque-speed curve up to a three-dimensional finite-element, computational fluid dynamics model. There is often tension between high-level and detailed models, with aspects of one informing aspects of the other, resulting in an iterative process of refinement.

The challenge in biology is that we tend to observe complex interactions and behaviors, from which we must deduce appropriate models that lead to understanding. There is no computer algorithm that will perform this task for us—it requires two manifestly human capabilities: pattern recognition and creativity. Here, *pattern recognition* means being able to relate observed biological phenomena to a "library of models" from experience in studying and designing feedback control systems. Creativity is needed to interpret, as well as to expand, this library of models in the face of new biological data. To illustrate, a couple of examples involving bat pursuit and echolocation are described below.

Bat Pursuit Example

Various pursuit problems have been explored both in nature and in engineering, and biologically motivated pursuit strategies include classical pursuit, constant-bearing pursuit,[16] and motion-camouflaged pursuit. In an engineering context, pursuit is exemplified by the field of missile guidance.[17] When the evader can sense and respond to the motion of the pursuer, the theory of differential games may be appropriate to properly understand the engagement.[18] Indeed, certain insects that bats prey upon have developed countermeasures that are remarkable in their own right, and the sensorimotor systems and biomechanics of, for example, moths have been explored in detail.[19] Moths flying at night, where visual processing is of limited utility, have developed sensitive gyroscopic force sensors (for roll, pitch, and yaw rates) to enable high-performance maneuvers in three dimensions to evade predators.

Recent work on pursuit laws in nature demonstrates how behavioral phenomena can be explained or modeled in terms of sensor-driven feedback laws.[20, 21] The phenomenon of motion camouflage was initially observed in the territorial maneuverings of insects—the hoverflies *Syritta pipiens*—by Srinivasan and Davey[22] (using data that had been obtained earlier by Collett and Land[23]). In motion-camouflaged pursuit, a pursuer approaches a target (the "evader") such that the pursuer always appears to the evader to either lie along a line of sight with the same fixed object, or to be at constant bearing. In the former case, the pursuit is stealthy because the evader cannot use motion parallax cues to distinguish the pursuer from the background. In the latter case, the pursuit is stealthy because the motion parallax of the pursuer is indistinguishable from that of a very distant object. (Compound eyes are much more sensitive to optic flow rather than looming, making this an effective strategy for insects.) Both types of motion-camouflaged pursuit are observed in insect maneuvers,[24, 25] but being at constant bearing, which can be considered motion camouflage "with respect to a point at infinity," is easier to analyze mathematically.[26, 27]

Although the *phenomenon* of motion camouflage had been documented in experiments with insects, a *feedback law* to achieve motion camouflage was not formulated until later.[28] It was also observed that trajectories of free-flying echolocating bats (*Eptesicus fuscus*) pursuing prey insects in an instrumented room were geometrically indistinguishable from motion-camouflaged pursuit (during a particular stage of the engagement).[29] For bat-insect engagements, the term *motion camouflage* is a bit awkward, because the engagements take place in the dark, and acoustic sensing predominates over visual sensing for both the pursuer and evader. (In fact, in these experiments, the insect's hearing organ is disabled to prevent the insect from executing stereotyped evasive maneuvers in response to the bat vocalizations.) Therefore, the term *constant absolute target direction* (CATD) is used to denote the bat's strategy (equivalent to motion camouflage with respect to a point at infinity). The missile guidance strategy known as *pure proportional navigation guidance* (PPNG) is very similar to the CATD strategy.

Roughly speaking, the feedback law associated with CATD takes the sensory information available to the bat and causes it to maneuver so as to approach the insect along a line of constant bearing. The sensor measurement required by the bat to achieve this is the transverse component of the relative velocity of the bat and insect—that is, the component of the relative velocity that is perpendicular to an imaginary line joining the bat and insect. The bat is able to measure this quantity using echolocation. An analogous sensing task occurs with PPNG for missile guidance: the rotation of the line-of-sight vector is the corresponding sensor measurement. The bat (or missile) then "steers" so as to null its sensor measurement. There is a versatile formulation of this pursuit model (and related models of interacting animals or vehicles) that allows precise mathematical conclusions to be reached.[30, 31] Modeling (in the sense used here) really means developing a mathematical model within a particular formulation that balances intuitive insight and physically meaningful representation of a phenomenon of interest. Therefore, both the mathematical (systems and control) expertise and natural science knowledge are prerequisites.

The power of the pursuit model and other behavioral models is that they occupy a status separate from the specific implementation platform (that is, the insects, bats, or missiles). The platform-dependent details, such as sensor capabilities and constraints, can be interpreted in the context of these models to yield insight as to why nature has made certain (evolutionary) choices. Thus, models help us to organize and prioritize biological data.

Bat Echolocation Example

Bat echolocation is of military interest both because of the analogies between radar and sonar processing and because ultrasound may prove useful in its own right as a complementary sensing modality for unmanned vehicles. Bats use ultrasound for both sensing and communication, although here we are focused on sensing. Many decades of research have led to an appreciation of how truly remarkable bats' acoustic sensing capabilities are, and we are still learning.[32] With advances in solid-state electronics, we are better positioned than ever before to take advantage of this body of knowledge in actual sensor systems.[33]

To give some idea of what bats have managed to accomplish, and to offer a comparison to radar, note that the speed of sound in air is on the order of 1,000 feet (ft)/second. Although there are many different types of bats with various ranges of echolocation frequencies, 40 kilohertz (kHz) to 100kHz is a typical frequency range. The corresponding wavelength range is then 0.01 ft to 0.025 ft, or 3 millimeters (mm) to 8mm. By analogy, the speed of light is roughly 1,000 ft/microsecond, so that for radar frequencies of 40 gigahertz (GHz) to 100GHz, the corresponding wavelength is also 3mm to 8mm. Bats are limited at the high-frequency end of their range by atmospheric attenuation and at the low-frequency end by the size of prey insects they need to be able to detect. They are able to achieve sub-wavelength resolution, detecting insects with diameters down to a millimeter, and avoiding wires with diameters less than 0.1mm.[34] Bats can detect insect wingbeats, distinguish surface textures, and compensate for Doppler shift resulting from their own flight.[35] They can also determine azimuth and elevation using binaural processing.

Detection of wires is a serious problem for UAVs flying in urban environments. Currently, the only reliable, commercially available on-board sensor for this purpose is lidar, and even the smallest lidar systems are on the order of 10 pounds, limiting their use to relatively large (greater than 10-foot wingspan), liquid-fuel–powered UAVs (for example, Tier II UAVs). Smaller UAVs, such as Dragon Eye, would seem to be more logical candidates for low-flying missions in urban environments, but wires (such as power lines) pose a significant challenge. This puts into perspective how a sensing task that a bat would consider trivial is nevertheless quite challenging for humans.

Although different types of bats use a variety of waveforms, an analogy can be drawn between a typical bat and a so-called frequency-modulated continuous wave (FMCW) radar. FMCW waveforms include pulses (at constant or varying center frequencies), chirped waveforms, and tones (continuous wave signals). "Automotive radars," which are starting to appear and could become ubiquitous if a low enough price point is achieved, use frequencies around 80GHz and FMCW waveforms. Automotive radars are assigned to this band and use FMCW waveforms so as to limit the propagation distance to a few hundred meters and to avoid interference among multiple radars operating in close proximity. Although the percent bandwidths and percent Doppler shifts are quite different for bat ultrasound and automotive radar, the analogy between them is still quite strong. In radar processing, the term *micro-Doppler* is used to describe the extraction of motion information (for slowly moving objects, such as humans walking) from radar returns.[36]

What we can learn from bats is how they use feedback control in their echolocation processing to extract information from their surroundings. As bats move through their environment and probe it acoustically, they are somehow able to build up, in ways we have yet to understand, a mental image of what is important to them. Further complicating the situation is the fact that even in low light, bats are generally able to incorporate visual sensing along with acoustic sensing. By studying the acoustic waveforms bats use, and observing the bats in carefully controlled experiments, the capabilities of their acoustic sensing systems can be determined. The next step is deducing what feedback mechanisms are present in their acoustic processing, and how they are used. For example, certain bats use intensity compensation: as a prey insect is approached, the vocalization intensity is reduced, so that a constant intensity reaches the bat's ears.[37] Another example is increasing pulse repetition rate as a target is approached. These adjustments made on the fly (that is, based on feedback) presumably serve to keep the return signals in the right "range" to interact with banks of neural filters that are "wired" for high sensitivity to particular features. What type of neural filtering is used and why certain feedback mechanisms are present are questions best tackled jointly by biologists and experts in the field of systems and control. The answers are likely to have a significant impact on design of sensor systems in general and sensor systems for unmanned vehicles in particular.

Conclusion

In the opening vignette, reference was made to futuristic battles between armies of UAVs. We believe that if these futuristic battles were to take place, they would extensively incorporate biologically motivated technology and processing. In the Tactical Electronic Warfare Division at the Naval Research Laboratory, a multivehicle simulation testbed for coordinated control of unmanned vehicles, including UAVs and unmanned surface vessels, is being built.[38, 39] Making contact with biology has already proved indispensable in communicating with the academic research community and in designing various components of this testbed effectively. On its face, this may seem surprising in the context of an engineered system with an engineering-oriented set of applications. In an earlier era, connecting such an effort to biology would indeed have been improbable. But today, our effort would be stifled if *not* for contact made with biology.

NOTES

1 J. Whitfield, "Portrait of a Serial Killer: A Roundup of the History and Biology of the Malaria Parasite," *Nature News* (2002).

2 M. Vespe, G. Jones, and C.J. Baker, "Biologically Inspired Waveform Diversity for Synthetic Autonomous Navigation Sensing," unpublished, 2007.

3 G.J. Vora et al., "Microarray-based Detection of Genetic Heterogeneity, Antimicrobial Resistance, and the Viable but Nonculturable State in Human Pathogenic *Vibrio spp*," *Proceedings of the National Academies of Science* 102, no. 52 (2005), 19109–19114.

4 Ibid.

5 E.D. Sontag, "Molecular Systems Biology and Control," *European Journal of Control* 11, no. 4–5 (2005), 396–435.

6 M.E. Csete and J.C. Doyle, "Reverse Engineering of Biological Complexity," *Science* 295, no. 5560 (2002), 1664–1669.

7 R. Ghaemi and D. Del Vecchio, "Evaluating the Robustness of a Biochemical Network Model," *Proceedings of the 46th IEEE Conference on Decision and Control* (2007), 615–620.

8 D. Angeli and E.D. Sontag, "Oscillations in I/O Monotone Systems under Negative Feedback," *IEEE Transactions on Circuits and Systems, Special Issue on Systems Biology*, 2008.

9 G. Jones and E.C. Teeling, "The Evolution of Echolocation in Bats," *Trends in Ecology and Evolution* 21, no. 3 (2006), 149–156.

10 H. Korn and D.S. Faber, "The Mauthner Cell Half a Century Later: A Neurobiological Model for Decision-making?" *Neuron* 47 (2005), 13–28.

11 M.V. Srinivasan and S. Zhang, "Visual Motor Computations in Insects," *Annual Review of Neuroscience* 27 (2004), 679–696.

12 J.R. Clay, "Excitability of the Squid Giant Axon Revisited," *Journal of Neurophysiology* 80 (1998), 903–913.

13 A.H. Cohen and P. Wallén, "The Neuronal Correlate of Locomotion in Fish," *Experimental Brain Research* 41, no. 1 (1980), 11–18.

14 D. Abramovitch, "The Outrigger: A Prehistoric Feedback Mechanism," *Control Systems Magazine* (2005), 57–72.

15 D.A. Mindell, *Between Human and Machine: Feedback, Control, and Computing before Cybernetics* (Baltimore: The Johns Hopkins University Press, 2004).

16 B.R. Fajen and W.H. Warren, "Visual Guidance of Intercepting a Moving Target on Foot," *Perception* 33 (2004), 689–715.

17 N.A. Shneydor, *Missile Guidance and Pursuit* (Chichester, UK: Horwood, 1998).

18 R. Isaacs, *Differential Games* (New York: John Wiley and Sons, 1965).

19 S.P. Sane et al., "Antennal Mechanosensors Mediate Flight Control in Moths," *Science* 315 (2007), 863–866.

20 E.W. Justh and P.S. Krishnaprasad, "Steering Laws for Motion Camouflage," *Proceedings of the Royal Society of London* 46 (2006), 3629–3643.

21 P.V. Reddy, E.W. Justh, and P.S. Krishnaprasad, "Motion Camouflage in Three Dimensions," *Proceedings of the 45th IEEE Conference on Decision and Control* (2006), 3327–3332.

22 M.V. Srinivasan and M. Davey, "Strategies for Active Camouflage of Motion," *Proceedings of the Royal Society of London* 259 (1995), 19–25.

23 T.S. Collett and M.F. Land, "Visual Control of Flight Behaviour in the Hoverfly, *Syritta pipiens*," *Journal of Comparative Physiology* 99 (1975), 1–66.

24 Srinivasan and Davey.

25 A.K. Mizutani, J.S. Chahl, and M.V. Srinivasan, "Motion Camouflage in Dragonflies," *Nature* 423, no. 604 (2003).

26 Justh and Krishnaprasad.

27 Reddy, Justh, and Krishnaprasad.

28 Justh and Krishnaprasad.

29 K. Ghose et al., "Echolocating Bats Use a Nearly Time-optimal Strategy to Intercept Prey," *Public Library of Science Biology* 4, no. 5 (2006), 865–873, e:108. These experiments were conducted in the Auditory Neuroethology Lab led by Professor Cynthia Moss at the University of Maryland; available at **www.bsos.umd.edu/psyc/batlab/**.

30 Justh and Krishnaprasad.

31 Reddy, Justh, and Krishnaprasad.

32 Vespe, Jones, and Baker.

33 C.J. Baker, M. Vespe, and G.J. Jones, "Target Classification by Echo Locating Animals," *Proceeding of the IEEE Conference on Waveform Diversity and Design* (2007), 348–352.

34 Vespe, Jones, and Baker.

35 Ibid.

36 V.C. Chen et al., "Micro-Doppler Effect in Radar: Phenomenon, Model, and Simulation Study," *IEEE Transactions on Aerospace and Electronic Systems*, 42, no. 1 (2006), 2–21.

37 Vespe, Jones, and Baker.

38 E.W. Justh and V. Kowtha, "Biologically Inspired Models for Swarming," in *Evolutionary and Bio-inspired Computation: Theory and Applications, Proceedings of the SPIE* 6563 (2007), 656302.

39 R.S. Cortesi, K.S. Galloway, and E.W. Justh, "A Biologically Inspired Approach to Modeling Unmanned Vehicle Teams," *Proceedings of the SPIE* 6964 (2008), 696405.

chapter 11

NEUROROBOTICS: NEUROBIOLOGICALLY INSPIRED ROBOTS

JEFFREY KRICHMAR

Neurorobots[1] are robotic devices that have control systems based on principles of the nervous system. These models operate on the premise that the "brain is embodied and the body is embedded in the environment." Therefore, neurorobots are grounded and situated in a real environment. The real environment is required for two reasons. First, simulating an environment can introduce unwanted and unintentional biases to the model. For example, a computer-generated object presented to a vision model has its shape and segmentation defined by the modeler and directly presented to the model, whereas a device that views an object hanging on a wall has to discern the shape and figure from ground segmentation based on its on active vision. Second, real environments are rich, multimodal, and noisy; an artificial design of such an environment would be computationally intensive and difficult to simulate. However, all these interesting features of the environment come for "free" when a neurorobot is placed in the real world.

A neurorobot has the following properties:

- It engages in a behavioral task.
- It is situated in a real-world environment.
- It has a means to sense environmental cues and act upon its environment.
- Its behavior is controlled by a simulated nervous system having a design that reflects, at some level, the brain's architecture and dynamics.

As a result of these properties, neurorobotic models provide heuristics for developing and testing theories of brain function in the context of phenotypic and environmental interactions.

Although there have been great advances in autonomous systems,[2, 3, 4, 5] the controllers of these machines are still very much tailored to specific missions and do not have the behavioral repertoire we normally associate with that of biological organisms. Behavior-based robotics[6] do not learn from experience and cannot adapt to environmental change. Probabilistic robot controllers[7] need an accurate model of their sensors and actuators. Robots controlled by reinforcement learning or machine learning[8] are driven by reward expectation and do not address attention, novelty, and threat assessment.

Neurorobotic models may provide a foundation for the development of more effective robots, based on an improved understanding of the biological bases of adaptive behavior. A robotic controller modeled after the vertebrate nervous system, in which the robot's behavior approaches the complexity and flexibility associated with higher order animals, would be a major step forward in the design of autonomous systems. Advances in computational models of the brain as well as computation power are making this a distinct possibility in the not too distant future. Neurally inspired robotic control would be flexible, experience-dependent, and autonomous—just like a biological organism.

Classes of Neurorobotic Models

There are too many examples of neurobiologically inspired robotic devices to exhaustively list in this brief review. However, the approach has been applied to several distinct areas of neuroscience research:

- motor control and locomotion
- learning and memory systems
- value systems and action selection.

The remainder of this article will briefly touch on a few representative examples.

Motor Control and Locomotion

Neurorobots have proved useful for investigating animal locomotion and motor control and for designing robot controllers. Neural models of central pattern generators, pools of motorneurons that drive a repetitive behavior, have been used to control locomotion in robots.[9, 10, 11] Kimura and colleagues have shown how neurorobotics can provide a bridge between neuroscience and biomechanics by demonstrating emergent four-legged locomotion based on central pattern generator mechanisms modulated by reflexes. Their group developed a model of a *learnable* pattern generator and demonstrated its viability using a series of synthetic and humanoid robotic examples. Ijspeert and colleagues constructed an amphibious salamander-like robot that is capable of both swimming and walking, and therefore represents a key stage in the evolution of vertebrate-legged locomotion. A neurorobotic implementation was found necessary for testing whether the models could produce locomotion both in water and on ground and investigating how sensory feedback affects dynamic pattern generation.

An intriguing neural inspiration for the design of robot controllers is the mirror neuron system found in primates. Mirror neurons in the premotor cortex are active both when a monkey grasps or manipulates objects and when it watches another animal performing similar actions.[12] Neuroroboticists, using this notion of mirror neurons, have suggested that complex movements such as reaching and locomotion may be achieved through imitation.[13, 14, 15, 16, 17]

Another strategy for motor control in neurally inspired robots is to use a predictive controller to convert awkward, error-prone movements into smooth, accurate ones. Recent theories of motor control suggest that the cerebellum learns to replace primitive reflexes with predictive motor signals. The idea is that the outcomes of reflexive motor commands provide error signals for a predictive controller, which then learns to produce a correct motor control signal prior to the less adaptive reflex response. Neurally inspired models have used these ideas in the design of robots that learn to avoid obstacles,[18, 19] produce accurate eye,[20] and generate adaptive arm movements.[21, 22, 23]

Learning and Memory Systems

A major theme in neurorobotics is neurally inspired models of learning and memory. One area of particular interest is navigation systems based on the rodent hippocampus. Rats have exquisite navigation capabilities in both the light and the dark. Moreover, the finding of place cells in the rodent hippocampus, which fire specifically at a spatial location, have been of theoretical interest for models of memory and route planning.[24] Robots with models of the hippocampal place cells have been shown to be viable for navigation in mazes and environments similar to those used in rat spatial memory studies.[25, 26, 27, 28] Recently, large-scale systems-level models of the hippocampus and its surrounding regions have been embedded on robots to investigate the role of these regions in the acquisition and recall of episodic memory.[29, 30, 31]

Another learning and memory property of importance to the development of neurorobotics is the ability to organize the unlabeled signals that robots receive from the environment into categories. This organization of signals, which in general depends on a combination of sensory modalities (for example, vision, sound, taste, or touch), is called *perceptual categorization.* Several neurorobots have been constructed that build up such categories, without instruction, by combining auditory, tactile, taste, and visual cues from the environment.[32, 33, 34]

Value Systems and Action Selection

Biological organisms adapt their behavior through value systems that provide nonspecific, modulatory signals to the rest of the brain that bias the outcome of local changes in synaptic efficacy in the direction needed to satisfy global needs. Examples of value systems in the brain include the dopaminergic, cholinergic, and noradrenergic systems.[35, 36, 37] Behavior that evokes positive responses in value systems biases synaptic change to make production of the same behavior more likely when the situation in the environment (and thus the local synaptic inputs) is similar; behavior that evokes negative value biases synaptic change in the opposite direction. The dopamine system and its role in shaping icrosys making has been explored in neurorobots and brain-based devices.[38, 39, 40] Doya's group has been investigating the effect of multiple neuromodulators in the "cyber-rodent," a two-wheeled robot that moves autonomously in an environment.[41] These robots have drives for self-preservation and self-reproduction exemplified by searching for and recharging from battery packs on the floor and then communicating this information to other robots nearby through their infrared communication ports. In addition to examining how neuromodulators such as dopamine can influence decisionmaking, neuroroboticists have been investigating the basal ganglia as a model that mediates action selection.[42] Based on the architecture of the basal ganglia, Prescott and colleagues embedded a model of it in a robot that had to select from several actions depending on the environmental context.

Conclusion

Higher brain functions depend on the cooperative activity of an entire nervous system, reflecting its morphology, its dynamics, and its interaction with the environment. Neurorobots are designed to incorporate these attributes such that they can test theories of brain function. The behavior of neurorobots and the activity of their simulated nervous systems allow for comparisons with experimental data acquired from animals. The comparison can be made at the behavioral level, the systems level, and the neuronal level. These comparisons serve two purposes: first, neurorobots can generate hypotheses and test theories of brain function. The construction of a complete behaving model forces the designer to specify theoretical and implementation details that can be easy to overlook in an ungrounded or disembodied theoretical model. Moreover, it forces these details to be consistent. Second, by using the animal nervous system as a metric, neurorobot designers can continually make their simulated nervous systems and resulting behavior closer to those of the model animal. This, in turn, allows the eventual creation of practical devices that may approach the sophistication of living organisms.

NOTES

1 A.K. Seth, O. Sporns, and J.L. Krichmar, "Neurorobotic Models in Neuroscience and Neuroinformatics," *Neuroinformatics* 3 (2005), 167–170.

2 A. Cho, "Robotics. Robotic Cars Tackle Crosstown Traffic—and Not One Another," *Science* 318 (2007), 1060–1061.

3 W.W. Gibbs, "From Finish to Start. Was the Grand Challenge Robot Race in March the Fiasco it Appeared To Be? Hardly, Argues William 'Red' Whittaker. The Annual Event is Pushing Mobile Robotics to Get Real," *Scientific American* 291 (2004), 33–34.

4 S. Squyres, *Roving Mars: Spirit, Opportunity, and the Exploration of the Red Planet* (New York: Hyperion, 2005).

5 B. Yenne, *Attack of the Drones* (St. Paul, MN: Zenith Press, 2004); J. Jones and D. Roth, *Robot Programming: A Practical Guide to Behavior-Based Robotics* (New York: McGraw-Hill, 2003).

6 Jones and Roth.

7 S. Thrun, W. Burgarde, and D. Fox, *Probabilistic Robotics* (Cambridge: MIT Press, 2005).

8 R.S. Sutton and A.G. Barto, *Reinforcement Learning: An Introduction* (Cambridge: MIT Press, 1998).

9 A.J. Ijspeert et al., "From Swimming to Walking with a Salamander Robot Driven by a Spinal Cord Model," *Science* 315 (2007), 1416–1420.

10 H. Kimura, Y. Fukuoka, and A.H. Cohen, "Biologically Inspired Adaptive Walking of a Quadruped Robot," *Philosophical Transactions Series A: Mathematical, Physical, and Engineering Sciences* 365 (2007), 153–170.

11 M. Lewis, F. Tenore, and R. Etienne-Cummings, "CPG Design using Inhibitory Networks," paper presented at IEEE Conference on Robotics and Automation, Barcelona, 2005.

12 G. Rizzolatti and M.A. Arbib, "Language within Our Grasp," *Trends in Neuroscience* 21 (1998), 188–194.

13 A. Billard and M.J. Mataric, "Learning Human Arm Movements by Imitation: Evaluation of a Biologically Inspired Connectionist Architecture," *Robotics and Autonomous Systems* 37 (2001), 145–160.

14 S. Schaal, "Is Imitation Learning the Route to Humanoid Robots?" *Trends in Cognitive Sciences* 3 (1999), 233–242.

15 S. Schaal, A. Ijspeert, and A. Billard, "Computational Approaches to Motor Learning by Imitation," *Philosophical Transactions Series B: Biological Sciences* 358 (2003), 537–547.

16 S. Schaal and N. Schweighofer, "Computational Motor Control in Humans and Robots," *Current Opinion in Neurobiology* 15 (2005), 675–682.

17 J. Tani, M. Ito, and Y. Sugita, "Self-organization of Distributedly Represented Multiple Behavior Schemata in a Mirror System: Reviews of Robot Experiments using RNNPB," *Neural Networks* 17 (2004), 1273–1289.

18 J.L. McKinstry, G.M. Edelman, and J.L. Krichmar, "A Cerebellar Model for Predictive Motor Control Tested in a Brain-based Device," *Proceedings of the National Academy of Sciences of the United States of America* 103 (2006), 3387–3392.

19 B. Porr and F. Worgotter, "Isotropic Sequence Order Learning," *Neural Computation* 15 (2003), 831–864.

20 P. Dean et al., "Saccade Control in a Simulated Robot Camera-head System: Neural Net Architectures for Efficient Learning of Inverse Kinematics," *Biological Cybernetics* 66 (1991), 27–36.

21 Ibid.

22 S. Eskiizmirliler et al., "A Model of the Cerebellar Pathways Applied to the Control of a Single-joint Robot Arm Actuated by McKibben Artificial Muscles," *Biological Cybernetics* 86 (2002), 379–394.

23 C. Hofstotter, M. Mintz, and P.F. Verschure, "The Cerebellum in Action: A Simulation and Robotics Study," *European Journal of Neuroscience* 16 (2002), 1361–1376.

24 J. O'Keefe and L. Nadel, *The Hippocampus as a Cognitive Map* (Oxford: Clarendon Press, 1978).

25 A. Arleo and W. Gerstner, "Modeling Rodent Head-direction Cells and Place Cells for Spatial Learning in Bio-mimetic Robotics," in *From Animals to Animats 6: Proceedings of the Sixth International Conference on Simulation of Adaptive Behavior* (Cambridge: MIT Press, 2001).

26 N. Burgess et al., "Robotic and Neural Simulation of the Hippocampus and Rat Navigation," *Biological Science* 352 (1997), 1535–1543.

27 M.J. Mataric, "Navigating with a Rat Brain: A Neurobiologically-inspired Model for Robot Spatial Representation," in *From Animals to Animats*, ed. J. Arcady Meyer, and S.W. Wilson (Cambridge: MIT Press, 1991), 169–175.

28 M.J. Milford, G.F. Wyeth, and D. Prasser, "RatSLAM: A Hippocampal Model for Simultaneous Localization and Mapping," paper presented at 2004 IEEE International Conference on Robotics and Automation, New Orleans, LA, 2004.

29 J.P. Banquet et al., "A Hierarchy of Associations in Hippocampo-cortical Systems: Cognitive Maps and Navigation Strategies," *Neural Computation* 17 (2005), 1339–1384.

30 J.G. Fleischer et al., "Retrospective and Prospective Responses Arising in a Modeled Hippocampus during Maze Navigation by a Brain-based Device," *Proceedings of the National Academy of Sciences of the United States of America* 104 (2007), 3556–3561.

31 J.L. Krichmar et al., "Spatial Navigation and Causal Analysis in a Brain-based Device Modeling Cortical-hippocampal Interactions," *Neuroinformatics* 3 (2005), 197–221.

32 J.L. Krichmar and G.M. Edelman, "Machine Psychology: Autonomous Behavior, Perceptual Categorization, and Conditioning in a Brain-Based Device," *Cerebral Cortex* 12 (2002), 818–830.

33 A.K. Seth et al., "Spatiotemporal Processing of Whisker Input Supports Texture Discrimination by a Brain-based Device," in *Animals to Animats 8: Proceedings of the Eighth International Conference on the Simulation of Adaptive Behavior*, ed. S. Schaal et al. (Cambridge: MIT Press, 2004), 130–139.

34 A.K. Seth et al., "Visual Binding Through Reentrant Connectivity and Dynamic Synchronization in a Brain-based Device," *Cerebral Cortex* 14 (2004), 1185–1199.

35 G. Aston-Jones and F.E. Bloom, "Nonrepinephrine-containing Locus Coeruleus Neurons in Behaving Rats Exhibit Pronounced Responses to Non-noxious Environmental Stimuli," *Journal of Neuroscience* 1 (1981), 887–900.

36 M.E. Hasselmo et al., "Neuromodulation, Theta Rhythm and Rat Spatial Navigation," *Neural Networks* 15 (2002), 689–707.

37 W. Schultz, P. Dayan, and P.R. Montague, "A Neural Substrate of Prediction and Reward," *Science* 275 (1997), 1593–1599.

38 A. Arleo, F. Smeraldi, and W. Gerstner, "Cognitive Navigation Based on Nonuniform Gabor Space Sampling, Unsupervised Growing Networks, and Reinforcement Learning," *IEEE Trans Neural Networks* 15 (2004), 639–652.

39 J.L. Krichmar and G.M. Edelman, "Machine Psychology: Autonomous Behavior, Perceptual Categorization, and Conditioning in a Brain-Based Device," *Cerebral Cortex* 12 (2002), 818–830.

40 O. Sporns and W.H. Alexander, "Neuromodulation and Plasticity in an Autonomous Robot," *Neural Networks* 15 (2002), 761–774.

41 K. Doya and E. Uchibe, "The Cyber Rodent Project: Exploration of Adaptive Mechanisms for Self-Preservation and Self-Reproduction," *Adaptive Behavior* 13 (2005), 149–160.

42 T.J. Prescott et al., "A Robot Model of the Basal Ganglia: Behavior and Intrinsic Processing," *Neural Networks* 19 (2006), 31–61.

chapter 12

BIOMIMETIC, SOCIABLE ROBOTS for HUMAN-ROBOT INTERACTION

Eleanore Edson, Judith Lytle, and Thomas McKenna

Ever since Isaac Asimov coined the term *robotics* in a short story published over 60 years ago,[1] science fiction has inundated the collective imagination with countless scenarios of people interacting with robots that strongly resemble humans in both form and behavior. Since the mid-20th century, science and engineering communities have labored to make the robots of reality attain the sophistication of their fictional counterparts. But the current state of the art is a far cry from the friendly and intelligent humanoid robots envisioned in popular culture.

Military science and technology (S&T) programs endeavor to bridge this gap between fiction and reality. An Office of Naval Research (ONR) S&T program in human-robot interaction (HRI) is working to develop biologically inspired, or *biomimetic*, robots that can naturally communicate and cooperate in human-robot teams during military operations. Fruition of these HRI projects will have numerous impacts on U.S. defense policy.

Why Biomimetic, Sociable Robots?

Currently, humans rely upon robots to perform a diverse array of functions.[2] Widespread use of robots in manufacturing and other physically laborious tasks precipitated economic and societal changes in developed countries during the 20th century. As robots became more advanced and specialized, humans began employing them in dangerous tasks related to scientific exploration,[3] search and rescue missions in urban environments,[4, 5] hazardous material handling,[6] and military operations, such as disarming improvised explosive devices (IEDs) in Iraq and Afghanistan.[7] For the past several decades, physically challenged populations, such as the elderly and the handicapped, have used robots for transportation and other types of assistance in navigating day-to-day activities. More recently, an interesting new application of robots has arisen: companionship for humans. The Japanese-designed AIBO, built to resemble a small dog, is the most famous autonomous entertainment robot. The makers of AIBO followed up with the creation of SDR–3X and SDR–4X II, humanoid robots that not only navigate, walk, and dance, but also feature facial recognition as well as speech comprehension and production capabilities.[8]

Designing sociable robots for companionship purposes is one thing. Less intuitive is the application of humanoid robots for military use. Which operational needs of the military, and specifically the U.S. Navy and Marine Corps, require biologically inspired, sociable robots?

Military Operational Needs

The military's predominant need is to find ways to reduce the risk of injury or mortality related to dangerous missions. A relatively mundane task such as driving a supply truck can be deadly if the driver must navigate through a hostile environment or treacherous terrain. Moreover, the current conflict in Iraq has posed an unprecedented level of threat of being harmed by IEDs or other explosives.

In addition to reducing casualties, the military wants to reduce the financial cost of operations. Productivity of military personnel can be extended by reducing the types of tasks humans are required to perform. For example, the personnel productivity of the Navy is limited by ship occupancy and expeditionary manpower.

To assess remote or dangerous environments and increase productivity, the military employs robots and—increasingly—autonomous ones. Current use of autonomous systems is limited by available manpower and skill types; at present, military personnel who interact with robots do so in a supervisory role that requires more attention and specialization than interacting with a peer. In order for robots to continue to meet the evolving needs of the military, these machines will need to require less human supervision and be more adaptable to the unpredictable nature of the real world: in essence, robots will need to achieve even greater degrees of autonomy.

Enhanced autonomy is necessary but not sufficient, however, for robots to help humans conduct complex tasks in increasingly uncertain environments. One potential solution to the military's sophisticated operational problems and needs is to expand the ability for military personnel to collaborate in peer-to-peer teams with unmanned systems. In order for robots to be teammates, they must be not only more autonomous but also capable of natural, "mixed-initiative" interactions with humans.[9]

Science and Technology Solutions

Natural interactions between humans and robots require autonomous systems to be as biologically realistic as possible. Biological systems, from basic organisms to complex organs such as the human brain, are characterized by several qualities, including adaptability, robustness, versatility, and agility.[10] Endowing autonomous systems with these desirable properties would markedly improve robots' task versatility as well as their ability to operate in dynamic, unpredictable real-world environments. Most importantly, biological qualities will enable robots to engage in efficient and natural teamwork.[11]

However, today's state of the art in autonomous systems is characterized by several major shortcomings. First, current autonomous systems are prone to communications loss. Second, present-day systems require high-level communication and reasoning to coordinate planning and execution in dynamic battle spaces. Finally, acute problems exist with sending autonomous vehicles into urban areas with mixed populations (friend, foe, unknown, casualties).

Modern-day warfighters' tasks require not only a range of skills but also a growing demand on their working memories. The lower a warfighter's cognitive load is, the easier it is to concentrate on the tasks at hand. However, proper control of an autonomous robot requires the supervision of one or more human operators. As a result, autonomous systems become another task consuming humans' time and attention.

In recent years, the Department of Defense (DOD) has come to recognize the significant importance of developing methods and technologies to enable autonomous systems to not only learn from humans but also collaborate naturally and effectively as peers. Such a dynamic between warfighters and robots could result in higher productivity, survivability, reliability, and efficacy of military missions. In response to this need, DOD has provided the Navy, Air Force, and Army with increased funding for accelerating the research and development (R&D) in the field of HRI.

For the past 3 years, ONR's program has invested in HRI projects at both military facilities (such as the Naval Research Laboratory) as well as private and public academic institutions. At present, a recent influx of DOD funding has allowed ONR to oversee an ambitious multidisciplinary effort to create cognitive models and dexterous mobile robots to enable direct HRI and team collaboration. These ONR-funded HRI research efforts are at the forefront of science, working to address fundamental scientific and engineering questions about intelligent systems, perception, communication, cognition, and psychology in order to create robots that will help humans meet the challenges of modern warfare.

Background on Biologically Inspired Robots

For the purposes of the military, robots need to collaborate effectively and naturally in human-robot teams. For these interactions to become a reality, robots must be as similar to humans as possible. Creation of a biomimetic robot—that is, an autonomous, intelligent, and sociable one—presents many scientific and technological challenges that will require a bridging of neuroscience, cognitive psychology, engineering, and robotics to overcome.

In order for robots to be capable "team players," they must possess an intelligent system that is based upon the types of representations and cognitive function found in the human brain.[12] Thus, autonomous robots must have various cognitive abilities, such as perception and situation assessment, learning and working memory, problem-solving, goal maintenance, language understanding, and communication skills. Further, these cognitive capabilities must be linked to sensory and motor systems that can collect information about the external environment and then allow the robot to respond to that environment.[13]

Human Brain Basics

The human brain is the most complex and powerful computational device on Earth. The cerebral cortex is the general region responsible for complex brain function and its higher level processing occurring in the neocortex, an area of the brain unique to mammals that composes the top-most layer of the cerebral cortex.

The basic working unit of the brain is the neuron, of which there are approximately 130 billion in the average adult human brain. The human neocortex alone contains an estimated 22 billion neurons.[14, 15] The basic components of a neuron are its cell body; extensions from the cell body that receive information called *dendrites*; and the long "output" process that sends information called an *axon*. The predominant site of transmission of information between neurons is called the *synapse*, typically formed between one neuron's axonal end and another neuron's dendrite; there are an estimated 150 trillion synapses in the human neocortex.[16]

Far from being a homogenous organ, the adult brain's billions of neurons and trillions of synapses are organized into different regional areas that perform one or more functions. Complex brain functions such as language, sensory processing, and abstract thinking are carried out in the four major sections of the cerebral cortex: the occipital, temporal, parietal, and frontal lobes. The association of information from different sensory areas into an organized understanding of the outside environment is carried out by the occipital, temporal, and parietal lobes. This complex is also responsible for the comprehension and creation of language, abilities that are separately governed by the Wernicke's and Broca's areas, respectively. The frontal lobe, also known as the prefrontal cortex, can be thought of as the center of intellect, allowing humans to plan behavior, reason, problem-solve, and think abstractly.

The foundation of computational modeling is biological realism,[17] with different types of cognitive architectures/neural networks being influenced by the brain's structure and function to varying degrees. Biologically, neurons are real, living processing elements and have powerful computational capabilities.[18] Their computational equivalents are simple, neuron-like processing elements that constitute the building blocks of neural networks.[19] The axonal and dendritic processes in a brain are simulated as network connections of neural networks, while synaptic strength is simulated as adjustable strengths of network connections between the processing elements. The human brain relies upon parallel processing in order to handle countless, simultaneous calculations, thoughts, observations, reasoning, and so forth. Accordingly, neural networks also function via parallel processes.[20]

However, there are some caveats to biologically realistic cognitive architectures; artificial neural networks cannot replicate the human brain with absolute fidelity for a few reasons. First, in order to function properly, neural networks need to have some built-in constraints that do not necessarily exist in the real brain.[21] Second, it is not currently possible to reduce certain biological phenomenon such as consciousness to a sum of its constituent parts. Even if it were, artificial neural networks are not yet sophisticated enough to accurately and thoroughly simulate all of the subtle inputs that produce complex functions of the brain.[22] Finally, there is a great deal to learn about the biological mechanisms that underlie human cognitive capabilities and behavior.

Crucial Component of Human Behavior: Mirror Neurons

When it comes to physical movement, whether it is a toddler being taught how to wave or eat with a utensil or a pitcher getting coached on throwing a curve ball, humans often learn new movements by watching and imitating others. Imitation is thought to be central not only for humans' ability to learn motor skills, but also for the development of communication skills.[23, 24, 25, 26, 27] Imitation is also thought to underlie understanding of the intention behind other's actions.[28] The paramount importance of imitation is revealed when one considers how social organization and, more fundamentally, survival are dependent on understanding the actions of others.[29]

What neurological mechanisms allow the human brain to perceive and interpret actions, and how does this ability relate to communication and empathy? Nearly two decades ago, researchers investigating the visual and motor areas of monkey brains discovered a special set of neurons that responded to both a monkey's movement as well as its observation of the same movement performed by another.[30, 31, 32] Accordingly, these types of neurons became known as *mirror neurons*.[33]

In monkeys, mirror neurons are part of a cortical circuit that involves the parietal lobe along with a section of the frontal lobe that governs sensory guided movement: the premotor cortex.[34, 35] This parietal-premotor cortical circuit links observation of a movement to the preparation to execute that same movement.[36] Motor neurons also selectively respond to performing and perceiving "goal-directed" actions, such as grabbing an object, and do not activate in response to more elementary movements.[37, 38] These findings from monkey experiments made researchers curious about whether human brains also feature mirror neurons.

To search for mirror neurons in human brains, researchers cannot record from single neurons. Instead, scientists use imaging techniques to view activity levels in specific regions of the human brain in response to performance and observation of motor acts. In the past dozen years, various neuroimaging techniques have revealed that the human premotor and parietal regions, just as in the monkey brain, also show activity in response to performing a certain action and perceiving that same action performed by others.[39, 40] A crucial difference between human and monkey brains that reflects human's sophistication compared to primates is that the core region for the human mirror system is Broca's area, the brain region credited with producing language.[41, 42, 43, 44, 45, 46, 47, 48, 49, 50, 51, 52, 53] The emergent theory from this work was that humans' use of visual and auditory systems to recognize others' actions eventually gave rise to language.[54, 55]

Moreover, mirror neurons in both monkeys and humans respond only to "biological actions"—that is, actions performed by a living being as opposed to a machine.[56, 57] Mirror neurons also respond to sounds associated with goal-directed actions.[58] These various discoveries led researchers to theorize that mirror neurons code for the "overall goal" and not just individual elements of an action.[59]

Researchers elucidated more crucial information about the functional role of mirror neurons by testing how much visual information about a goal-directed action is required to active mirror neurons. In this set of experiments, various goal-directed movements (combinations of grasping, holding, and placing an object) were performed but the final stages of the actions (hand-object contact) were obscured from the monkeys' view. Recordings from premotor cortical mirror neurons demonstrated that the monkeys' mirror neurons still fired in response to the goal-directed movements. This implied that the monkeys' premotor cortex had internally generated an understanding of the full action and that mirror neurons were involved in inference of the action's goals.[60]

As for human mirror neurons' role in prediction, neuroimaging of human subjects found that their observation of others' movements coupled with the subjects' intention to imitate those actions resulted in greater activity in premotor and parietal areas compared to observation coupled with plans to recognize those same actions in the future.[61, 62] Based on these findings, scientists proposed that mirror neurons constitute the basic units of "a more sophisticated predictive system that allows humans to infer goals from the observation of actions."[63, 64, 65]

This hypothesis was tested in human brain imaging studies, which found that parietal neurons had different responses to identical actions that had different goals. These findings provided evidence that mirror neurons are involved in prediction of an action's intention.[66] Thus, researchers postulated that mirror neurons constitute a basic key element in social communication and understanding of others, also known as empathy.[67, 68, 69, 70, 71]

Neural Networks Model the Brain

For the past several years, the HRI field has sought to create artificial models of cognition that are "human compatible."[72] Such models, also referred to as *cognitive architectures* or *neural networks*, are products of a multidisciplinary effort of neurobiology, computational neuroscience, cognitive science, engineering, computer science, robotics, and psychology. These models, or neural networks, endow robots with abilities that range from object recognition to decisionmaking to "mind-reading." At the most rudimentary level, basic mechanisms and fundamental principles of synaptic transmission can be the basis for modeling spatial and temporal pattern recognition processes. Because of their abilities to solve a wide range of complex tasks such as cognitive skills while also being biologically relevant, neural networks constitute extremely important feats of engineering.[73]

The neural network field is diverse and can be classified into several major categories based on how and to what degree the brain is modeled. As discussed in the book *Artificial Intelligence and Neural Networks*,[74] the major subfields are the following:

- *Computational neuroscience*. This field consists of modeling the brain at various levels of sophistication—from single neurons to entire functional regions—in great detail. Simulations of brain activity in these model systems are performed, resulting in predictions that can be empirically tested. The purpose of this subfield is to simulate the brain as realistically as possible, so the biological realism of new algorithms is of paramount importance.

- *Artificial neural networks.* Neural networks are leveraged in this engineering and industrial field to solve engineering problems and build electronic devices for certain applications. Successful and simple performance of the devices' function(s) is the priority, whereas biological realism is irrelevant.
- *Connectionism.* This is the field of modeling cognitive architectures that are inspired by biological neurons and neural connections. The emphasis is on developing networks that have simple, elegant representations.
- *Biological neural networks.* This represents a small but important emerging subfield that reverse-engineers the nervous system in order to generate new neural architectures. These novel neural networks are then incorporated into general technological applications.

Designing Mirror Neuron Function in Robots

Given mirror neurons' crucial role in human learning, "emotional" intelligence, and behavior, the neural network field has aspired to replicate mirror neurons in robots in order to allow them to emulate human emotion and understand human intent.[75]

In order to reach this goal, an interdisciplinary research community consisting of neuroscientists, computational neuroscientists, computer engineers, and robotic engineers is working on two long-term goals that have yet to be realized. First, robots must be able to both understand the goals of the human they are observing and mimic those goal-directed actions. Second, once the robot has successfully observed, understood, and mimicked the human's goals, it must be capable of adapting those goals under different circumstances and in different environments.[76]

What is the current state of the art? A distilled and simplified explanation of the fundamental elements of computational mirror systems is described here. The first stage must involve the robot perceiving and mapping an observed action and then converting that information to its own internal frame of reference. "Perception algorithms" may be written to allow the robot to observe and measure human movement. The next phase will involve the robot learning to generate motions based on what is observed. This may involve scaling and converting the observed information into the robot movement that most closely resembles the human movement. Part of the challenge of translating movement is determining how to abstract the observed motion so that only the crucial information about goals is retained while the irrelevant information is discarded. In some cases, a neural network mirror system will contain a "closed-loop mode," where the outputs of movement prediction are fed back as inputs, as opposed to outside observations of the human's movement, to the robot's learning system. This internal, closed-loop, feedback system is the computational equivalent of the robot visualizing the observed movements.

Another method for removing the need for a human to repeatedly demonstrate a motion is to program reinforcement learning, whereby a reward function motivates the robot to learn how to execute the motion by practicing the action. Although this reward function is effective for teaching the robot how to perform various movements, it is ineffective in inferring the human's goals. Finally, some systems will enable the robot to recognize a motion. In the recognition phase, the mirror neuron system relies on what has been learned and generated in order to predict human actions. This prediction ability allows the robot to recognize the same motion again. Some neural network models of a mirror system, such as a connectionist neural network program, have a dynamic quality to them that allows real-time interaction and synchronization between "top-down behavior generation" and "bottom-up recognition" processes: a parallel behavior recognizer and generator system.[77, 78]

Advances in Human-Robot Interaction R&D

A major S&T goal at ONR is to develop human-centric robots that can work in partnership with humans in a peer-to-peer dynamic, as well as in remote teams. For this to be possible, robots must be compatible with human cognitive, attentive, and communications capabilities and limitations. This compatibility requires robots to have several basic capabilities, including the ability to recognize faces and individuals, communicate in natural language styles of humans, as well as understand and perform communicative gestures.[79] A more sophisticated requirement for successful HRIs in real-world scenarios is the integration of cognitive models, action schemes, and statistical estimations. Furthermore, integration of behavior models and distributed control capabilities is essential for effective tasking across human-robot teams. These capabilities will allow robots to cope with and react to their changing environments in real time, as well as interact with their human counterparts accordingly.

However, as discussed earlier, the current state of the art in HRI is several stages behind such peer-to-peer interactions, due to several inherent challenges in human-robot collaborations: noisy signals due to visual or auditory recognition problems, lack of spatial awareness due to communication restraints, lack of shared knowledge or misunderstanding of capabilities and limitation, linguistic ambiguities, and interpretations of perceived intent (including situational awareness, local tasking, and so forth).

To overcome these challenges and realize the goal of biomimetic, human-compatible robots, ONR has invested in an ambitious multi-university research initiative (MURI). What follows is a brief overview of the MURI-based advances in engineering artificial sensory, language, and cognitive abilities. Further, this summary will describe interface designs that will allow trust to increase between humans and robots as well as enable them to share the same goal structure.

Visual and Auditory Recognition

Humans are highly visual and communicate through a wide variety of visual cues including eye gaze direction, body language, and facial expressions. Accordingly, there has been considerable work on modeling the human visual system for robotics applications. Since there is a large amount of visual data in a given field of view, attention systems are also incorporated. Habituation filters, which encourage robots to continually search for interesting things, are also included. Bottom-up, or stimulus-driven, processing draws attention toward specific objects in the visual field, whereas top-down processing is task-based.

Photoelectric devices and charge-coupled device cameras have given way to biologically inspired vision sensors, which include artificial foveas and circuitry that mimic the neuronal connectivity of the human retina. These artificial retinas have some of the same visual processing capabilities as the human retina, including lateral inhibition (a method by which nerve cells can determine the origin of a stimulus) and temporal sensitivity.[80] Currently, a novel retinal circuit is being developed that focuses on the ganglion cell, one of several cell types involved in retinal processing. This novel circuit enhances the response of cells to approaching objects, which may signal danger, in the context of a demanding visual environment.[81] Visual recognition of human activity, or kinematic tracking, is also being undertaken.[82]

Biomimetic robots also generally include a microphone sound sensor. Processing of incoming sound waves is permitted with software models, or with hardware models mimicking the human hearing organ, the cochlea.[83] This model mimics hippocampal function, with cochlea-like preprocessing. Current work focuses on biologically based pattern recognition systems in space and time for acoustic signal identification. The goal is to transition this application to security-breaching noise detection systems using progressively elaborate models with complex connectivity patterns and network architectures.[84]

Other research being done on multimodal sensor fusion with neuromorphic sensors for both visual and auditory processing is based on sensory systems of the barn owl. A silicon retina can respond to relative changes in intensity, and helps shape the auditory spatial map. For auditory processing, a silicon cochlea models responses as a set of bandpass filters. In each modeling paradigm, signal processing occurs similarly to biological processes of vision or hearing.

Somatosensory Function

The human somatosensory system includes the sensations of light touch, pain, pressure, temperature, and joint and muscle position sense, or proprioception. In robotics, artificial touch sensors are generally composed of grids of pressure sensors that contain resistive or capacitive components. Proprioceptive feedback is provided by position sensors, which include potentiometers, magnetic sensors, linear displacement sensors, and digital encoders, among other sensors. Velocity is measured with tachometers, force is measured with stretch sensors coupled to position sensors, and position gauges register small movements in space.[85]

Language

If peer-to-peer interaction between humans and robots is to occur naturally, it makes sense to program robots with an understanding of human communication styles.[86] Humans largely communicate through verbal language. Human-machine language interaction consists of automated speech recognition and synthesis, language processing, dialogue and discourse, and affect recognition and expression. Currently, learning-by-doing algorithms create composite methods that result in interleaved perception and action responses.[87] One thrust is perception-motor learning by imitation, where a robot learns a sound and attempts to repeat it. This will evolve from a sensorimotor grammar capability to linguistic grammar capabilities, resulting in elementary language interaction, elementary reasoning, and learning. More intense study into robust natural language and dialogue systems, affect recognition, and effective conveyance of critical information is also being undertaken to improve HRI.[88]

Learning

Robots must continually learn about themselves, about the humans with whom they interact, and about their environments. Algorithms have historically been based on single-robot, single-human interactions, and have used simple models of human intentional states, not learned models. More recently, planning algorithms and models have been generalized to multiperson teams and multiple objectives. Ongoing research suggests that the human brain uses Bayesian principles for inferring actions and subsequent behavior.[89] Bayesian inference provides a mathematically rigorous platform for information processing given uncertain situations. Bayesian modeling is being applied to design anthropomorphic robots that can learn from observation and experience. Additionally, a noninvasive brain-robot interface is being developed that will allow for control of virtual or physical objects directly through brain activity.

In order to learn and communicate effectively, robot partners require flexible memory models similar to those of their human counterparts. A vast amount of literature already exists on the neuropsychology of memory reconsolidation. Ongoing research into the neural circuitry that supports episodic memory (that is, the memory of events, times, places, and so forth, and its associated emotions) and spatial navigation seeks to demonstrate a causal relationship between specific neural activity patterns and memory performance.[90] Current work on information flow focuses on reconsolidation at the computational or behavioral level, as well as at the functional or algorithmic level.[91] The goal is a human-like working memory system that involves adaptive recognition and new memory formation.

Mind Reading

Optimal peer-to-peer HRI requires interactive training in human-robot teams. This type of learning includes acquainting human partners with both the capabilities and limitations of robot partners, establishing common knowledge about strategies and procedures, negotiating roles for joint activities, and mastering joint activities through practice. Such robust and adaptable collaboration in dynamic environments requires "mind-reading" skills.

Research into mirror neuron function is being applied to mind-reading skills in robotics. Biologically speaking, mirror neurons are key elements in a predictive system that allows for inference of intent, goals, and the like from observation. In robotics and neural network modeling, emulation of mirror neuron activity in prediction tasks and action recognition is currently being applied to learning.[92] The immediate goal is to improve robot learning by implementing algorithms based on imitation—learn by doing. The long-term goal is the establishment of highly functioning human-robot teams that can participate effectively in collaborative decisionmaking processes.

Collaborative decisionmaking is useful for understanding and predicting human behavior by the robot. In turn, humans can apply their own mental models to understand and predict the behavior of the robot. Prior work in this area has included ambiguity resolution and visual and mental perspective-taking for collaborative efforts with multiple human partners. This research is currently being extended with studies to incorporate language and nonverbal behavior in estimating mental models. Perspective-taking is also being integrated with models of spatial cognition to model human beliefs and goals.[93, 94, 95] This architecture is organized around the ability to understand the environment from the perspective of another individual.

Increase Trust

Trust between peers is a major foundation of effective teamwork, especially in the dangerous environments encountered during military operations. In order to increase the level of trust between humans and robots, a human must understand a robot's limitations and abilities. The consequent need is optimal human-robot visualizations and interfaces. Such interfaces will provide scalable visualizations, support situational awareness, and allow transitions with an understanding of historical context. Technical advances in this area have included the development of an integrated HRI system for multiple human users and mixed-type robot systems. For example, research is currently being performed to develop a robotic platform for security and service applications.[96] The goal of this project is to develop intelligent systems that can work in self-organizing teams able to respond robustly to changing inputs. This work will result in a fully configurable multirobot network that can self-organize in task allocation and can execute practical task requests by human collaborators.

In order for robots to operate safely alongside humans, the status of shared goals and plans must be continually monitored. Sensing capabilities, multiagent cooperation, plan learning and execution, and sensory channels (speech, gestures, actions) all present real challenges to safety. Advances in gesture recognition, personal identification, visual tracking and activity recognition, and dialogue capabilities technologies are being developed to improve trust and increase effectiveness of human-robot collaboration.[97, 98, 99]

Conclusion

Reducing casualties and the financial cost of operations as well as increasing the productivity of military personnel are the primary driving forces behind defense S&T investments. To help meet these objectives, autonomous robots will need to be more adaptable to the unpredictable nature of the real-world environment as well as be capable of collaborating in peer-to-peer teams with military personnel. Natural interactions between robots and human teammates require robots to be compatible with human cognitive, attentive, and communications capabilities and limitations. Further, these human-robot collaborations are dependent on trust, which requires robots to understand the goals of their human teammates and necessitates that humans understand the limitations and abilities of the robot. Thanks to advances in the fields of neurobiology, computational neuroscience, cognitive science, engineering, computer science, robotics, and psychology, it is possible to engineer robots with artificial sensory, language, and cognitive abilities in order to make autonomous systems more biologically realistic. The ONR is at the forefront of this ambitious, multidisciplinary effort to create biomimetic robots that are capable of more natural and efficient human-robot interaction. Fruition of these HRI projects will have numerous impacts on U.S. defense policy.

NOTES

1 I. Asimov, "Runaround," *Astounding Science Fiction* (1942), 94–103.

2 H.A. Yanco and J.L. Drury, "A Taxonomy for Human-robot Interaction," in *Proceedings of the AAAI Fall Symposium on Human-Robot Interaction*, AAAI Technical Report FS-02-03 (2002), 111–119.

3 J.G. Bellingham and K. Rajan, "Robotics in Remote and Hostile Environments," *Science* 318, no. 5853 (2007), 1098–1102.

4 J.L. Casper and R.R. Murphy, "Workflow Study on Human-robot Interaction in USAR," *Proceedings of IEEE International Conference on Robotics and Automation* (2002), 1997–2003.

5 D.J. Bruemmer, J.L. Marble, and D.D. Dudenhoffer, "Intelligent Robots for Use in Hazardous DOE Environments," *Proceedings of the Workshop on Measuring the Performance of Intelligence Systems* (2002).

6 Ibid.

7 A. Klein, "The Army's $200 Billion Makeover: March to Modernize Proves Ambitious and Controversial," *The Washington Post*, December 7, 2007, A1.

8 M. Fujita, "How to Make an Autonomous Robot as a Partner with Humans: Design Approach Versus Emergent Approach," *Philosophical Transactions of the Royal Society A* 365, no. 1850 (2007), 21–47.

9 C. Breazeal et al., "Tutelage and Collaboration for Humanoid Robots," *International Journal of Humanoid Robots* 1, no. 2 (2004), 315–348.

10 R. Pfeifer, M. Lungarella, and F. Iida, "Self-Organization, Embodiment, and Biologically Inspired Robotics," *Science* 16 (2007), 1088–1093.

11 Breazeal et al., 315–348.

12 D. Perzanowski et al., "Building a Multimodal Human-robot Interface," *IEEE Intelligent Systems* 16, no. 1 (2001), 16–21.

13 R. Hecht-Nielsen and T. McKenna, "The Development of Cortical Models to Enable Neural-Based Cognitive Architectures," in *Computational Models for Neuroscience: Human Cortical Information Processing* (Berlin: Springer-Verlag, 2002), 171–204.

14 B. Pakkenberg and H.J. Gundersen, "Neocortical Number in Humans: Effect of Sex and Age," *The Journal of Comparative Neurology* 384, no. 2 (1997), 312–320.

15 J.L. Saver, "Time is Brain-Quantified," *Stroke* 37 (2006), 263–266.

16 B. Pakkenberg et al., "Aging and the Human Neocortex," *Experimental Gerontology* 38, no. 1/2 (2003), 95–99.

17 R.C. O'Reilly, "Six Principles for Biologically-based Computational Models of Cortical Cognition," *Trends in Cognitive Sciences* 2 (1998), 455–462.

18 T.M. McKenna, "The Role of Interdisciplinary Research Involving Neuroscience in the Development of Intelligent Systems," in *Artificial Intelligence and Neural Networks: Steps toward Principled Integration*, ed. V. Honavar and L. Uhr (San Diego: Academic Press, Inc., 1994), 75–92.

19 S.J. Hanson and C.R. Olson, eds., *Connectionist Modeling and Brain Function: The Developing Interface* (Cambridge: The MIT Press, 1990).

20 V. Honavar and L. Uhr, eds., *Artificial Intelligence and Neural Networks: Steps toward Principled Integration* (San Diego: Academic Press, Inc., 1994), xvii–xxxii.

21 R.C. O'Reilly, "Biologically Based Computational Models of High-level Cognition," *Science* 314 (2006), 91–94.

22 Ibid.

23 M. Tomasello, A. Kruger, and H. Ratner, "Cultural Learning," *Behavioral and Brain Sciences* 16, no. 3 (1993), 495–552.

24 V. Gallese and A. Goldman, "Mirror Neurons and the Simulation Theory of Mind-reading," *Trends in Cognitive Sciences* 2, no. 12 (1998), 493–501.

25 R.W. Byrne and A.E. Russon, *Behavioral and Brain Sciences* 21, no. 5 (1998), 667–684.

26 M. Iacoboni et al., "Cortical Mechanisms of Human Imitation," *Science* 286 (1999), 2526–2528.

27 M. Jeannerod, "Neural Simulation of Action: A Unifying Mechanism for Motor Cognition," *NeuroImage* 14, no. 1 (2001), S103–109.

28 Iacoboni et al., 2526–2528.

29 G. Rizzolatti and L. Craighero, "The Mirror-Neuron System," *Annual Review of Neuroscience* 27 (2004), 169–192.

30 G. Rizzolatti et al., "Functional Organization of Inferior Area 6 in the Macaque Monkey. II. Area F5 and the Control of Distal Movements," *Experimental Brain Research* 71 (1988), 491–507.

31 M. Taira et al., "Parietal Cortex Neurons of the Monkey Related to the Visual Guidance of Hand Movement," *Experimental Brain Research* 83 (1990), 29–36.

32 V. Gallese et al., "Action Recognition in the Premotor Cortex," *Brain* 119 (1996), 593–609.

33 Ibid.

34 Rizzolatti et al., 491–507.

35 Taira et al., 29–36.

36 L. Fadiga et al., "Motor Facilitation during Action Observation: A Magnetic Stimulation Study," *Journal of Neurophysiology* 73, 2608–2611.

37 Rizzolatti et al., 491–507.

38 G. Rizzolatti, L. Fogassi, and V. Gallese, "Neurophysiological Mechanisms Underlying the Understanding and Imitation of Action," *Nature Review Neuroscience* 2 (2001), 661–670.

39 Fadiga et al., 2608–2611.

40 Iacoboni et al., 2526–2528.

41 S.T. Grafton et al., "Localization of Grasp Representations in Humans by Positron Emission Tomography. 2. Observation Compared with Imagination," *Experimental Brain Research* 112 (1996), 103–111.

42 G. Rizzolatti et al., "Localization of Grasp Representations in Humans by PET: 1. Observation Versus Execution," *Experimental Brain Research* 111 (1996), 246–252.

43 J. Decety et al., "Brain Activity during Observation of Actions. Influence of Action Content and Subject's Strategy," *Brain* 120 (1997), 1763–1777.

44 J. Grezes, N. Costes, and J. Decety, "Top-down Effect of the Strategy on the Perception of Human Biological Motion: A PET Investigation," *Cognitive Neuropsychology* 15 (1998), 553–582.

45 F. Hamzei et al., "The Human Action Recognition System and its Relationship to Broca's Area: An fMRI Study," *NeuroImage* 19 (2003), 637–644.

46 S.H. Johnson-Frey et al., "Actions or Hand-object Interactions? Human Inferior Frontal Cortex and Action Observation," *Neuron* 39 (2003), 1053–1058.

47 M. Petrides and D.N. Pandya, "Comparative architectonic analysis of the human and the macaque frontal cortex," in *Handbook of Neuropsychology*, ed. F. Boller, H. Spinnler, and J.A. Hendler (Amsterdam: Elsevier, 1994).

48 K. Nelissen et al., "Observing Others: Multiple Action Representation in the Frontal Lobe," *Science* 310 (2005), 332–336.

49 M. Petrides, "Broca's Area in the Human and the Nonhuman Primate Brain," in *Broca's Region*, ed. Y. Grodzinsky and K. Amunts (New York: Oxford University Press, 2006), 31–46.

50 F. Binkofski et al., "A Fronto-parietal Circuit for Object Manipulation in Man: Evidence from an fMRI-Study," *European Journal of Neuroscience* 11 (1999), 3276–3286.

51 L. Fadiga et al., "Speech Listening Specifically Modulates the Excitability of Tongue Muscles: A TMS study," *European Journal of Neuroscience* 15 (2002), 399–402.

52 K.E. Watkins, A.P. Strafella, and T. Paus, "Seeing and Hearing Speech Excites the Motor System Involved in Speech Production," *Neuropsychologia* 41 (2003), 989–994.

53 L. Metta et al., "The Mirror-neurons System: Data and Models," *Progress in Brain Research* 164 (2007), 39–59.

54 G. Rizzolatti and M.A. Arbib, "Language within our Grasp," *Trends in Neurosciences* 21 (1998), 188–194.

55 L. Craighero et al., "The Mirror-neurons System: Data and Models," *Progress in Brain Research* 164 (2007), 39–59.

56 J. Grèzes, N. Costes, and J. Decety, "The Effects of Learning and Intention on the Neural Network involved in the Perception of Meaningless Actions," *Brain* 122 (1999), 1875–1887.

57 Rizzolatti, Fogassi, and Gallese, 661–670.

58 E. Kohler et al., "Hearing Sounds, Understanding Actions: Action Representation in Mirror Neurons," *Science* 297 (2002), 846–848.

59 S.J. Blakemore and C. Frith, "The Role of Motor Contagion in the Prediction of Action," *Neuropsychologia* 43 (2005), 260–267.

60 M.A. Umiltà et al., "I Know What You Are Doing. A Neurophysiological Study," *Neuron* 31 (2001), 155–165.

61 Grèzes, Costes, and Decety, 1875–1887.

62 Blakemore and Frith, 260–267.

63 V. Gallese and A. Goldman, "Mirror Neurons and the Simulation Theory of Mind-reading," *Trends in Cognitive Sciences* 2, no. 12 (1998), 493–501.

64 Jeannerod, S103–109.

65 Blakemore and Frith, 260–267.

66 L. Fogassi et al., "Parietal Lobe: From Action Organization to Intention Understanding," *Science* 308 (2005), 662–667.

67 Gallese and Goldman, 493–501.

68 Jeannerod, S103–109.

69 P.L. Jackson and J. Decety, "Motor Cognition: A New Paradigm to Study Self-Other Interactions," *Current Opinion in Neurobiology* 14, no. 2 (2004), 259–263.

70 Fogassi, 662–667.

71 K. Nakahara and Y. Miyashita, "Understanding Intentions: Through the Looking Glass," *Science* 308 (2005), 644–645.

72 Breazeal et al., 315–348.

73 Hanson and Olson, eds.

74 McKenna, 75–92.

75 M. Arbib, "Robot Brains and Emotions. Interview on the Talking Robots Postcast," available at **http://lis.epfl.ch/index.html?content=resources/podcast/**.

76 Ibid.

77 C.G. Atkeson et al., "Using Humanoid Robots to Study Human Behaviour," *IEEE Intelligent Systems* 15 (2000), 46–56.

78 J. Tani, M. Ito, and Y. Sugita, "Self-organization of Distributedly Represented Multiple Behavior Schemata in a Mirror System: Reviews of Robot Experiments Using RNNPB," *Neural Networks* 17, no. 8/9 (2004), 1273–1289.

79 A. Schultz and G. Trafton, "Toward Cognitively Plausible Human Robot Interaction," presentation, Office of Naval Research (ONR) Program Review: Human Robot Interaction and Human Activity Recognition, May 3–4, 2007, Arlington, VA.

80 B. Roska, "A Mammalian Looming Detector and Its Neural Circuit," presentation, ONR Program Review: Computational Neuroscience, Sensory Augmentation, and Brain Machine Interface, April 25–26, 2007, Arlington, VA.

81 Ibid.

82 G. Mori and J. Malik, "Recovering 3D Human Body Configurations Using Shape Contexts," *IEEE Transactions on Pattern Analysis and Machine Intelligence* 28, no. 7 (2006), 1052–1062.

83 T. Berger, "Biologically-based Temporal Pattern Recognition Systems for Acoustic Signal Identification," presentation, ONR Program Review: Computational Neuroscience, Sensory Augmentation, and Brain Machine Interface, April 25–26, 2007, Arlington, VA.

84 Ibid.

85 G.A. Bekey, *Autonomous Robots: From Biological Inspiration to Implementation and Control* (Cambridge: MIT Press, 2005).

86 Perzanowski et al., 16–21.

87 R. Granger, "End to End Brain-Circuit Framework for Interactive Cognitive Robotics," presentation, ONR Program Review: Human Robot Interaction and Human Activity Recognition, May 3–4, 2007, Arlington, VA.

88 M. Scheutz, "Effective Human-Robot Interaction Under Time Pressure Through Robust Natural Language Dialogue and Dynamic Autonomy," presentation, ONR Program Review: Human Robot Interaction and Human Activity Recognition, May 3–4, 2007, Arlington, VA.

89 R. Rao, "Probablistic Neuromorphic Systems," presentation, ONR Program Review: Computational Neuroscience, Sensory Augmentation, and Brain Machine Interface, April 25–26, 2007, Arlington, VA.

90 P. Lipton, "Real Time Recording and Stimulation of Brain Cell Assemblies," presentation, ONR Program Review: Computational Neuroscience, Sensory Augmentation, and Brain Machine Interface, April 25–26, 2007, Arlington, VA.

91 H. Siegelmann, "Memory Reconsolidation and Computational Learning," presentation, ONR Program Review: Computational Neuroscience, Sensory Augmentation, and Brain Machine Interface, April 25–26, 2007, Arlington, VA.

92 O. Jenkins, "Dynamical Motion Vocabularies for Human-Robot Interactions," presentation, ONR Program Review: Human Robot Interaction and Human Activity Recognition, May 3–4, 2007, Arlington, VA.

93 C. Breazeal, "Three Projects on Human Robot Teaming," Presentation, ONR Program Review: Human Robot Interaction and Human Activity Recognition, May 3–4, 2007, Arlington, VA.

94 Jenkins.

95 Schultz and Trafton.

96 M. Nicolescu, "Understanding Intent Using a Novel Hidden Markov Model Representation Robotic Platform for Security and Service Application," presentation, ONR Program Review: Human Robot Interaction and Human Activity Recognition, May 3–4, 2007, Arlington, VA.

97 C. Jones, "Peer-to-Peer Embedded Human Robot Interaction," presentation, ONR Program Review: Human Robot Interaction and Human Activity Recognition, May 3–4, 2007, Arlington, VA.

98 P. Rybski, "Peer-to-Peer Embedded Human Robot Interaction," presentation, ONR Program Review: Human Robot Interaction and Human Activity Recognition, May 3–4, 2007, Arlington, VA.

99 Schultz and Trafton.

chapter 13

BIOMECHANICALLY INSPIRED ROBOTICS

JOHN SOCHA AND JUSTIN GRUBICH

The tens of millions of species in existence on Earth today are the products of 3.8 billion years of evolution. Some of these organisms survive and flourish at astounding physiological extremes of temperature, pressure, and pH, and exist in the most inhospitable habitats, from the depths of the oceans to arid deserts and remote frozen glaciers of Antarctica. Beyond their innate abilities to grow, heal, and reproduce, the great diversity of organisms that have evolved can, from an engineering perspective, perform some remarkable feats. Nature has produced creatures that are capable of a vast range of functional abilities. On dry land, they can walk, run, or jump over complex terrain and climb, slide, or crawl over slippery surfaces. Some take to the air with active flight or glide through the trees to bypass the terrestrial clutter below. Still others have conquered the aquatic realm with unrivaled swimming and maneuvering abilities, while some remain hidden using serpentine bodies to burrow and slither between cracks and small spaces in the earth. How do our current robotics stack up to the wide-ranging and impressive mechanical capabilities of nature? These questions are currently being answered by integrative research initiatives that combine multiple disciplines in engineering, physics, chemistry, and more recently biology.

While the reproduction of behaviors such as running and climbing in robots may seem simple on the surface, it is in actuality an immensely complex feat that requires an understanding of multiple layers of mechanics and control: force productions, counterbalances, kinematic motions, and even material properties of soft, hard, and compliant tissues. Hundreds of parameters and many thousands of measurements might be required to accurately describe the dynamics of such behaviors through three-dimensional space. Building a robotic device to mimic these types of highly complex behaviors requires interdisciplinary teams of researchers that integrate expertise from fields of science such as computer and mechanical engineering, materials science, biomechanics, and neurobiology. The goal of this chapter is to introduce concepts of biomimetics (that is, the use of biological organisms as models for new robotic designs) and describe some of the cutting edge robots with uses for military as well as civilian applications that have resulted from these types of research programs. Finally, we will advocate that the promise of future unforeseen breakthroughs in biologically inspired robotics will continue to rely upon support for basic research in biodiversity and zoology that can lead to the discovery of new organismal capabilities and subsequently, novel biomechanical principles for engineers to mimic.

Why Look to Nature?

Biomimetics is a field that looks to nature for mimicry or inspiration in engineering design. "Why reinvent the wheel when nature has done it already?" is the rationale that motivates its research goals (although, interestingly, nature has never invented a wheeled organism).[1] However, before diving into a discussion of bio-inspired robotics, we start with a caveat. Although the biological world is extremely diverse and ripe with fodder for inspiration, literally copying biology is not recommended or even practical for many applications. If one endeavors to design robotics that can perform as well as or better than their biological counterparts, one must be aware of two major realities of biology: first, organisms are products of their evolutionary history, where multiple, often antagonistic selection pressures can result in suboptimal designs; and second, evolution has shaped organisms using materials and designs that are not necessarily superior to human technologies. To put it another way, organisms are historical beings, products of a continuous line of ancestry and the shaping forces of evolution. Any changes in a new generation must build upon or alter the previous genetic stock, severely constraining the realm of what is currently possible. In human engineering, inventors are free to change whatever aspect of design they choose; they can use new materials, add or take away parts, or even scrap everything wholesale and start afresh. Organisms cannot do that, being inexorably limited by their past.

Furthermore, the primary driving forces of evolution—natural and sexual selection—do not act on individual traits (for example, the ability to jump high) but instead favor success in one simple endeavor: the production of viable offspring. Because of these forces, biological systems are rarely optimized for one particular function. Furthermore, not every component of a system is vital to its function, as evolution does not always weed out redundant or unnecessary parts as commonly believed. For example, consider the magnificent Emperor penguins made famous in the documentary, "March of the Penguins." Emperor penguins spend months each year at sea, swiftly swimming through the frigid ice-capped waters, hunting and being hunted, until that time comes in the fall when they make landfall on Antarctica to begin their breeding migration. They march hundreds of miles inland over icy crevices and crumbling glaciers to mate, lay eggs, and incubate their offspring through the long dark winter where temperatures can reach more than -50°C. Now consider the many functions of the "marching" feet of these icrosystem birds. They use them to steer like rudders while swimming, to walk and climb for hundreds of miles over slippery craggy terrain, and finally, to hold, incubate, and protect the eggs that will produce the next generation. Hence, the feet of Emperor penguins have over time evolved to perform multiple tasks that are at odds with each other. In addition, their feet must do each of these functions adequately enough to ensure their survival and that of their offspring. Thus, is it equally unfair to critique the penguin foot solely for being an imperfect rowing device, as it is unwise to blindly copy it as a propulsor. Organisms are multifaceted, and biology is not always the best answer for a particular human design goal.

Second, this difference in the processes of design (evolution versus human engineering), combined with differential availability of usable material and resources, has resulted in profoundly different technologies. Although fundamental physics and chemistry are the same for nature and human engineering, the results of their gestalt designs are often profoundly different. As pointed out by Vogel,[2] nature is wet, soft, and squishy; engineered structures are dry, hard, and stiff. Nature has more far more curved surfaces than flat, and rarely do you see a sharp corner or right angles. Our designs tend to focus on stiffness, whereas strength matters more in nature. Engineers make joints that slide, whereas natural structures bend, twist, or stretch, and can often reconfigure to accommodate forces. Metals are widely used in engineered structures but are entirely absent as a material in biology. Organisms experience more tensile forces; compression is more common in engineered structures. Rolling and rotation are common and important in society's technology, but almost completely absent in biology. The rotation at the heart of many engines, in particular, is entirely different from the engine that drives animal movement, muscle, which can only shorten. Lastly, engineered devices are designed for minimal, modular repair, but most organisms have continuous cell replacement—you literally are not the same person you were a year ago.

Recent Advances in Biologically Inspired Animal Robotics

The primary collaboration that drives modern robotics projects is between engineers and biologists. For engineers, the goal is to use biology to improve or create new robotics. Informed by the contrasts and caveats discussed above, "bio-inspiration" is more common than mimicry per se. This means that many designers take biological principles and run with them and do not adhere strictly to the rules of biology. For biologists, robots provide a way to test hypotheses of function and control.[3] Although computational modeling and simulations are possible, only physical modeling will work for many real-world situations.[4]

In the spirit of both goals—to answer biological questions and to push the boundaries of engineering—research teams have developed a multitude of bio-inspired robots based on a broad array of organisms, including humans, dogs, salamanders, snakes, fish, insects, spiders, lobsters, and octopi. Below we highlight some novel findings in current and past projects as an entrée into the bio-robotics world. For each, we discuss the sources of biological inspiration, examples of current robotics, challenges and goals for future improvements, and application for military usage.

Conquering Complex Terrain: Challenges for Land-based Locomotors

Existing robotic platforms that use wheels and tracks are limited in their mobility in the rugged environments and complex urban terrain where military personnel and first responders operate. The unique load-carrying and climbing capabilities and agility that terrestrial four-legged animals such as horses, rocky mountain goats, and antelopes have evolved are being mimicked in the robotic platforms known as Big Dog, a four-legged platform designed to carry heavy loads over rough or uneven terrain.

Biomechanical research of legged animals was key to establishing the behavioral dynamics of motion, force, balance, and control principles that have led to dramatic increases in the capabilities of robotic technologies in the Big Dog program. In particular, high speed video and force platforms were used to describe the climbing and running ability of billy goats and guinea fowl over rough rocky terrain and steep grades.[5] By integrating live animal biomechanics data as blueprints of performance, the Big Dog robots can replace traditional tracked or wheeled systems with legged mobility that can move through rocky terrain, travel efficiently, and even recover from stumbles. Indeed, the latter ability was a major breakthrough for legged robotic designs. The findings from biomechanical research of how legged animals kinematically and neuromuscularly control their balance and velocity in rough terrain gave engineers insights into how simple mechanical models of walking or climbing extend to unsteady conditions, and thereby provided inspiration for the design of the dynamically stable Big Dog legged robot. The ultimate goal of the Big Dog effort that has emerged from the former Biodynotics program is to develop, on a commercial scale, a biologically inspired robotic mule that can navigate difficult rugged environments while carrying additional payloads of supplies, weapons, tools, or sensors to facilitate military and disaster relief missions.

A challenge in designing many-legged robots is control and actuation of the multiple legs and joints while maintaining balance. With simplicity in walking in mind, Dennis Hong and colleagues at Virginia Tech have created an unusual three-legged robot called STriDER (Self-excited Tripedal Dynamic Experimental Robot).[6] This bizarre-looking creation is all legs and hips with no body—imagine a tripod of bending legs connected by rotating joints. The inspiration came from a principle of human walking: during the swing phase, in which the lifted leg swings forward like a pendulum, there is minimal neural control. Unlike many other legged robots that actively control leg movement throughout all phases, STriDER uses a similar passive swing phase, simplifying the required control.

The key design features of STriDER include a single DC motor and gear set that can move each leg independently, and three joints that allow this centrally located motor to rotate with each step. The robot takes a step by moving one leg at a time. First, the robot shifts its center of mass forward on two legs; as it falls forward, the remaining third leg swings between the stationary two and lands in front. The central section containing the motor is flipped 180 degrees during the step, which is enabled by the multiple joints at the leg ends. Depending on the sequence of which leg swings through, the robot either moves straight forward, albeit in a zig-zag fashion, or turns.

STriDER, like Big Dog, is a tall robot. Because of its height and inherent stability when all three legs are static, it may be best used as a platform for cameras or other remote sensing in hard-to-reach areas. Of all the robots discussed in this chapter, this is probably the most bio-inspired rather than biomimetic per se, but it clearly shows the benefits of borrowing choice features of natural design.

Beneath the Surface: Swimming with Four Limbs

Underwater swimmers come in many forms, ranging from animals that use the whole body for propulsion, such as eels and jellyfish, to those with rigid bodies and flexible fins, such as boxfish. Multiple types of bio-inspired swimmers have been developed, the most prominent of which is Robotuna, a flexible-bodied robot based on the blue-fin tuna, one of the fastest cruising swimmers in the ocean. One purpose of such robots is to create new autonomous underwater vehicles (AUVs), which have traditionally involved use of rotary propellers. However, current propeller-based and fish-based designs alike suffer in their ability to hold position or to accelerate quickly, making them unsuited for some applications.

Turtles provide an alternative inspiration for underwater locomotion. With four flippers and central rigid body, they are able to easily hover in place and can create quick movements for acceleration and adjustment to perturbances. Two robotic turtles take advantage of these features: Madeleine, a collaborative project led by John Long at Vassar College, and Roboturtle, led by Michael Triantafyllou at MIT. Both robots consist of a sealed, rigid body packed with sensors and controllers and with room for other potential payload and four bilaterally symmetrical flippers that can be actuated in any pattern.

Madeleine's flippers have an airfoil shape and are flexible, matching the material properties of real turtle appendages, and they pitch up and down to produce propulsive vortices. Roboturtle's flippers add an extra degree of freedom, able both to pitch along the fore-aft axis and flap to the side. The mobility and coordination of these flippers gives the robots unprecedented control, allowing them to easily hold position in the water and to generate large forces quickly for acceleration and maneuverability. Roboturtle's extra degree of freedom of movement in the flippers makes it especially good at reacting to transient forces in unsteady environments, such as choppy seas or changing currents.

In addition to aims for AUV development, robots such as Madeleine nicely illustrate how biologists can learn from engineered devices. One longstanding question in functional morphology concerns why it is that modern aquatic tetrapods—animals with four appendages like sea turtles, penguins, seals, sea lions, dolphins, and whales—employ only the front or rear limbs for steady, straight swimming.[7] Why not take advantage of the propulsion ability of all four limbs? Puzzlingly, extinct species such as plesiosaurs used all four, as inferred by their anatomy. By using the robot Madeleine, it was possible to isolate specific patterns of flipper movement to directly test this question. Long and colleagues found that during cruising, two flippers produced the same top speed but at half the power cost as with four flippers. However, when starting from a complete stop, the four-flippered configuration produced the highest acceleration, lending evidence to previous hypotheses that plesiosaurs were acceleration specialists.

Up the Wall: Robotic Climbers

Scansorial movement—travel up vertical surfaces—faces its main challenge from gravity. Insects, spiders, lizards, and frogs are expert animal climbers that can, for example, nimbly attach to surfaces, run up and down, and jump off if needed. They can even do this upside-down, suspended from a horizontal surface such as a ceiling. They can move over smooth or rough surfaces, and have no problems returning to normal locomotion on the ground. How do animals stick to the side of a building, tree, or rock face with the force of body weight pulling downward, and how do they move up or down without becoming unbalanced and tipping over?

The key physical factor in the building of robots that climb is adhesion. The issue is not producing enough adhesive force—this is easy to do with magnets, suction, or pressure sensitive adhesives (such as sticky tape)—but *controlling* the adhesion. Consider a limbed animal climbing a wall. To move upward, the animal must lift at least one appendage off the surface, and in doing so, must produce enough force to overcome adhesion, advance the appendage, and "restart" the adhesive force. In the meantime, the lifted appendage changes the force balance on the animal, such that the other appendages have to support the weight and counter any torques created when the limb is lifted off. Despite these issues, animals use a range of mechanisms to adhere and climb. Generally, larger animals such as cats use claws to penetrate into a compliant surface such as bark; these animals cannot scale smooth, hard surfaces. Tree frogs and some insects use sticky pads, and other insects use small spines that can catch on tiny surface irregularities. Geckos and some spiders use fine hair-like structures that adhere to almost any surface using intermolecular forces.

Multiple climbing robots have been produced following the inspiration of these mechanisms. The RiSE (Robots in Scansorial Environments) project, a large multi-institution collaboration, is a particularly nice example of how a detailed knowledge of biological systems can be integrated with engineering to produce impressive climbing robots.[8] The most advanced climbers produced by this Defense Advanced Research Projects Agency (DARPA)-funded team include Spinybot, inspired by insects that use small, claw-like hooks, and Stickybot, modeled after geckos, which have highly specialized feet and toes. Here we highlight Stickybot for its unusual and highly biologically based components—peeling toes, directional adhesion, and differential control of toe placement.

A robot such as Stickybot never would have been made without an understanding of the gecko's specialized adaptations for climbing. Geckos are amazing climbers, able to stick to almost any surface on Earth, even under water or in a vacuum.[9] Although extremely sticky, a gecko can attach and detach its toes in milliseconds; furthermore, the toes do not stick to each other. Even more remarkably, gecko toes are self-cleaning—dirt is whisked away after a few steps of walking. Unlike sticky tape, the bottom of the toes does not degrade into a gooey and less effective mess over time.

The way that gecko toes stick is a combination of the complex toe pad anatomy and the particular direction and timing of forces during locomotion. Gecko toes are hierarchical structures, with key features from the macro to nano scales. Toe pads have ridges called lamellae that run perpendicularly along the long axis of the toe, and these lamellae are composed of hair-like stalks called setae. Each seta branches into multiple split ends, with each end being tipped by an extremely small plate called a spatula. Spatulae, which are less than a micron in width, are the sites of contact and adhesion between the gecko and the substrate. Kellar Autumn and colleagues have shown in a series of experiments how this system functions as a programmable adhesive, dependent on both the anatomical features and the gecko's movement patterns. When the lizard pulls its foot across the surface, it preloads the setae and angles them properly to set up the attachment force. This force is extremely high—a tokay gecko can hang from a single toe with no problem. However, the force is dependent on the angle of the setal stalk with the substrate. If angled more than 30 degrees from upright, the adhesion force disappears, allowing detachment. The way that the gecko unsticks while moving is by curling its toes backward, a feat termed *digital hyperextension*, which increases the setal angle enough to detach. Unlike pulling sticky tape from a surface, the gecko uses very little force for this detachment, contributing to a very energy-efficient system. The toes then uncurl when placed back on the surface, and reattach.

Stickybot is a climbing machine that mimics the gecko's mechanisms of climbing from the macro to micro scale. The general gecko-inspired features include a flexible body, toes that curl, and hierarchical toe pads with hair-like adhesives. Starting at the smallest scale, the toe pads are composed of directional polymer stalks that mimic the setae and spatulae together. The stalks are made of polyurethane from a mold, which imbues the stalks with slanted tips, an important feature that gives the robot a form of controllable adhesion. When the stalks are loaded (that is, when the foot presses down and pulls across the surface), the tip face has maximum surface area in contact and adheres to the surface. Detachment in the robot's foot is also inspired by the gecko. To release the toes, steel cables connected to their tips cause them to curl upward when pulled back, changing the angle of the stalks and thus decreasing the attachment force. At the macro level, Spinybot has a differential rocker system that actuates the toes, which allows them to move independently and conform to uneven surfaces. Lastly, Stickybot has a flexible spine and sensors in the shoulders to help distribute forces more evenly among the legs.

The net effect of these features is a robot that can vertically climb a variety of surfaces including glass, glossy tile, and granite. Though extremely impressive, Stickybot (and other biomimetic climbers) are still far less capable than their natural counterparts. For example, although Stickybot can deal with small surface irregularities, it cannot maneuver over larger bumps or accommodate larger obstacles such as a window ledge. Most climbing robots have difficulties negotiating the vertical to horizontal transition (such as reaching the top of a cliff), but the use of large, flexible tails to create favorable force balance is ameliorating that problem as well. A further challenge includes issues of scale: as the robots become bigger, their self-weight or added payload increases disproportionately compared to the adhesive forces created by the surface area of contact. This means that larger climbing robots may need to use different design principles, or look elsewhere for inspiration.

Manipulation: Grasping with Agility in Three Dimensions

Mammal tongues, elephant trunks, and cephalopod tentacles can perform an amazing variety of functions that allow these animals to deftly interact with and manipulate their environment. For instance, giraffes can wrap their tongues around high tree branches to strip off leaves using gross motions, or they can more deftly and delicately pick off leaves while avoiding the many spiny thorns found in acacia trees. Elephant trunks are amazing appendages that are strong enough to pull down trees and yet agile enough to pick a peanut out of your hand. Squid have eight arms and two specialized tentacles, while octopi have only eight multitalented arms. Squid tentacles are specialized for quick-strike prey capture. The arms of squid and octopi perform a myriad of more general manipulation and locomotion behaviors. These biological manipulators have inspired the design and application of soft robotic manipulators.[10] There are many potential applications that can be imagined for an octopus-like robotic manipulator. For example, search-and-rescue robots equipped with "octopus arms" could efficiently navigate a rubble pile, searching for victims in crevices, analyzing structural integrity in a collapsed building, and grasping and handling payloads of varying dimensions, threats, and fragility (for example, rocks, unexploded ordnance, and injured humans).

Inspiration from these fantastically flexible limbs led a multidisciplinary team in the DARPA Biodynotics program to conduct research on the design and use of an innovative style of robotic limb known as OctArm. The OctArm is a soft robotic manipulator based on the biological model of an octopus arm. Octopus arms, like mammal tongues and elephant trunks, work as muscular hydrostats. The anatomical design of these types of appendages is devoid of bones and instead uses a pressurized chamber that is controlled by a range of unique muscle architectures. The function is similar to a balloon that, when squeezed at one end, elongates as the other end extends away from the increase in pressure. Because of this design, muscular hydrostats have extreme flexibility translating into a vast range of motor capabilities. Octopus arms can shorten, elongate, twist, and bend in any plane at any point along their length, making them excellent animal models to mimic for dexterous grasping and manipulating robotic devices. In contrast to rigid links, levers, and rotating joints, which most conventional robot limbs are based upon, OctArm has three pneumatic tube sections that can create continuous bending motions. Thus, the OctArm is a highly flexible robotic limb that is able to achieve similar performance capabilities to a living octopus arm in that it can pivot, extend, shorten, and bend at almost any point along its length.

This enhanced functional ability over conventional robotics allows for compliant whole-arm manipulation of objects. Instead of using a hand or claw to grasp via a pinch grip, the OctArm is able to encircle an object and grasp with distributed pressure. This allows the arm to lift objects that are many times larger than those graspable by traditional manipulators. The soft, compliant nature of OctArm allows it to firmly grasp uniquely shaped objects and successfully handle fragile materials. In addition, the arm's ability to wrap around an object eliminates the potential of piercing it. The arm has been designed to attach to a small, unmanned ground vehicle, which allows the vehicle's operator to enter a hazardous environment and manipulate objects while remaining at a safe distance.

The OctArm has been tested in several battlefield tasks including mock scenarios where it successfully grasped and removed heavy mannequin soldiers from a battlefield without further "injuring" them and while providing safety to the remote operator. In addition, OctArm has been successful in search and recovery of mock destroyed buildings. With its tip-mounted camera and octopus-inspired suction cup devices, it is able to navigate its flexible limb into irregularly rocky rubble, sense objects, and retrieve them.

Squeezing through Cracks: Inspiration from Single-celled Locomotion

Many kinds of cells can move on their own. Unicellular organisms such as bacteria or protists use rotating or undulating flagella, or wave with rows of smaller cilia. Individual amoebae, as well as many types of animal cells such as leucocytes, keratocytes, and fibroblasts, move not with an engine such as a flapping flagellum, but by extending a protrusion, attaching it to the substrate, and then pulling themselves forward, a sort of "bootstraps" way of movement.

The particulars of how one unicellular species moves have served as inspiration for an unusual type of cylindrical robot conceived by Dennis Hong and his colleagues. Their "whole-skin locomotion" (WSL) device is loosely based on movement of the "giant" amoeba *Amoeba proteus*.[11] The body form of the amoeba can be thought of as a membraneous bag surrounding a rigid gel and a liquid inner core. To move, the giant amoeba extends a protrusion (a pseudopod, or "false foot") forward, with the inner liquid moving to the tip. The bottom surface contacts the substrate and remains stationary, and the back end contracts. Amoebae can even advance forward in fine capillary tubes, with all surfaces in contact circumferentially.

Hong's amoeba-inspired WSL device is an elongated torus, a sausage-shaped robot with a hollow middle. It works in icrosy by turning itself inside out in the following manner. The outer skin is used for traction, and as it rolls forward the device as a whole moves too. As the outer skin moves forward and reaches the forward end, it rolls over the edge and back down the middle and becomes the inner skin, marching backward as the device moves forward as a whole. Another way of thinking of this is as a three-dimensional tank tread that rolls in on itself—the tread itself is the vehicle.

Because of its ability to squeeze between cracks and obstacles, this device has tremendous potential for uses such as search and rescue, where it could follow along through spaces such as rubble piles. Additionally, it can be used to travel through piping and could be especially valuable in systems with variable diameter or surface roughness, and flexible tubing; it could even be used to search the human digestive system. One challenge with development and deployment of a WSL robot for military usage is how to take advantage of the design; because the device as a whole constantly rotates around itself, there is no stationary platform for objects such as cameras. However, this is likely to be a minor technical issue, as it is possible to embed sensors or other material into the skin itself. WSL may ultimately prove to be a valuable tool for crawling and searching through uneven terrain.

Continuing Sources of Biological Inspiration

The root of these realized robotics projects is biological inspiration. Engineers are perfectly capable of identifying principles in plain view—the Wright brothers' bird watching led to their use of wings in design for flight—but for basic biological research to flourish, it is critical to provide continuing sources of inspiration. A large proportion of research and funding today centers around model systems such as fruit flies, mice, and zebrafish, and for good reason—with many basic questions worked out, they are expedient for addressing higher level questions. But the biological world is extremely diverse and complex, and these model systems are not always so model. Different organisms have solved life's problems in their own particular ways, something that we are still only beginning to appreciate. How can we mimic or be inspired by phenomena that have not been discovered or explained? In the past few years alone, we have learned of ants that explosively propel themselves with their jaws, a new species of shark that instead of swimming actually crawls on modified fins, an octopus that walks uniquely across the seafloor by sequentially rolling only two tentacles, and snakes and ants that glide through the air. The research that uncovered these stories was motivated not by promises of future practical applications such as robotics, but by desires to understand the rich spectrum of functional mechanisms among all organisms. Can we extract useful robotics principles from these diverse phenomena? Whether or not it is possible, we cannot know until we have first identified and understand what is out there to begin with. To do this we must continue to support basic biological research. What we have left to understand in areas such as biomechanics, physiology, and neural control, in species known and yet to be discovered, is immense. It is imperative that we preserve and discover the Earth's richness in diversity and continue supporting the research that brings this to light.

NOTES

1 M. LaBarbera, "Why Wheels Won't Go," *The American Naturalist* 121, no. 3 (March 1983).

2 Steven Vogel, *Cat's Paws and Catapults: Mechanical Worlds of Nature and People* (New York: W.W. Norton and Company, May 1998).

3 R.E. Ritzman et al., "Insect Walking and Biorobotics: A Relationship with Mutual Benefits," *BioScience* (January 2000).

4 R. Pfiefer, M. Lungarella, and L. Fumiya, "Self-organization, Embodiment, and Biologically-inspired Robotics," *Science* 318, no. 5853 (November 16, 2007).

5 M.A. Daley and A.A. Biewener, "Running over Rough Terrain Reveals Limb Control for Intrinsic Stability," *Proceedings of the National Academy of Sciences* 103, no. 42 (2006), 15681–15686.

6 I. Morazzani et al., "Novel Tripedal Mobile Robot and Considerations for Gait Planning Strategies Based on Kinematics," in *Recent Progress in Robotics: Viable Robotic Service to Human*, ed. Sukhan Lee, Il Hong Suh, and Mun Sang Kim (Berlin: Springer, 2009).

7 J.H. Long et al., "Four Flippers or Two? Tetrapodal Swimming with an Aquatic Robot," *Bioinspiration and Biomimetics* 1 (March 2006), 20–29.

8 K. Autumn et al., "Robotics in Scansorial Environments," *Proceedings of SPIE* 5804 (2005).

9 K. Autumn, "How Gecko Toes Stick," *American Scientist* 94 (2006), 124–132.

10 W. McMahan et al., "Robotic Manipulators Inspired by Cephalopod Limbs," 2004, funded by DARPA BioDynotics Program, Contract N66001-03-C-8043.

11 M. Ingram, "Whole Skin Locomotion Inspired by Amoeboid Motility Mechanisms: Mechanics of the Concentric Solid Tube Model," thesis, Virginia Polytechnic Institute and State University, August 10, 2006, available at **http://scholar.lib.vt.edu/theses/available/etd-09182006-152023/unrestricted/Ingram_WSL_Thesis_ETD.pdf**.

chapter 14

BIOLOGICAL AUTOMATA and NATIONAL SECURITY

Yaakov Benenson

Automata are associated with two rather distinct notions: a commonly used one of any "machine that is relatively self-operating," and a specialized, computer science concept of finite state machines, in particular finite automata. Biological automata referred to in this chapter follow the former notion; specifically, they resemble the autonomous, self-operating systems best described in Norbert Wiener's book *Cybernetics*:

> automata effectively coupled to the external world . . . by flow of incoming messages and of the actions of outgoing messages. . . . Between the receptor or sense organ and the effector stands an intermediate set of elements, whose function is to recombine the incoming impression into such form as to produce a desired type of response in the effectors. . . . The information fed into the central control system will very often contain information concerning the functioning of the effectors themselves. . . . Moreover, the information received by the automaton need not be used at once but may be delayed or stored so as to become available at some future time. This is the analogue of memory. Finally, its very rules of operation are susceptible to some change on the basis of the data which have passed through its receptors in the past, and this is not unlike the process of learning.[1]

Wiener discussed at length the similarities between mechanical and "natural" automata—a discussion whose origins date back to Rene Descartes, who described animals as "biological automata."

In the natural world, single cells were for a long time considered the smallest "automata" with a nontrivial behavior. An aggregation of billions of cells in higher organisms endows these organisms with irreducibly complex high-level *behavior* that determines how they respond as a whole to changing environments and in particular how they make decisions under uncertain conditions and with incomplete information. This complexity reaches its peak in humans who excel in nondeterministic behavioral patterns that are light years away from the simple reflexes of lower organisms. However, all the way down to organs and individual cells we find molecular-scale processes, otherwise called *pathways* or *networks*, that themselves act as autonomous information processors and that respond to environmental information in a programmed, mostly deterministic fashion. For example, hormones are secreted in our body and are sensed by individual cells via special receptor molecules on their surface. These receptors then report about the presence of a hormone to the intracellular components; ultimately, multiple data from a variety of external stimuli are integrated by a cell to take an appropriate action (for example, to activate a stress response). These natural molecular systems sense, process, and respond to environmental molecular signals; therefore, they fit Wiener's definition of an automaton. Intuitively, we may not regard them as automata because they have no physical boundary and are not separated from the environment in which they operate. Instead, the whole molecular ensemble, when placed in a certain volume, represents one "distributed" automaton that has a certain emergent behavior. A useful analogy may be found in the world of chemical engineering: there, a control system of a plant has sensors and actuators distributed over large distances. The separation between the system and its environment (the plant itself) is done by listing all the elements that directly belong to the system. Similarly, a specification of a molecular automaton is a list of all molecules that comprise the automaton, their concentrations, and the parameters of their interactions with the environment and among themselves.

Interpretation of biological processes in these and similar terms (for example, as signal-processing and information-processing systems, control systems, circuits, and so forth) has a rich history. It has recently inspired an effort in a synthetic direction—namely, design and construction of artificial molecular systems with novel sensory, computation, and actuator properties. Success in this endeavor will supplant natural automata with engineered ones and afford a wide range of applications from sophisticated intracellular sensing to single-cell diagnostic tools and "designer" organisms with novel functionalities.

Design Paradigm and Potential Applications

Development of biological automata is still in its infancy. Eventual success in this endeavor will require massive integration of cutting-edge results from a few tangential technologies still under active development. Apart from that, real ingenuity will be required on the developers' part to solve both conceptual problems pertaining to the systems' architecture and technical issues that may impede delivery of working applications. However, the anticipated benefits of the technology could be substantial. It is therefore instructive to consider these benefits even at this early stage.

We can predict the impact of biological automata by examining the impact made by the advent of digital computing machines. Conceptually, both technologies deal with the processing of information, but in practice a lot depends on what generates this information and how it is relayed to a "computer." One kind of data is generated by humans. These data by definition must be stored and transmitted by ways accessible to other humans and therefore should be formulated as a sequence of symbols, sounds, and images. It can be relatively easily digitized and fed into a digital computer for further processing. Another kind of information does not result from an intelligent human activity but has its origins in the natural world. Its availability to digital computer processing depends on our ability to collect these data via sensing. In essence, anything that can be measured by a physical device and converted into an electric current can be digitized and fed as an input to a computer. The combination of sensors and computers is very powerful. Meteorology stations use measurements from a large number of environmental sensors to feed predicting models that run on a supercomputer, making highly reliable weather forecasts that have become indispensable for seamless functioning of a modern society. Yet certain events in nature can never be observed, due either to the lack of sensors or our inability to gather the data because of the temporal or spatial features of an event.

As shown by the weather forecast example, a capability to rapidly process large amounts of data has significantly advanced many aspects of human existence. It is important to note that it is a combination of powerful sensing and computation technologies that is most useful. In the absence of practical methods with which to gather and relay data, computer programs would largely remain worthless. All human-generated data can be fed into a digital computer due to either its inherently digital character (symbols) or its easy digitations (sound and images). In parallel, increasing amounts of data from natural sources are becoming accessible as more and more sensors are being developed. However, not all information from natural sources can be extracted and processed using existing tools. One such kind of information is generated by a living organism to coordinate the activity of its parts, organs, and cells. These are normally chemicals or specialized cells that circulate in an organism. We may expect that these organism-wide indicators will eventually be rendered accessible by miniature conventional sensors introduced in a human body. Even today, tiny cameras can transmit images from the depths of the digestive tract, while nuclear imaging can uncover abnormalities with high spatial precision.

On an even smaller scale, an individual cell generates vast amounts of molecular information to regulate the activity of its various pathways. These and other molecular indicators can tell a lot about the physiology of a cell and, more importantly, report on problems. A problem with a single cell in an organism can eventually become a problem for the entire organism if not checked in time. For example, a cancer usually develops from a single abnormal cell. Moreover, many early molecular signals precede an emergence of a problem, and they potentially could be used to trigger a preventive action. Sensing multiple molecular features in a single cell could both tell us about the current state in this cell and help us predict its future. However, as of today, we have neither the required sensors that would operate inside individual live cells nor the computers capable of processing this information. Considering the sensory component, we contend that there are basic physical limitations to developing sensors that would convert molecular signals in single cells to an electrical current or other comparable outputs. Even if these sensors could be developed, there is hardly a conceivable way to transmit the sensory output to a computer for processing. What is required here is a change in paradigm, which may come from carefully studying biology. In fact, biology solves similar tasks using very different engineering principles. Biological sensors convert molecules to other molecules instead of converting molecules to an electrical current. Biological processors are built from molecules that interact with the molecular outputs of the sensors to generate a response. These processors operate inside single cells. We can use these principles to design artificial systems with novel and powerful capabilities. Molecular automata technology will combine the power of sensing of multiple molecular indicators in individual live cells with their programmable processing inside these cells. They will be biology-friendly so that they can function in individual cells and will be built much like the natural molecular pathways, with sensors, computers, and actuators all composed of biological molecules. In fact, in certain cases the operation of these automata could go completely unnoticed by an outside observer, much like the action of myriad biological transformations inside our bodies go unobserved, at least as long as they function properly.

To recapitulate, molecular automata comprising a set of versatile molecular sensors and a general-purpose, programmable molecular computing module will be used in a broad sense to collect and distill molecular information from individual cells and to use the result of a molecular computation to trigger a desired response. Specifically, the applications will include detection of normal complex cell phenotypes in an organism in real time; detection of diseased cells and their labeling or destruction; making predictions about a future development of a disease; creating "designer" cells with novel properties; and many more. Unlike digital computers that are only useful for complex problems that cannot be solved by an unaided human brain, molecular automata will be useful even if their computational power is modest, as their utility stems from the capacity to gather and process information at a place where existing approaches fail. Even a trivial automaton with a single sensor could be extremely useful.

Successes and Challenges

The work on biological automata has just begun. However, some encouraging results have emerged in recent years (theoretical work on these issues dates back to the 1960s). A large effort has focused on biochemical systems—that is, systems composed of biological components that operate outside of the native biological environment. These systems sometimes approximate real biological processes, although care must be taken when making any conclusions due to many complicating factors peculiar to the real-life biological milieu. These biochemical systems focused on two complementary models of computation, those of logic circuits and a state machine. The latter system in its most advanced form represented a fully functional molecular automaton that worked outside of a live cell. Nevertheless, it demonstrated the complete sequence of steps, beginning with sensing molecular signals, a decisionmaking computation and, ultimately, actuation. In parallel, the work on purely biological systems has been focused on the development of simple logic circuits as precursors for biological computers and, separately, on the development of biosensors. Recently, new work has looked at a framework for a scalable, general-purpose automaton that may function inside human cells. This automaton utilizes a computation core based on logic evaluation.

To summarize, development of biological automata has progressed significantly along both theoretical and practical axes. From a theoretical standpoint, we have moved from a few vague ideas toward concrete feasible designs whose implementation is only a matter of time. We have also identified a few key challenges we must tackle in order to deliver real-life applications. From a practical perspective, we have tested a variety of approaches and have a good understanding of their relative merits and drawbacks. If in the past we faced conceptual challenges, they have largely been overcome, and we are in a position to advance to real-life applications.

This transition raises a number of issues, one of which is the delivery of bio-automata into the cells of an organism. According to the common wisdom in the field, the automata will be built as a set of genes, with each gene coding for one structural component of the system. Ideally, the genes are to be delivered transiently into all cells of an organism and function there for a period of time just long enough to gather and process the relevant information. Delivering a collection of genes to all cells of an organism is a technical challenge that is being addressed by active research in the field of gene therapy, which also requires similarly efficient delivery. Besides, the timing of an automaton action has to be precisely controlled. Moreover, as the genes are successfully delivered to a cell, we must rely on cellular processes to generate the gene products. Unfortunately, a cell is not a completely predictable machine, as illuminated by the current research in systems biology. Instead, a cell is characterized by a great deal of randomness; for example, the exact amount of a gene product generated in a unit time cannot be predicted precisely. Therefore, an automaton design should anticipate and counter this randomness by a variety of compensatory and stabilizing elements, and engineering these elements is a challenge in its own right. Nevertheless, we expect that these and other challenges will be overcome as a few tangential technologies mature. For example, automata delivery to cells will be solved by methods developed in gene therapy. Similarly, development of robust architectures will be informed by the studies in systems biology and insights from natural systems that have similar performance requirements.

Implications for National Security

Being both an emerging and an enabling technology, the exact impact of biological automata on the issues of national security is hard to predict. In a very broad sense, these systems will open ways to entirely new approaches to a medical treatment, preventive or therapeutic. In particular, these automata will be capable of detecting the very early onset of disease symptoms with highest possible spatial precision, that of a single cell. As such, they could potentially be used to fight biological and chemical warfare agents, especially in their preventive capacity. Specifically, we may expect this new technology to supplant immunization as a way of fighting bio-agents and supplant antidotes in fighting chemical agents. As a matter of fact, our body is under constant attack by foreign agents; these attacks are normally dwarfed by the immune system or other response mechanisms in our body. Biological agents, by definition, are the ones that have no appropriate built-in response in an organism. In other words, our biology has not evolved to resist these agents, probably because their occurrence has been rare or nonexistent and thus has not warranted an "invention" of such response. Currently, the best known protection against biological agents is immunization using a vaccine. However, immunization often fails because it responds directly to the attacking agent; mutating the agent or slightly modifying it in other ways may render an immune response useless. Bio-automata technology will enable us to develop new types of responses, based on those characteristics of the biological or chemical agents that are not easily modified. For example, if a certain toxin blocks ion channels, the sensors of a bio-automaton will detect changes in ion-transport properties and other metabolic parameters inside the cell rather than the toxin itself. This sensing will generate a similar response to all toxins that have this particular mechanism. Moreover, the automata residing in individual cells may respond by synthesizing a novel type of channel not blocked by a toxin, or simply by boosting the synthesis of the native ion channel in order to overcome the impact of the toxin. When the toxin will have been removed from an organism, the automaton will self-inactivate in order to maintain a normal number of ion channels per cell. Given that the number of mechanisms by which different agents may cause damage is much more limited than the number of agents, we could in principle envision a general-purpose automaton that would fight all major agents in the same time. To reiterate, this new approach will focus on mechanistic consequences of an agent action, rather than on the agent itself, and thus will significantly decrease the degrees of freedom of agents' developers, as it is much harder to devise a new action mechanism than to design a new compound whose mechanism is established already.

Launching automata to probe the physiological state of a live cell is only one possible level of operation for these systems. A higher level of application complexity will result from the development of "designer cells" harboring these automata that receive their input information from the larger environment. The sensors therefore will reside on the cell membrane, but the computational module and the actuation could still function inside the cell. As a result, designer cells will act as individual agents interacting with each other and their environment and perform organism- and environment-level tasks. Within the medical realm, designer cells harboring biological automata could revolutionize emergency medical treatment in the battlefield. In particular, they could substantially enhance current efforts to deliver new cell-based regenerative treatments. For example, stem cells and their derivatives are currently viewed as a potential source of biological material to replace or replenish damaged or absent tissue, respectively. However, using nonmodified stem cells might result in poor control over the regeneration process and, potentially, tumor formation as demonstrated in a few cases. Moreover, the regeneration process, even if successful, could take weeks or months to complete.

It is becoming increasingly clear that these cell therapies will operate best once the therapeutic cells are enhanced with a variety of automata-like mechanisms. These mechanisms should include an ability to discern the location within the organism and the organ/tissue to be regenerated; safety mechanisms to sense and prevent potential tumorigenesis; and mechanisms that accelerate growth when needed but slow it down once the regeneration process is completed. A number of specialized treatments could be developed, for example, to fight the loss of eyesight due to an eye injury. This case could be one of the first areas where regenerative treatments will be implemented in an emergency mode, because the actual damaged area is small, while the consequences are devastating and a fast response is crucial. The complexity of an eye, combined with its relatively small size, warrants inclusion of highly sophisticated automata-like mechanisms into the therapeutic cell agents. Similar considerations apply for certain brain and/or nervous system injuries that could be limited in size but extremely dangerous in their consequences. It is fair to say that our level of basic understanding of both the eye and the nervous systems has to improve significantly to make possible emergency regenerative therapies, but the development of basic enabling technologies has to continue hand in hand with the basic biomedical research so that both can converge at some point in the future.

Apart from battlefield-specific medical applications, molecular automata embedded in designer cells may address a variety of needs for situations that require interacting with and affecting chemical and biological agents on the battlefield. The very nature of these agents makes them amenable to detection and treatment using antidotes in the form of engineered biological entities. Ideally, an area contaminated with the nonconventional agents will be treated with biological antidotes such as microorganisms or even primitive multicellular organisms such as worms or flies. These antidotes will be programmed to detect and be attracted to high concentrations of the harmful agents, decontaminate them by digestion or chemical neutralization, and self-destruct once their mission is accomplished. Once again, the built-in automata will be used not only to enable the basic antidote function of the organism, but also to ensure its safety for the environment by preventing its escape into the wild.

More mundane tasks, such as decontamination of the environment from common toxic wastes that remain on a battlefield and its surroundings, could also be accomplished by similar means. Given an anticipated complexity of the automata, we could envision a single generic "bug" that would manage the majority of common contaminations, as opposed to specialized bugs that fight individual agents separately.

Apart from the above applications, we believe that many new uses of the technology for national security will emerge. Digital computers, initially intended to serve as fast calculating machines, have transformed all aspects of our daily life as well as the practices of modern warfare in all its aspects. The capacity to process biological information in situ, either in individual cells, organisms, or the natural environment, will lead to multiple ideas that as of now are completely unanticipated.

Conclusion

Biological automata comprising molecular sensors, computers, and actuators have the potential to access biological signals in individual cells of an organism in real time, process these signals in a programmable fashion, and generate an appropriate response. Although these ideas have been around for some time, only recently have we reached a point when actual technical solutions are becoming feasible and may be tested in live cells in culture. There is a long way from a cell culture to an organism, but once feasibility has been demonstrated, we may expect that increased investment will drive the technology forward at a much faster pace, resulting in real-life applications in a relatively short period of time. The anticipated impact of biological automata may be compared to that of digital computing machines. Digital computers fed with real-life sensory data have proven indispensable in monitoring and controlling large-scale manmade systems as well as in the analysis and forecast of natural phenomena. Similarly, biological automata may become indispensable in monitoring, diagnosing, and ultimately controlling molecular processes at the level of a single cell and, consequently, an entire organism. Like other transformative technologies of this kind, it will undoubtedly have an impact on various aspects of national security, in particular in bio-defense. However, past experience has shown that most applications arise after an enabling technology is developed. We expect that this will be the case with biological automata.

NOTE

1 Norbert Wiener, *Cybernetics: or Control and Communication in the Animal and the Machine* (Cambridge: MIT Press, 1948).

part four

HUMAN APPLICATIONS

chapter 15

ENHANCED HUMAN PERFORMANCE and METABOLIC ENGINEERING

James J. Valdes and James P. Chambers

The transformation of the current legacy U.S. Army to the Future Force will require a degree of doctrinal and technical innovation surpassing that of any past revolutions in military affairs. The Future Force can best be described as smaller, lighter, faster, and more agile, flexible, and lethal than any army today. The Future Force will also possess an unprecedented degree of autonomy, with small units operating with limited support for lengthy periods. These conditions will necessarily place enormous physical and mental demands on the Soldier.

The biological sciences have been largely treated with benign neglect by military planners, at best relegated to medical and biodefense applications. The physical sciences and information technology have played a dominant role in the transformation to date, making spectacular and devastating enhancements to the effectiveness of the current force. With rapid advances in biotechnology within the past decade culminating in the sequencing of the human genome, the biological sciences have crossed a technological threshold and are poised to play a pivotal role in shaping the Future Force. From biosensors to new energy sources, biofuels to lightweight materials, molecular computing to new therapeutics, the biological revolution will have an enormous impact on materiel, training, and the Soldier.[1, 2]

Athletes and Performance

The key to any army is the individual soldier. Training and materiel are critical, but monitoring soldiers' physiological status and readiness and possibly enhancing their metabolic efficiency and performance are within our technological grasp. Certainly, there are precedents, crude as they are, in the sports world. Current approaches can be reduced to nutritional supplements (often called nutriceuticals),[3] drugs (such as anabolic steroids), and gene doping.[4]

Juhn's[5] excellent article on supplements and ergogenic (or performance enhancing) aids reviews the supplements of choice for athletes and points out the many methodological flaws in much of the research on them. Also noteworthy are the wide variety of individual responses to supplements as well as observable deficiencies in responses of trained versus untrained subjects. For example, the popular supplement creatine, a primary supplier of the energy source adenosine triphosphate (ATP), appears to be ergogenic in repetitive cycling and possibly weight lifting, but not in running, swimming, or other activities requiring endurance. Protein, amino acid, and antioxidant supplements appear to have no ergogenic effects at all, although they cause no harm as dietary supplements. Interestingly, caffeine is one of the most reliably ergogenic supplements, while having negligible side effects.

The most well-known performance enhancing drugs used by athletes comprise the class known as anabolic steroids. That these drugs are highly effective as ergogenics while having deleterious effects on general health is beyond dispute. Kicman and Gower's[6] fine review article on the use of anabolic steroids in sports examines both positive and negative aspects of these drugs. The body of evidence supports the conclusion that anabolic steroids increase muscle mass in women and children, with smaller effects in adult healthy males, and may also increase performance by stimulating the production of icrosystems (erythropoietin, or EPO) and erthyrocytes[7] (red blood cells); the negative effects include liver disease,[8] psychological effects including aggression,[9] and possible cardiovascular disease.[10]

Gene therapy has the potential to correct metabolic deficiencies caused by genetic mutations but can also be used to artificially enhance metabolic capacity. The Finnish cross-country skier and Olympic gold medalist Eero Mantyranta had a genetic mutation that increased his red blood cell mass and hence his oxygen carrying capacity.[11] This effect can now be accomplished artificially by taking the drug EPO or by delivering the appropriate EPO gene using gene therapy techniques. This constitutes only the first step in performance enhancement using genomic approaches, as a map of health- and performance-related genes published by Rankinen et al. is continually being expanded.[12] Notwithstanding the corrective and self-regulating mechanisms of metabolic pathways and unresolved problems with gene therapy, it is not fantasy to suggest the possibility of "tailoring" the human genome and metabolic pathways for optimal performance purposes.

The Physiology of Energy

ATP is the immediate source of energy for all cellular processes and must be present in living systems at all times to drive energy-dependent processes. The two major sources of ATP are the tricarboxylic acid cycle and the electron transport chain, and glycolysis. The respective metabolic pathways are all interconnected via glycogen, fat, and protein reserves for storage and ATP retrieval, and these pathways are exquisitely fine-tuned and regulated. Indeed, they are so important to life and so conserved evolutionarily that many known mutations are lethal.

Mitochondria are dynamic intracellular organelles with an interesting history. They are thought to have been ancient aerobic bacteria that were assimilated into (taken up by) our ancestral cells, known as pro-eukaryotes, to form a symbiotic relationship that allowed the evolution of multicellular aerobic organisms.[13] Mitochondria produce roughly 80 percent of the cell's energy, although this amount differs from tissue to tissue and in response to metabolic challenge. Theoretical modeling of the metabolic pathways involved in energy production supports the commonsense notion that millions of years of evolution have resulted in highly efficient pathways with similar structured features.[14] That said, individual variations and data from clinical studies suggest that there is room for improvement. Put another way, there is no generic mitochondrion, and cell, tissue-specific, and individual differences remain to be studied and exploited.[15]

A few words on the structure of mitochondria are important to demonstrate the experimental accessibility of these organelles. They have both an inner and outer membrane. The inner membrane is impervious to essentially all ions and most uncharged particles, possessing specific protein carriers—transporters—for a number of important constituents in the energy production process. Some key features of mitochondria include a respiratory assembly (also known as a respirasome), which is an integral component of the inner membrane, and enzymes critical to the citric acid cycle and fatty acid oxidation. These coupled enzymatic reactions occur on the matrix side of the inner mitochondrial membrane, are all critical in finality to production of ATP, and can be manipulated experimentally. Additionally, respirasomes are now known to be highly organized multi-subunit protein electron transporting complexes.[16]

Mitochondria are not simply passive energy factories that produce ATP at a maximal rate. Their efficiency, expressed as a ratio of ATP produced to oxygen consumed, varies across tissues. This variation is determined by the physiological "mission" of the mitochondria, which, from a global metabolic point of view, could be maximizing ATP production and cellular phosphate potential while minimizing the cost of ATP production, or performing any of these processes either individually or in combination,[17] depending on the tissue and metabolic demands.

Oxidative phosphorylation (OP) is a key element in this metabolic balance. OP, the enzymatic phosphorylation of ATP precursor adenosine diphosphate (ADP) to ATP coupled to election transport from substrate to molecular oxygen, is a key controlling element in this balance. It is an aerobic (hence, an oxidative) process, as distinguished from glycolytic phosphorylation, which can occur anaerobically. Approximately three molecules of ATP are formed from ADP and phosphate for each atom of oxygen, as shown below:

$$NADH + H^{+} 1/2 + 3ADP + 3P_{2} + NAD^{+} + 4H_{2}O + 3ATP$$

The key feature of OP is that protons (H^+) are moved via the Fi/Fo ATPase complex from the mitochondrial matrix to the intermitochondrial space, establishing a proton, that is, pH, a gradient. The respirasomes and Fi/Fo ATPase complex that synthesizes ATP within the inner mitochondrial membrane are independent systems linked by this proton gradient. Thus, OP can be "uncoupled" by dissipating the gradient. The degree or tightness of coupling of OP is critical to the efficiency of the process, and aberrant metabolic states arising from the effects of various toxins that uncouple OP result in the generation of heat at the expense of energy. A convenient measurement, which can be performed experimentally, is determination of the ratio of inorganic phosphate per atom of oxygen consumed.

To summarize, oxidation of NADH is coupled to phosphorylation of ADP, arising from proton movement from the intermitochondrial space back through the inner mitochondrial membrane to the matrix side. Three proton-moving complexes have been experimentally reconstituted, and electron transport at these sites induces conformational changes in respirasomes. Finally, ADP and ATP transport into and out of the mitochondria is coupled (that is, ADP enters only when ATP exits), and this is mediated by an ATP-ADP translocase. Any interference of translocase function will uncouple OP. Uncoupling OP results in thermogenesis (heat generation). Excluding naturally occurring hibernation in many animals, such uncoupling can be seen in metabolic diseases, hormonal control via norepinephine, or toxicosis with plant glycosides such as atractyloside. The key point is that despite numerous feedback controls and a highly evolutionarily conserved system, individual and tissue differences exist and the entire system is accessible to nutritional, pharmaceutical, and genomic interventions.

Metabolic Interventions

The term *genomics*, which refers to the study of genes and their expression, has become widely used in mass media. Less well known is the emerging field of *metabolomics*, the study of the many metabolites in the body that represent the sum of human physiological activity. It is now possible to quickly measure a broad array of metabolites from a small blood sample and obtain a metabolic "snapshot" of an individual. Physicians currently assess a few such biomarkers, triglycerides and cholesterol being the primary ones, but the ultimate aim of metabolomics would be to build integrated databases across populations and to use this information to design nutritional, pharmaceutical, and/or training interventions to enhance performance.[18] More to the point, these interventions could be tailored to the individual.

The focus of metabolic assessment in today's health care system is disease oriented. That is, the physician uses particular markers such as cholesterol ratios or blood glucose levels to diagnose conditions such as cardiovascular disease and diabetes. A future science of metabolomics will focus on health and will look at a broad array of lipid and protein metabolites with respect to health status, drug effects, stress, and other environmental challenges such as heat, cold, and altitude. Since lipid metabolic pathways are well characterized, any changes induced by artificial interventions would yield immediate clues about approaches to either ameliorate deleterious effects or enhance performance.

The ultimate goal is to tailor diet and training to an individual Soldier's personal metabolism and to do so with reference to mission requirements and environmental conditions. An individual's metabolism could be fine tuned to the desired task for optimal performance under any mission conditions: cold, heat, high altitude, sleep deprivation, and so forth. It is now known, for example, that after conditions of prolonged strenuous activity, it is necessary to provide both carbohydrates and fats, rather than carbohydrates alone, to replenish the cell's energy stores. Individual metabolic profiling would take this to the next level of sophistication.

Other nutritional approaches are also possible. The carnitine transport system moves long-chain fatty acids (LCFA) into the mitochondria. Fat is a rich source of energy, and LCFAs are required for ATP production. The carnitine transporter can be up- or down-regulated in a number of ways. The stress experienced and strenuous activity undertaken during military missions would likely deplete levels of ATP, resulting in relatively higher levels of ADP and NAD^+. The result would be to up-regulate the electron transport chain, and this could be exploited by increasing the availability of nutritional substrates such as LCFAs.

Adipocytes (fat cells), like glial cells in the brain, were once thought to be inert. They are now known to be intimately involved in metabolism and may be manipulated to good effect. Adipocytes are known to secrete many cytokines, which are involved in energy storage, mobilization, and metabolism, and to participate in a network of adipose tissue signaling pathways allowing adaptation to metabolic challenges.[19]

White adipocytes secrete leptin, a protein that is involved in the regulation of food intake, energy expenditure, insulin regulation, and energy balance.[20] The gene for leptin has now been cloned, and administration of leptin reduces food intake. Leptin is regulated by neuropeptide Y and its secretion can be regulated by, among others, cortisol, insulin, and beta-adrenergic agonists. Thus, this system, key to energy regulation, is subject to manipulation.

Finally, we come to direct genetic engineering approaches. While it is no trivial task to alter one, let alone multiple, key protein in metabolic pathways, the potential payoff could be substantial. Gene therapy, the introduction of healthy genes into cells to compensate for defective genes, relies on safe and efficient gene delivery systems and the regulation and control of gene expression, techniques that are being developed and refined today. Metabolic engineering, the directed improvement of cellular properties through the modification of specific biochemical reactions or the introduction of new ones, will afford the systematic elucidation of metabolic control via a top-down control analysis.

There are more physiologically invasive ways by which one might increase ATP synthesis. Under normal physiological conditions, if ATP is not turned over as a result of biological work, ADP is a rate-limiting factor. Similarly, if the turnover rate of ATP is high, the ADP transporter may not be able to keep pace with the demands for ADP. Genetic manipulation of the ADP transporter has the potential to increase its efficiency, hence making more ADP available for phosphorylation to ATP.

Perhaps the area with the greatest potential to enhance metabolic efficiency is with the respiratory chain supercomplexes known as respirasomes.[21] There are five such supercomplexes, each composed of multiple subunits. The functional roles of respirasomes are now largely known. One strategy that could result in increased ATP production is to alter the subunits via genetic engineering. Respirasomes are involved in electron transport, and altering the affinity of the subunits for transported electrons could result in the generation of four, rather than three, ATP molecules for each electron pair transported. Metabolic engineering of mammalian respirasomes to either increase their affinity for electron transport or to produce megacomplexes is clearly testable with current gene therapy technology and experimental models.

Conclusion

The mitochondrial pathways involved in energy production are highly conserved genetically and possessed of complex regulatory feedback mechanisms. Mutations are often lethal. That said, the fact that considerable individual phenotypic differences exist at all levels of the systems, coupled with our evolving understanding of mitochondrial diseases, suggests that genetic manipulation of key components is possible. Further, the ability of drugs and toxins such as atractyloside and bongkrekic acid and of neurohormonal systems to alter energy metabolism indicates the possibility of nutritional and pharmaceutical interventions. Finally, while the ability to selectively enhance human metabolic performance is speculative, it is clearly a testable concept using cell culture, animal models, and genomic techniques available today.

NOTES

1 Board on Army Science and Technology, *Opportunities in Biotechnology for Future Army Applications* (Washington, DC: National Academy Press, 2001).

2 Office of Net Assessment, *Exploring Biotechnology: Opportunities for the Department of Defense* (Washington, DC: Information Assurance Technology Assessment Center, 2002).

3 M.S. Juhn, "Popular Sports Supplements and Ergogenic Aids," *Sports Medicine* 30, no. 12 (2003), 921–939.

4 P. McCrory, "Super Athletes or Gene Cheats?" *British Journal of Sports Medicine* 37 (2003), 192–193.

5 Juhn, 921–939.

6 A.T. Kicman and D.B. Gower, "Anabolic Steroids in Sport: Biochemical, Clinical, and Analytical Perspectives," *Annals of Clinical Biochemistry* 40 (2003), 321–356.

7 A.D. Mooradian, J.E. Morley, and S.G. Korenman, "Biological Actions of Androgens," *Endocrine Reviews* 8 (1987), 1–28.

8 M.G. DiPasquale, *Anabolic Steroid Side Effects: Facts, Fiction and Treatment* (Ontario, Canada: MGD Press, 1990), 63.

9 H.G. Pope, Jr., and D.L. Katz, "Homocide and Near Icrosys by Anabolic Steroid Users," *Journal of Clinical Psychiatry* 51 (1950), 28–31.

10 B.F. Hurley et al., "Effects of High Intensity Strength Training on Cardiovascular Function," *Medicine and Science in Sports and Exercise* 16 (1984), 483–488.

11 C. Aschwanden, "Gene Cheats," *New Scientist* (2000), 24–29.

12 T. Rankinen et al., "The Human Gene Map for Performance and Health Related Fitness Phenotypes," *Medicine and Science in Sports and Exercise* 34 (2002), 1219–1233.

13 L. Margulis, *Symbiosis in Cell Evolution* (San Francisco: W.H. Freeman, 1981).

14 O. Ebenhoh and R. Heinrich, "Evolutionary Optimization of Metabolic Pathways. Theoretical Reconstruction of the Stoichiometry of ATP and NADH Producing Systems," *Bulletin of Mathematical Biology* 63, no. 1 (2001), 25–55.

15 I.E. Scheffler, "A Century of Mitochondrial Research: Achievements and Perspectives," *Mitochondrion* 1 (2000), 3–31.

16 H. Schagger and K. Pfeiffer, "The Ratio of Oxidative Phosphorylation Complexes I-V in Bovine Heart Mitochondria and the Composition of Respiratory Chain Super Complexes," *Journal of Biological Chemistry* 276, no. 41 (2001), 37861–37867.

17 C.B. Cairns et al., "Mitochondrial Oxidative Phosphorylation Thermodynamic Efficiencies Reflect Physiological Organ Roles," *American Journal of Physiology* 274, no. 5 (1998), R1376–R1383.

18 S.M. Watkins and J.B. German, "Towards the Implementation of Metabolic Assessment of Human Health and Nutrition," *Current Opinions in Biotechnology* 13 (2002), 512–516.

19 G. Fruhbeck et al., "The Adipocyte: A Model for Integration of Endocrine and Metabolic Signaling in Energy Metabolism Regulation," *American Journal of Physiology Endocrinology and Metabolism* 280, no. 6 (2001), E827–847.

20 K.L. Houseknecht, "The Biology of Leptin: A Review," *Journal of Animal Science* 76, no. 5 (1998), 1405–1420.

21 H. Schagger and K. Pfeiffer, "Supercomplexes in the Respiratory Chains of Yeast and Mammalian Mitochondria," *The EMBO Journal* 19, no. 8 (2000), 1777–1783.

chapter 16

FUNCTIONAL NEUROIMAGING in DEFENSE POLICY

Richard Genik III, Christopher Green, and David C. Peters II

Functional neuroimaging is the use of technology to measure and monitor aspects of the working brain in real time. This is an important part of the biotechnology revolution that is affecting many parts of our daily lives and will play an important role in defense policy decisions in the near future. Several devices perform functions that are capable of providing pieces of this immensely complicated puzzle, including electroencephalography, computerized tomography, functional magnetic resonance imaging, positron emission tomography, magnetoencephalography, functional transcranial doppler sonography, and functional near-infrared spectroscopy.

It is critical to understand that future applications will depend on multimodal methodologies and miniaturization technologies that are still under development. Scanning devices will become smaller and more portable. The techniques will increasingly become cheaper and easier to use. Ultimately, disparate devices will merge into comprehensive brain scanning tools that provide extensive, integrated, interactive information about the brain in real time.

The capability of measuring the working brain in real time has potentially important implications for a number of practical applications relevant to defense and national policy issues. Predictions about future applications of technology are always speculative, but we are definitely far enough down the road of this emergent technology to reliably predict likely advances. These will include, but will not be limited to, quick and secure data transmission, reliable intelligence acquisition from captured unlawful combatants, enhanced training techniques, augmented cognition and memory enhancement of soldiers and intelligence operatives, screening terrorism suspects at checkpoints or ports of entry not covered by constitutional protections (such as airports), and furthering the goal of soldier-machine interface devices such as remotely piloted vehicles and prosthetics. Many important and fast-advancing contributions in this field have already been made to clinical and battlefield medicine.

Functional Specialization

The human brain evolved in layers, with evolutionarily newer structures covering older ones. The developing brain in a human embryo grows outward in a pattern that largely mirrors human evolution. The brain is divided into discrete regions controlling vital functions such as sensory transduction and cognition for the cortex. This functional specialization is a critical element in brain research, and it arises mainly from layered development. These functionally specific regions activate when a person is presented with a stimulus, such as a scent exciting an area of the temporal lobe, or a vision exciting the occipital lobe.

Mapping these areas of activation and the circuits that control specific behaviors and cognitive processes is the goal of functional neuroimaging. It is an enormous goal, and creating even a crude map of the functionality of the brain is daunting. The brain involves many, many more connections than even the most advanced computer. This is because, unlike a computer that utilizes binary connections, brain circuits involve thousands of neurons and millions of permutative connections to perform even the simplest task of cognition or behavior. Indeed, one can think of gray matter neurons as transistors (or the "cell" of a computer processing unit), and white matter fibers as the wires that connect the transistors—although in the brain, each neuron can have up to 10,000 or more connections.

To provide some perspective on the enormity of the task of mapping connected areas, consider that a typical human infant's brain has approximately 200 billion gray matter neurons at birth. With the exception of some interesting findings suggesting ongoing neurogenesis in the dentate gyrus (a portion of the "reptilian" or mid-brain), the newborn will never grow any more neurons. Yet by age 1, the child's brain has filled the skull and tripled in size from approximately 350 grams to a near-adult size of 1,000 grams. If almost no new neurons are developing, where does all this extra mass come from? Although the number of neurons is largely fixed from birth, the brain continually forms complex white matter connections between neurons at an almost inconceivable rate. The human infant may form 20,000 distinct neural connections every second for the first year of life. These are merely individual connections. If we factor in circuits that involve thousands or tens of thousands or even millions of permutated reciprocal connections,[1] it becomes clear that there are far more combinations of neural circuits in the cerebral cortex than atoms in the known universe.[2]

Undaunted, scientists are cautiously but busily mapping the connectivity of the human brain, publishing thousands of new papers on the topic each year. Many studies are being published investigating correlative classifications between traditional psychological testing, clinical observations, and brain scans to establish biomarkers for pathological states, cognitive and behavioral tasks, task-specific aptitude measures, detecting deception, and even predicting neuropathological propensity.

Electroencephalography

The oldest device used to assess brain function in real time is the electroencephalograph (EEG), which works by placing electrodes on the scalp and recording the summated electrical signal from nominally 50,000 local neurons. Early clinical electroencephalography was used primarily to detect and diagnose epilepsy, but with advances in computer technology, informative new experimental paradigms and techniques are being developed. EEG recordings can be separated into two major types: continuous and discrete. Continuous recordings are the traditional multitrace waveforms recorded since EEGs began in the late 19th century, and activity is classified by the frequency of the dominant waveform (0–30 hertz) on any given channel—for example, alpha waves constitute frequencies between 8 and 13 hertz. Discrete recordings are triggered by an event, such as an external flash of a light, after which the next 1 to 2 seconds of activity are recorded. In discrete recordings, the "normal" EEG waves are considered background.

Discrete recordings, the current main area of basic research, are called *evoked potentials* when the brain response peak above background is unconscious, nominally classified as any signal before 100 microseconds from an external stimulation. *Event-related potentials* are generally any other peak in the recording, the *evoked* nomenclature being dropped due to the possibility that conscious thought intervened in a decision process.

Quantitative EEG (QEEG) uses postrecording computer analysis common to other functional imaging techniques to analyze the relationship between each of the electrodes placed at the scalp. By comparing the relative signal strength of each electrode, and precisely measuring the latency (time between stimulus onset and onset of the electrical response) and amplitude (strength of the electrical signal) at each location on the scalp, the likely location of the neuronal source can be inferred.[3] However, localization methodology continues to be a matter of intense scientific investigation.[4]

Neurotherapy is an experimental technique that uses a QEEG brain map to analyze psychiatric problems from attention deficit disorder to depression to schizophrenia. Patients are then subjected to a conditioning protocol to train the abnormal brain activity toward a statistically more normal pattern. Neurotherapy has been shown to reduce aberrant symptoms of many conditions in a number of studies.[5]

EEG has several advantages over other functional neuroimaging techniques:

- equipment is relatively cheap (around $15,000)
- reliable recordings only require a single technician
- unmatched temporal resolution is measured in milliseconds.

The main drawback of EEG is that the source localization has only 1 centimeter resolution at present.

Positron Emission Tomography

Harvard University psychologist Stephen Kosslyn spent many years of his life attempting to solve the source localization problem in EEG. In 1986, after several years of equivocal research findings and no publications, he gave a dejected talk about his "dismal failure." As he was leaving the stage, two young medical physicists approached to tell him about something new they were working on that they called positron emission tomography (PET).[6]

The physicists had been experimenting by injecting radioactive bioassays (biologically active substances that are absorbed by cellular metabolism) directly into their carotid arteries. When the short-lived radioactive bioassay decays, it emits a positron. Positrons quickly annihilate with a local electron to produce two or three high-energy photons. These photons are detected by the PET imager, and their origin is determined by projecting their path back to intersection. Since the substances were metabolized and incorporated into the working brains of the subjects, areas of greater positron emission were assumed to be regions that were recently more "active."

However, PET techniques for observing the functional brain in real time are more limited for several reasons:

- temporal resolution (function of time) is extremely low, 30–90 minutes
- results depend on scanning speed and how quickly data can be collected
- spatial resolution (function of distance) is substandard, within a few centimeters
- results depend on the precision of the recording method and whether data has a specifically identifiable location
- equipment cost is enormous
- huge number of team members is required
- technique is invasive
- technique is already outdated in many circumstances.

There are several important clinical uses of PET, particularly for imaging blood flow and metabolic activity.[7] The "ground truth" of PET will remain extremely important in better understanding and assessing other neuroscience tools. Dr. Kosslyn continues his work attempting to identify the circuits and functional brain regions involved in image generation and formation using the full array of neuroscience tools, including PET.[8]

Computerized Tomography

Computerized tomography (CT) relies on x-rays and computer technology. The original systems were dedicated to head imaging only, but whole body systems with larger patient openings became available in 1976. CT works by directing a tightly focused x-ray beam at the target from many different angles around a single axis of rotation. A computer combines the multiple "slices"[9] into an integrated, three-dimensional picture of the brain and surrounding hard tissue structures.

Since CT is an x-ray–based technique, it best images hard tissues and has limited application for functional analysis of soft tissue. However, CT is unmatched in structural imaging. CT neuroscans are used clinically primarily to diagnose aneurisms and other cerebrovascular events because blood outside a vessel is imaged quite well by x-ray. The use of contrast agents in the blood allows better, albeit indirect, imaging of functional brain activity at the price of increased invasiveness.

Current multislice CT systems can image the entire brain and skull in 5 seconds, reconstruct the images, and send them to a radiologist at a remote site before the patient has time to stand up. As CT scan times have grown faster, more anatomy can be scanned in less time. Faster scanning helps to eliminate artifacts from patient motion caused by physiology such as normal respiration, heartbeat, or peristalsis.

The main drawback of CT is its use of ionizing radiation. One CT scan can deliver the same radiation dose as dozens of chest x-rays.[10] Beginning in the early 1990s, research efforts began to focus more on imaging techniques that did not rely on carcinogenic radiation.

Functional Magnetic Resonance Imaging

Along with CT, magnetic resonance imaging (MRI) is widely accepted as a gold standard in medical imaging. Functional MRI (fMRI), however, has technical challenges that have slowed its introduction into common clinical use.[11] The first clinical fMRI diagnostic codes were approved in 2006.[12] The technique is used in pre-neurosurgical mapping of functional brain regions before excision procedures. For example, one option for intractable epilepsy is neurosurgery to remove the excitable brain regions causing the seizures.

Using functional imaging, doctors can avoid removing regions of the brain that are functionally important and should not be removed.

Functional MRI relies on the blood oxygen level–dependent (BOLD) effect. Scientists have known for over 100 years that changes in blood flow and blood oxygenation in the brain are linked to neural activity. A complicated computer bank using technology developed with CT scanners creates a series of pictures that can identify the regions of the brain as they activate. A magnetic movie of blood oxygenation patterns in the working brain lets the researcher infer brain activity in real time as it thinks and responds. Even with such amazing technologies, we are at the very beginning of understanding the brain.

fMRI has many advantages over other functional imaging techniques:

- spatial resolution of the activation patterns is incredible, measured in millimeters
- temporal resolution is good, measured in a few seconds
- there are no known risk factors in healthy subjects.[13]

Despite the advantages of fMRI, these downsides remain:

- technique is still enormously expensive, with equipment costing millions of dollars
- scan typically takes a full hour per subject
- team of computer programmers is needed to adjust and interpret the data over several days.

Magnetoencephalography

Magnetoencephalography (MEG) is a completely noninvasive, nonhazardous technology for functional brain mapping that localizes and characterizes the electrical activity of the central nervous system by measuring the associated magnetic fields emanating from the brain. Every electrical current generates a magnetic field. However, unlike an electrical signal, magnetic fields are not distorted by traveling through the skull. Thus, with MEG, the source of the summated magnetic fields can be triangulated within a couple of millimeters.

Modern MEG scanners use as many as 300 superconducting quantum interference device[14] detectors, allowing very fast acquisition and extremely high localization of the source of the electromagnetic signal. The information MEG provides is entirely different from that provided by structural imaging techniques like CT or MRI. MEG is a dedicated functional scanner that provides functional mapping information of the working brain and measures actual neurological function. This provides complementary information to the excellent anatomical imaging capabilities of MRI and

CT. Using MEG, we are finally able to measure brain activity in real time with excellent spatial resolution.

MEG is also superior in many respects to other functional imaging modalities. It:

- offers vastly superior temporal and spatial resolution
- can be used to localize brain tumors or identify seizure activity
- provides presurgical mapping target for excision.

However, MEG has its disadvantages. It:

- costs more than MRI
- requires a team of technicians and engineers to keep it running
- needs huge levels of refrigeration and several layers of insulation
- requires a specially constructed scan room to limit measurement interference.

Transcranial Sonography

The common ultrasound provides stunning pictures of a fetus through the abdominal wall of a pregnant female. Unfortunately, transcranial sonography is considerably more limited because it is much more difficult to image structures hidden beneath the cranium. However, the skull is thin enough in a few "sonographic windows"[15] to provide a path for the ultrasonic signal. The transorbital window located above the zygomatic arch (the temple) is used to image the posterior, anterior, and medial cerebral arteries along with a few of the branches. Since portions of these arteries provide blood flow to various parts of the brain, doppler sonography can provide highly accurate real-time measurements of blood flow velocity changes. For example, the posterior cerebral artery innervates regions of the brain associated with visual interpretation.

Although both rely on blood flow, sonography is very different from fMRI, which measures blood oxygenation level changes with a spatial resolution of a couple millimeters. In functional transcranial doppler sonography (fTDS), the spatial resolution is determined by the size of the brain innervated by the blood vessel under study. These areas can be quite large, making the spatial resolution of fTDS extremely limited. Changes in blood velocity, which are presumed to directly measure changes in resistance of the artery, occur nearly instantaneously in an event-related experimental paradigm giving exceptional temporal resolution.

There remain several technical problems with fTDS. Only a limited number of large arteries can be imaged. Even in the arteries that are large enough and located within sight of the few available ultrasonic windows, the angle of the ultrasonic beam can make it very difficult to accurately measure blood flow changes.

However, fTDS has several advantages over other functional neuroimaging techniques. It:

- is very cost effective
- is portable
- offers continuous monitoring of blood flow activity
- provides excellent temporal resolution.

A significant amount of useful information about the working brain can be learned from sonography, particularly if one were willing to increase the number or size of the available ultrasonic windows in the skull.

Near-infrared Spectroscopy

Infrared spectroscopy depends on measurements of energy outside the visible spectrum. Near-infrared spectroscopy (NIRS) was first used during World War I to monitor blood oxygenation of bomber crews, a critical measurement prior to the B–29 innovation of pressurized cabins. Similar devices are employed today to measure pulse rates and blood oxygen saturation using a fingertip sensor. Water (the primary component by weight of most biological tissue) is almost transparent to the NIRS spectrum (700- to 900-nanometer wavelength). This makes BOLD signals in the first few millimeters of cerebral cortex through an intact skull accessible to monitoring using a light source and sensors.[16] Although NIRS functional research has been ongoing since the late 1930s, the recent breakthroughs in fMRI have renewed interest in this technology.

There are no approved clinical neuroimaging uses for this technique, but the potential experimental uses are very exciting, and it has the advantages of:

- costing between $25,000 and $300,000[17]
- being portable (even wireless) and completely noninvasive
- offering temporal resolution similar to fMRI.[18]

NIRS can theoretically be combined with EEG, transcranial sonography, and other functional neuroimaging sensors. Unlike fMRI, where the subject is confined to the bore of the magnet, NIRS movement artifacts can be limited by proper affixation of sensors to the scalp. An experimental subject can sit upright, walk around, and perform tasks while his or her brain is being monitored. The only drawback is that NIRS measures best the first few 2 to 3 centimeters of cortex, so deep brain imaging, at least through an intact skull, is very challenging. According to the biogenetic theory, the outermost layer of the cortex should be the most functionally advanced, and there is little doubt that a significant amount of measureable brain function occurs there.

Screening and Detecting Deception

There is already a considerable amount of research on the objective of detecting deception using neuroimaging instruments. The traditional polygraph has been in use for many decades, but this instrument relies on galvanic skin response, pulse, and respiration measures. The assumption underlying the polygraph is that the nearly universal cultural prohibitions against telling an untruth for personal gain will result in increased sympathetic activity when a subject tells a lie. In plain terms, telling a lie should make a person nervous, and this can be picked up by measuring autonomic responses. The problem is that trained intelligence or military persons, psychopathic or antisocial subjects, and others can effectively beat the polygraph by suppressing the expected autonomic response to the telling of an untruth. Functional neuroimaging instruments rely on intrinsic responses of the brain and therefore do not depend on this cultural assumption.

Much of the recent deception research has used fMRI to image patterns of activation in the brain during tasks that require the two states of deception and truthfulness. When a subject answers a question, he or she must either recall an actual memory or recall a false memory and thus confabulate. Theoretically, the brain activation patterns should be quite different when a subject is activating the circuits of an actual memory than when activating the circuits of false memories because actual memories are stored in the hippocampus, while the recollection of false memories (and the ability to *create* new false memories) is largely stored in the cortex.

Lie or truth events are both mediated in the frontoparietal network, but a lie event is characterized by increased working memory and intensive activity in the inferolateral cortex involved in task and response selection inhibition.[19] Several fMRI studies using a number of different paradigms have found that activations in the dorsal, medial, and ventral lateral prefrontal cortex are specific neural correlates of committing deception.[20] However, the use of fMRI is limited by the tremendous cost and size of the equipment, the difficulty in processing the data, and the fact that subjects must hold their head absolutely still inside the bore of a magnet during the scanning session.

In contrast, since NIRS scanners can be wireless and completely portable, they can be used practically in field experiments. In the first published NIRS study on detecting deception, 21 right-handed persons wore a functional NIRS device while playing a poker-like card game. Twenty of the 21 subjects showed greater activation in the ventral lateral prefrontal cortex and mid-frontal gyrus (a portion of the medial prefrontal cortex) when they were bluffing than when they were telling the truth.[21]

There are many applications of this technology in screening suspected security threats. For example, NIRS could be used covertly without any knowledge on the part of the subject at airports or security checkpoints. When homeland security personnel, soldiers, or other officials detect a possible security threat or believe that a person moving through a checkpoint is acting suspiciously, the subject could be taken to a private room where a covert NIRS scanner is located. The subject could be questioned and scanned without even knowing that his or her brain is being assessed. A reliable, although preliminary, determination of the truthfulness of his or her statements to security officials could be made.

The pace of neuroscience research in the area of detecting deception is increasing exponentially. In the near future, the primary barrier to implementing this technology is likely to come not from technical or scientific challenges, but rather from the constitutional and legal arenas that are well outside the scope of this chapter.

Multimodal Imaging

The pinnacle of neuroscience research would be combining the exquisite temporal resolution of EEG/MEG data with the incredible spatial resolution of fMRI/NIRS data. One of many problems with doing so is that these devices measure different aspects of functional neuroimaging. EEG measures the electrical forces of the brain at the scalp, while MEG measures the magnetic force generated by those electrical field potentials. Both fMRI and NIRS measure the hemodynamic response, which cannot yield a better temporal resolution than the few seconds it takes for the contiguous blood vessels to dilate and deliver the increased levels of oxygenated blood to the parts of the brain that have been recently active. Transcranial doppler sonography permits near-instantaneous temporal resolution by measuring the blood flow velocity rather than oxygenation, but there are several practical limitations. Combining these different data streams into one image of the brain is the challenge.

There are essentially three ways that merging multimodal data has been attempted.[22] Converging evidence is the most common method and relies on either taking joint or time-locked consecutive measurements using multiple scanning devices. The two modalities can then be fused to create a composite image of function and anatomy. Currently, several multimodal scanners are available. CT/fMRI and CT/PET scanners can locate structural anomalies such as brain tumors much more readily than either technique used alone. MEG/MRI scanners also have considerable clinical potential. The estimated source location from MEG is combined with MRI images to create magnetic source images. The two sets of data are combined by measuring the location of a common set of fiducial points marked during MRI with lipid markers and during MEG with electrified coils of wire that give off the magnetic fields. The locations of the fiducial points in each data set are then used to define a common coordinate system so that superimposing (or coregistering) the functional MEG data onto the structural MRI data is possible. Several other multimodal scanners in development include nonferrous EEG devices for use in functional MRI scanners, NIRS/EEG, and fMRI/PET scanners. The problem is that many of these dual-use scanners create significant confounds in the data stream, and the resolution of these artifacts remains an active area of research.[23]

The other two methods for coregistering data from multimodal measurements are direct data fusion and computational modeling.[24] Direct data fusion attempts to combine the multimodal data streams using a mathematical/statistical algorithm.[25] Currently, the most commonly used program is the low-resolution electric tomography algorithm, which attempts to collapse the electrophysiological data from EEG into known spatial/temporal points. However, important nodes in the data network can be missed because of where two conditions are measured during two different activation conditions; a correspondence between them could represent one node as active as another node even though there are significant changes in interregional connectivity.

Finally, the multimodal problem could be addressed by developing a neural model that incorporates hypotheses about how the brain activates in response to a particular cognitive or behavioral operation. Unfortunately, although work continues in this area, nobody has developed a reliable computational model that predicts activation patterns for both electrophysiological (EEG/MEG) and hemodynamic (fMRI/NIRS) data.

In order to fully accomplish the goal of integrating multiple scanners, the electronic sensors need to become cheaper, smaller, and more efficient. This is extremely likely to happen, as a commonsense review of the development of virtually any electronic equipment demonstrates. For example, scientists predict that the multi-million-dollar fMRI scanners that weigh thousands of pounds could be reduced to a couple hundred pounds within just a few years.[26] This capability could even be mostly obsolete by then because we will have already coregistered NIRS (which is already small and portable) with the deep brain scans of fMRI and the instantaneous measurements of EEG and MEG.

Communication and Data Transfer

When a reliable computational model can predict activation patterns in the hippocampus (an important region of the brain responsible for learning and memory formation), it could be possible to effectively "beam" memories into the brain.[27] Several research groups are attempting to understand how images and sound signals are transducted in the hippocampus into memories. Neurons encode information by forming complicated circuits with other neurons. This web of connectivity occurs in multiple neurons reciprocally linked together, encoding and linking information. Most, and probably almost all, of these memory circuits link through the hippocampus.

In parallel, other groups are developing a neurochip, a silicon chip containing actual living neurons designed to mimic the human brain. The neurochip could, theoretically, be implanted into a brain, grow functional connections, and then interface directly with a computer.[28] At least one group has already developed a chip based on a portion of the hippocampus called the CA3, a very thin layer that receives input from the cerebral cortex and sends it back again. For various reasons, the CA3 layer of the hippocampus is thought to be where memory formation actually occurs and where this information is stored and linked.[29] Studies are already well under way that involve implanting the neurochip into the hippocampus of animals and transmitting simple memories directly into their brains. When this is accomplished, it might be possible to begin the process of laying down more complex memories.

This development has staggering implications for future defense policy. Instant and secure data transmission directly to the brains of soldiers on the battlefield would be possible. Augmented cognition of intelligence operatives would be a goal. Perhaps learning how to operate an aircraft could be accelerated from years of technical training to a few days. Perhaps reasonably proficient language skills could be implanted within just a few hours; the ability to reproduce technical schematics within a few minutes; the entire family and tribal history of a high-value target or a mission update report within a few seconds. We can only begin to imagine the potential applications of this new emergent technology.

Brain-Computer Interface

Functional neuroimaging devices detect various forms of energy emissions from the working brain, but these emissions are not static. The electrical energy of working neurons is detected by EEG and MEG, and the hemodynamic response of functional regions of the brain is measured using light (NIRS) and the distortions of a powerful magnetic field (fMRI). The metabolism of these neurons in the functional regions of the brain can be measured by a gamma ray detector using bioactive radioactive assays injected into the blood (PET).

A critical feature of these various energy emissions that is not very well understood is the concept of neuroplasticity. The working brain adapts function very quickly and very readily. Under an operant conditioning paradigm (that is, with training), the brain can quickly learn and adapt functioning. This remarkable organ is so flexible that a subject can learn to activate and deactivate functional regions as well as the electrical distribution, metabolic activity, and brain wave patterns throughout the brain after just few hours of training.

A brain-computer interface (BCI) takes advantage of the important feature of neuroplasticity by activating and controlling electronic or mechanical devices based only on brain activity.[30] There is a surprising amount of recent work in the area of connecting the brain directly to prosthetics. EEG and MEG scanners can record oscillation signals from the whole brain or from functional regions of it and activate a device when the subject specifically controls this activity. Slow cortical potentials and sensorimotor rhythm have both been used to activate electronic devices. Evoked potentials recorded by EEG, especially the P300 wave, have been used to activate and even operate communications equipment.[31] The BOLD signal and NIRS instruments measuring cortical blood flow have also been used as a BCI.[32] The main goal of research so far has usually been to exert some degree of control over a prosthetic or communication device.

However, we would pose a question: Is the operation of a human (or at least a human-like) hand or arm more complicated than the operation of an aircraft? The level of detail in the human organ can lead to the conclusion that operating an aircraft by thought alone might actually be easier than operating a hand or arm by thought alone. While it is far beyond the scope of this chapter to attempt a quantitative calculation of the number of functions in the operation of an aircraft versus the number of functions in the operation of a hand or arm, we could reasonably speculate that the number of functions in each, and the degree to which they could be automated, probably is not staggeringly different.

Brain-Computer Interface for Weapons Systems

If an aircraft can be controlled solely by the energy emissions of a working brain, then almost any conceivable weapons system could be similarly controlled. Unmanned tanks, artillery pieces controlled from thousands of miles away, and even search-and-destroy killer robots that discriminate between combatants and noncombatants are much closer than most of us can imagine. Combining the very fast operations of largely automated weapons systems with the unique judgment of a human brain would be especially helpful in an urban environment where a mission-critical goal is selectivity. Multimodal scanning technology with integrated EEG, NIRS, and possibly fMRI, PET, fTDS, and MEG components could operate many types of weapons systems with a very high degree of precision.

Functional neuroimaging technology can already reliably discriminate between men and women. It would not require exceptional advances in technology to have the capability to reliably discriminate between combatants with hostile intent and noncombatants without hostile intent. One early military application of this idea may be the use of NIRS technology on gun scopes, once sensors advance beyond the requirement of scalp contact. A sniper could take aim from a considerable distance and reliably determine if the target is male or female, a hostile combatant, or a civilian, simply from the infrared activation patterns in the first few millimeters of cerebral cortex. A weapons system, perhaps completely controlled via BCI, could reliably roll through a heavily defended urban center, selectively shooting and killing only the men with aggressive thoughts while sparing the women and children and men who are behaving and *thinking* peacefully. Full-scale production of killing machines based on these controversial ideas is just a few years away. The question is no longer *what*, but *when*. Given the speed at which these automated determinations could be made, and the possible uncertainty in targets of hostile thoughts, this intent identification technology is best implemented in combination with nonlethal weapons systems.

Conclusion

Any attempt to use known information to extrapolate a potential trajectory into the future is highly speculative. Some of the theoretical applications proposed here will be shown to be technically impossible, impractical, or even infeasible. Some applications that cannot even be imagined today will become common in 20 to 40 years. Currently, the technology—the numerous scanning devices that have been developed—is far in front of the science—the interpretation of the functional imaging data collected from these devices. The technology is now focused on making the scanning devices smaller, easier to use, and cheaper each year. Whatever the ultimate applications of this technology, we can be certain that the experimental paradigms will become easier to implement, and the pace of scientific discovery will continue to increase for the foreseeable future. However, within the next 20 to 40 years, there is no suggestion that the science will catch up to the technology. The only certainty is that functional neuroimaging research will provide unimagined benefits and applications in every area of our lives, including defense policy, but the human brain will remain largely a mystery.

NOTES

1 For example, if there are 1,000 possible combinations in the first link of a neural circuit, then the number of permutations or combinations with a second reciprocal link would be 1,000 x 999 x 998 all the way down to 1, or 1,000! ~ 10^{2568}.

2 See **www.holisticeducator.com**. There are only around 10^{80} atoms in the visible universe and perhaps 10^{82} including nonluminous matter.

3 K.L. Coburn et al., "The Value of Quantitative Electroencephalography in Clinical Psychiatry: A Report by the Committee on Research of the American Neuropsychiatric Association," *Journal of Neuropsychiatry and Clinical Neurosciences* 18, no. 4 (Fall 2006), 460–500.

4 J. Polich and J.R. Criado, "Neuropsychology and Neuropharmacology of P3a and P3b," *International Journal of Psychophysiology* 60, no. 2 (2006), 172–185.

5 D.J. Fox, D.F. Tharp, and L.C. Fox, "Neurofeedback: An Alternative and Efficacious Treatment for Attention Deficit Hyperactivity Disorder," *Applied Psychophysiology and Biofeedback* 30, no. 4 (December 2005), 365–373.

6 John McCrone, *Going Inside: A Tour Round a Single Moment of Consciousness* (New York: Fromm International, 2001), 28.

7 For example, brain tumors often provide a distinctive vasculature (angiogenesis) that is easily imaged by bloodborne radioactive assay techniques.

8 S.M. Kosslyn et al., "Two Types of Image Generation: Evidence from PET," *Cognitive, Affective, and Behavioral Neuroscience* 5, no. 1 (March 2005), 41–53.

9 The word *tomography* derives from the Greek word *tomos*, which means *slice*. These slices are axial to the body. Computerized tomography was originally known as computed axial tomography, terminology that survives among laypersons and seasoned professionals.

10 This is about same radiation exposure an airplane crew experiences in an entire year of flight operations.

11 "Official Position of the Division of Clinical Neuropsychology (APA Division 40) on the Role of Neuropsychologists in Clinical Use of fMRI," *The Clinical Neuropsychologist* 18 (2004), 349–351.

12 American Medical Association, *CPT Changes 2007: An Insider's View*, November 2006.

13 The primary reports of injury during magnetic resonance imaging resulted from the magnetic field attraction of ferrous substances within the body. Proper screening of subjects by attending personnel eliminates this risk.

14 Superconducting quantum interference device is a super-cooled electronic component designed to detect extremely small changes in magnetic fields.

15 S. Duscheck and R. Schandry, "Functional Transcranial Doppler Sonography as a Tool in Psychophysiological Research," *Psychophysiology* 40 (2003), 436–454.

16 S.C. Bunce et al., "Functional Near-Infrared Spectroscopy: An Emerging Neuroimaging Modality," *IEEE Engineering in Medicine and Biology Magazine* (July-August 2006), 54–62.

17 Ibid.

18 Currently, NIRS can localize hemodynamic changes within about 1 centimeter, while the best fMRI scanner can localize changes within a few millimeters.

19 D. Langleben and J. Loughead, "Telling Truth from Lie in Individual Subjects with Fast Event Related fMRI," *Human Brain Mapping* 26 (2005), 262–272.

20 L. Phan et al., "Neural Correlates of Telling Lies: A Functional Magnetic Resonance Imaging Study at 4 Tesla," *Academic Radiology* 12 (2005), 164–172.

21 S.C. Bunce et al., "Detecting Deception in the Brain: A Functional Near-infrared Spectroscopy Study of Neural Correlates of Intentional Deception," *Proceedings of the SPIE* 5769 (2005), 24–32.

22 B. Horwitz and D. Poeppel, "How Can EEG/MEG and fMRI/PET Data Be Combined?" *Human Brain Mapping* 17 (2002), 1–3.

23 K. Krakow et al., "EEG Recording during MRI Recording," *Human Brain Mapping* 10, no. 1 (April 2000), 10–15.

24 Horwitz and Poeppel.

25 Ibid.

26 R. Genik and C. Green, "Cognitive Avionics and Watching Space Crews Think: Generation After Next Applications in Functional Neuroimaging," *Aviation, Space, and Environmental Medicine* 76 (June 2005), B208–212.

27 H. Hoag, "Neuroengineering: Remote Control," *Nature* 423 (2003), 796–798.

28 E. Singer, "Silicon Brains," *Technology Review* (May-June 2007), 94–96.

29 Ibid.

30 N. Birbaumer, "Breaking the Silence: Brain-Computer Interfaces (BCI) for Communication and Motor Control," *Psychophysiology* 43 (2006), 517–532.

31 N. Birbaumer and L. Cohen, "Brain-Computer Interfaces: Communication and Restoration of Movement in Paralysis," *Journal of Physiology* 579, no. 3 (2007), 621–636.

32 Ibid.

chapter 17

FORGING STRESS RESILIENCE: BUILDING PSYCHOLOGICAL HARDINESS

Paul Bartone

Today's military forces must operate in environments of increasing complexity, uncertainty, and change, a fact that has increased stress levels along with the challenge to adapt. For some people, such stressful conditions can lead to a range of health problems and performance decrements. But this is not true for everyone. Many, indeed most, people remain healthy and continue to perform effectively despite high stress levels. While much attention is devoted to understanding and treating stress-related breakdowns, relatively little has gone toward understanding healthy, resilient response patterns and identifying the factors that protect people from the ill effects of stress. With better knowledge about resilience, future selection and training programs can focus more on maximizing stress-resilient response patterns and enhancing human health and performance. Extensive research over the last 25 years has determined that psychological hardiness is a key stress-resilience factor for many people.

Modern life is inherently stressful and is getting more so as the pace of technological change continues to increase. This is even truer for U.S. military forces, who are deploying more and for longer periods of time, on missions that are multifaceted, changeable, and ambiguous. While much attention has been devoted to studying those who break down under stress, the majority of people appear to respond with remarkable resilience even to severe or traumatic stress.[1] If the factors or pathways that lead to human resilience under stress were better understood, perhaps some of these factors could be developed or amplified in those individuals who are low in resilience and more vulnerable to stress.

This chapter focuses attention on psychological hardiness, one of several potential "pathways to resilience" posited by Bonanno.[2] It will first delineate the major psychological stress factors salient in modern military operations. Next, the concept of psychological hardiness is described, including theoretical background and representative research findings. Finally will be a discussion of how the hardy cognitive style can be built up in organizations through leader actions and policies. By focusing more attention on increasing psychological hardiness, military organizations can enhance human health and performance, while also preventing many stress-related problems before they occur. Based on both theoretical and empirical grounds, I argue that leaders in organizations can foster increases in the kinds of cognitions and behaviors that typify the high hardy person's response to stressful circumstances.

Psychological Stress Factors in Modern Military Operations

Military operations entail stressors of various kinds for the troops involved. Combat-related stressors are the most obvious ones and have garnered the most attention, but military operations in the post–Cold War era entail additional challenges and stress factors. The number of peacekeeping, peacemaking, humanitarian, and other kinds of operations has increased, even as military organizations shrink in size with the shift to all-volunteer forces, causing units to deploy more frequently. Increased deployments bring other stressful changes in military units as well, such as higher frequency and intensity of training exercises, planning sessions, and equipment inspections, all of which add to the workload and pace of operations.[3] Also, more frequent deployments require more family separations, a well-recognized stressor for Servicemembers.[4]

One possible avenue for reducing the stress associated with military operations is to lessen the frequency and duration of deployments. Unfortunately, political and strategic imperatives and troop shortages may prevent such an approach. The military is not alone in this regard; the same is true (at least at times) in other occupations and contexts. For example, following the September 11, 2001, terrorist attack on the World Trade Center, fire, police, and other emergency personnel necessarily maintained operations around the clock with the goal of locating possible survivors, as well as restoring essential services to the affected areas. Thousands of disaster response workers were involved in rescuing victims and restoring basic services in New Orleans following Hurricane Katrina in August 2005. In such crisis situations, continuous operations and extreme efforts are necessary to save lives; easing the pace of work may be considered unacceptable or even unethical.

When reducing stressful operations or activities is not a policy option, what can be done to minimize or counter the stressors associated with such operations? Answering this question with respect to the military case must begin with a clear understanding of the nature of the stressors personnel encounter on modern military deployments. Extensive field research with U.S. military units deployed to Croatia, Bosnia, Kuwait, and Saudi Arabia from 1993 through 1996, including interviews, observations, and survey data, identified five primary psychological stress dimensions in modern military operations: isolation, ambiguity, powerlessness, boredom, and danger.[5, 6] Today, with the greatly increased frequency and pace of deployments for U.S. forces and the long work periods involved,[7] an additional significant stress factor should be added to the list: workload or operations tempo. These dimensions are summarized in table 17–1. Although these stress dimensions are discussed as six distinct factors, in practice they overlap and interact in multiple ways.[8, 9]

table 17–1. PRIMARY STRESSOR DIMENSIONS IN MODERN MILITARY OPERATIONS

Stressor	Characteristics
Isolation	Remote location Foreign culture and language Far from family/friends Unreliable communication tools Newly configured units with unfamiliar coworkers
Ambiguity	Unclear/changing mission Unclear rules of engagement (ROE) Unclear command/leadership structure Role confusion Unclear norms, standards of behavior
Powerlessness	Movement restrictions ROE constraints on response options Policies prevent intervening, providing help Forced separation from local culture, people, events, places Unresponsive supply chain—trouble getting needed supplies/repair parts Differing standards of pay, movement, behavior for different units in area Indeterminate deployment length Do not know/cannot influence what is happening with family at home
Boredom (alienation)	Long periods of repetitive work activities without variety Lack of work that can be construed as meaningful, important Overall mission/purpose not understood as worthwhile or important Few options for play, entertainment
Danger (threat)	Real risk of serious injury or death from: • enemy fire, bullets, mortars, mines, explosive devices • accidents, including "friendly fire" • disease, infection, toxins in the environment • chemical, biological, or nuclear materials used as weapons
Workload	High frequency, duration, and pace of deployments Long work hours/days during the deployments Long work hours/days before and after deployments

Isolation

Military personnel typically deploy to remote areas, far from home and their families, and frequently lacking reliable methods for communicating. Troops are in a strange land and culture, away from familiar surroundings. Furthering the sense of psychological isolation is the fact that personnel may not know each other, since deployed units are typically task forces specially constituted for a particular mission.

Ambiguity

In modern military operations, a unit's mission and rules of engagement are often unclear, multiple missions might be in conflict, or the missions may change over time. The role and purpose of the military person may be similarly unclear. Confusion and mystery in the command structure often add to the uncertainty (Who is in charge of what?). Lack of understanding of host nation language and cultural practices, and how these may affect deployed forces, also increases the uncertainty (Which norms and practices are acceptable in the host culture, and which are not?). These ambiguities can also affect other military contingents as well as contractors in a multinational coalition force. All of these factors generate a highly ambiguous social environment.

Powerlessness

Security and operational concerns often lead to movement restrictions as, for example, when troops are restricted from leaving their base camp. Troops may also be banned from any interaction with the local populace and prevented from participating in familiar activities such as jogging for exercise or displaying the flag. There are frequently multiple constraints on dress and activities, and troops have few choices in their daily existence. Movement and communication restrictions also impede them from learning about local culture and language and resources that might be available in the area. All of this adds to a sense of powerlessness and lack of control over the surrounding environment. Troops may also see military personnel from other Services or countries operating under different rules and privileges in the same environment, but have no explanation for these alternate standards. Soldiers may observe local people who are wounded, ill, hungry, and in need of help but be unable to proffer assistance due to movement and contact restrictions and rules of engagement.[10]

Boredom

Modern military missions frequently involve long periods of "staying in place," often without much real work to do. As the weeks and months crawl by, troops start to get bored. To some degree, this boredom can be countered by providing more entertainment and sports activities. But the real problem of boredom is traceable to the lack of meaningful work or constructive activities to engage in. Daily tasks often take on a repetitive dullness, with a sense that nothing important is being accomplished.

Danger

This dimension encompasses the real physical hazards and threats that can result in injury or death that are often present in the deployed environment. Things such as bullets, mines, bombs, or other hazards in the deployed setting are included, as well as the risk of accidents, disease, and exposure to toxic substances. Current U.S. and coalition operations in Iraq and Afghanistan include many hidden dangers such as suicide bombers, snipers, and improvised explosive devices. This source of stress can be direct, posing a threat to the individual soldier, or indirect, representing threats to his or her comrades. Exposure to severely injured or dead people, and the psychological stress this can entail, also adds to the sense of danger for troops.

Workload

This factor represents the increasing frequency and duration of deployments that many military units are experiencing. Also, most deployments are characterized by a 24/7 work schedule in which soldiers are always on duty, with no time off. Work-related sleep deprivation is often a related feature. Training and preparation activities in the period leading up to a deployment also usually entail a heavy workload and extremely long days. The same is generally true for military units returning home from a deployment, who must work overtime to assure that all vehicles and equipment are properly cleaned, maintained, and accounted for.

What tools, strategies, or coping mechanisms can be applied in order to increase resilience or resistance to these stressors, at both the individual and unit levels? I focus next on the cognitive style of hardiness and suggest how leaders might utilize this construct to increase individual and group resilience under stress. It is also important to remember, however, that many other factors at various levels also influence how individuals behave and respond to work-related stress. Figure 17–1 lists some of these factors at the individual, organizational policy, and organizational structure levels.

Psychological Hardiness

The "hardiness" theoretical model first presented by Kobasa[11] provides insight for understanding highly resilient stress response patterns in individuals and groups. Conceptually, hardiness was originally seen as a personality trait or figure 17–1. FACTORS THAT INFLUENCE RESILIENCY

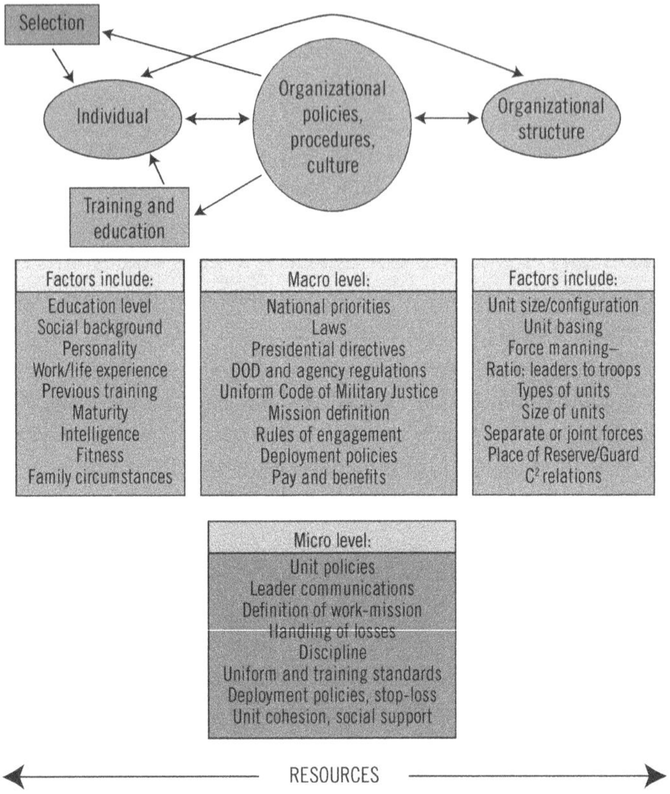

style that distinguishes people who remain healthy under stress from those who develop symptoms and health problems.[12, 13] Hardy persons have a high sense of life and work commitment and a greater feeling of control, and are more open to change and challenges in life. They tend to interpret stressful and painful experiences as a normal aspect of existence, part of life that is overall interesting and worthwhile.

Rather than a personality trait, psychological hardiness may be more accurately described as a generalized style of functioning that includes cognitive, emotional, and behavioral features, and characterizes people who stay healthy under stress in contrast to those who develop stress-related problems. The hardy-style person is also courageous in the face of new experiences as well as disappointments, and tends to be highly competent. The high hardy person, while not impervious to the ill effects of stress, is strongly resilient in responding to stressful conditions.

figure 17–2. MENTAL HARDINESS AND THE EFFECTS OF COMBAT STRESS*

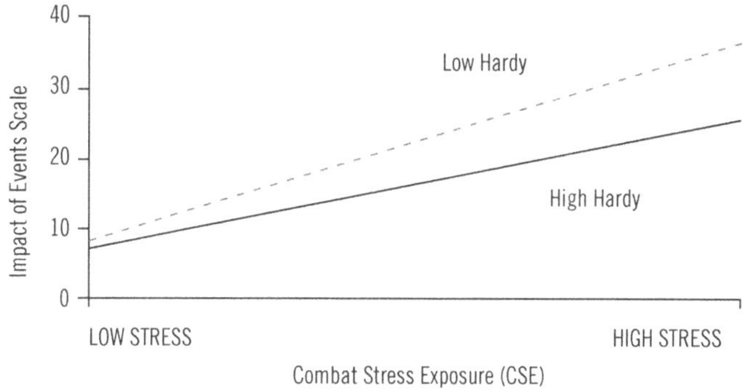

* Displays Hardy x CSE interaction (p. <.001) in regression model, N=824 Active duty, unstandardized betas used to map regression lines

The concept of hardiness is theoretically grounded in the work of existential philosophers and psychologists such as Heidegger,[14] Frankl,[15] and Binswanger[16] and involves the creation of meaning in life, even life that is sometimes painful or absurd, and having the courage to live life fully despite its inherent pain and futility. It is a global perspective that affects how one views the self, others, work, and even the physical world (in existential terms, *Umwelt*, the "around" or physical world; *Mitwelt*, the "with" or social world, and *Eigenwelt*, the "world of the self" or me). As early as 1967, using somewhat different terms, Maddi outlined the hardy personality type and contrasted it with the non-hardy "existential neurotic."[17] He used the term *ideal identity* to describe the person who lives a vigorous and proactive life, with an abiding sense of meaning and purpose and a belief in his own ability to influence things.

Since Kobasa's original 1979 report on hardiness and health in executives,[18] an extensive body of research has accumulated showing that hardiness protects against the ill effects of stress on health and performance. Studies with a variety of occupational groups have found that hardiness operates as a significant moderator or buffer of stress.[19] Hardiness has also been identified as a moderator of combat exposure stress in Gulf War soldiers.[20] Psychological hardiness has emerged as a stress buffer in other military groups as well, including U.S. Army casualty assistance workers,[21] peacekeeping soldiers,[22] Israeli soldiers in combat training,[23] Israeli officer candidates,[24] and Norwegian navy cadets.[25] Studies have found that troops who develop post-traumatic stress disorder (PTSD) symptoms following exposure to combat stressors are significantly lower in hardiness than those who do not exhibit PTSD.[26] Using data from this study of U.S. soldiers in the Gulf War, figure 17–2 shows the typical, and rather robust, interaction effect of hardiness and stress. Under low-stress conditions, those high in hardiness are fairly similar to those low in hardiness in terms of health—in this case, PTSD symptoms. However, under high-stress conditions, the resiliency effects of hardiness are most apparent. Here, those high in hardiness report significantly fewer PTSD symptoms than those low in hardiness (PTSD symptoms measured by the Impact of Events Scale).[27]

Psychological Hardiness as a Framework for Understanding Positive Leader Influence

How does hardiness increase resilience to stress? While the underlying mechanisms are still not fully understood, a critical aspect of the hardiness resilience mechanism likely involves the meaning that people attach to events around them and their own place in the experiential world. High hardy people typically interpret experience as overall interesting and worthwhile, something they can exert control over, and challenging, presenting opportunities to learn and grow. In organized work groups such as military units, this "meaning-making" process is something that can be influenced by leader actions and policies. Military units by their nature are group-oriented and highly interdependent. Common tasks and missions are group actions, and the hierarchical authority structure frequently puts leaders in a position to exercise substantial control and influence over subordinates. By the policies and priorities they establish, the directives they give, the advice and counsel they offer, the stories they tell, and perhaps most importantly the examples they set, leaders may alter the manner in which their subordinates interpret and make sense of experiences. Some empirical support for this notion comes from a study by Britt, Adler, and Bartone,[28] who found (using structural equation modeling) that hardiness increases the perception of meaningful work, which in turn increases the perception of positive benefits associated with a stressful military deployment to Bosnia.

Many authors have commented on how social processes can influence the creation of meaning by individuals. For example, Janis[29] used the term *groupthink* to describe how people in groups can come to premature closure on issues, with multiple individuals conforming to whatever is the dominant viewpoint in the group. Berger and Luckmann[30] argue that "reality" or perceptions of individuals reflect "social constructions," an incorporation into the individual mind of social definitions of the world. Karl Weik[31] discusses the process by which organizational policies and programs can influence how individuals within the organization "make sense" of or interpret their experiences, particularly at work. Even Gordon Allport,[32] a distinguished American personality psychologist, viewed individual meaning as often largely the result of social influence processes. Peers, leaders, and entire work units or organizational cultures can influence how experiences get interpreted. In particular, leaders who are high in hardiness themselves can exert substantial influence on those around them to process stressful experiences in ways characteristic of high hardy persons.

Thus, the operative power of psychological hardiness to defuse stressful experiences is related to the particular interpretations of such experiences that are typically made by the hardy person. If a stressful or painful experience can be cognitively framed and made sense of within a broader perspective that holds that all of existence is essentially interesting, worthwhile, fun, a matter of personal choice, and a learning experience, then the stressful event can have beneficial psychological effects instead of harmful ones. In a small group context, leaders are in a unique position to shape how stressful experiences are understood by members of the group. The leader who, through example and discussion, communicates a positive construction or reconstruction of shared stressful experiences may exert an influence on the entire group in the direction of his/her interpretation of experience. Leaders who are high in hardiness likely have a greater impact in their groups under high-stress conditions when, by their example as well as by the explanations they give to the group, they encourage others to interpret stressful events as interesting challenges to be met. This process itself, as well as the positive result (a shared understanding of the stressful event as something worthwhile and beneficial), could be expected to also generate an increased sense of shared values, mutual respect, and cohesion. Further support for this interpretation comes from a study showing that hardiness and leadership interact to affect small group cohesion levels following a rigorous military training exercise.[33] This interaction effect signifies that the positive influence of leaders on the growth of unit cohesion is greater when hardiness levels in the unit are high. This suggests that effective leaders encourage positive interpretations of stressful events and increase group solidarity, especially in a context of higher psychological hardiness levels.

As an example of the kind of leadership that can foster hardy-resilient response patterns throughout a unit, consider the case of a U.S. Army company commander who led a Patriot missile air defense artillery unit approaching the end of a 6-month deployment to Southeast Asia. The unit mission, which was to be prepared to shoot down any Scud missiles that might be launched from Iraq, had become rather dull and boring. Morale and cohesion were found to be quite low throughout the battalion. But surprisingly, one company/battery stood out as different from the rest, reporting very high morale and cohesion levels: the Headquarters and Maintenance Company. Further investigation revealed that shortly after arriving in theater, the company commander had set a major task for the unit, one that provided a common goal and a tangible mission to work on during their 6 months in the desert. He had heard about a large area nearby that had been used as an equipment dump after the Gulf War. There, several tons of discarded equipment and parts were buried in the sand, rusted and dirty. The commander set for his unit the task of excavating the dump and recovering, cleaning, and repairing as much equipment as possible over the course of the deployment. Five months later, they had salvaged over $1 million worth of equipment, which was returned to the Army supply system in good working order. Proudly displayed on the walls of the company work area and meeting room were large before-and-after photographs of the equipment dump, which the Soldiers had also rebuilt into a multisport athletic field. In interviews, all unit members spoke with great pride about this feat and of having saved considerable taxpayer dollars.

This case illustrates how a proactive, committed, high hardy leader can mobilize an entire work group in the direction of greater hardiness and stress resilience. The company commander asserted creative control under ambiguous conditions and sought out and found a meaningful, albeit secondary, mission for his unit. Without compromising the primary mission (maintaining the equipment and resources of the battalion), he created a major, challenging task that they could "get their arms around" and exercise control over. He was enthusiastic and worked alongside his Soldiers in the dump, helping them develop a shared sense of commitment to the task. Soldiers were also involved in planning the job, building a sense of commitment and control. The task was challenging but had a clear goal that could be accomplished within the 6-month deployment period. The commander also understood the importance of recognition and made sure that Soldiers received awards as well as military media attention for their accomplishments. Recognition in media such as the *Stars and Stripes* newspaper, as well as from senior leaders such as the Sergeant Major of the Army, further reinforced the sense of commitment and positive meaning within the unit, a shared belief that what they had done was important and valuable. While other units in the same battalion grew more alienated, bored, and powerless over the course of the deployment, one insightful and resilient leader showed how psychological hardiness—commitment, control, and challenge—could be increased within his part of the organization. This example shows how a high hardy leader may be able to influence an entire unit toward more hardy interpretations of experience, and the positive, resilient reactions that can follow.

Although more research is needed on this issue, sufficient evidence exists to support the view that leaders can increase high hardiness response patterns within their organizations and to provide a preliminary sketch of how the high hardy leader behaves in order to influence hardiness and stress resilience in the organization. The prototypical hardy leader:

- leads by example, providing subordinates with a role model of the hardy approach to life, work, and reactions to stressful experiences. Through actions and words, he or she demonstrates the strong sense of commitment, control, and challenge, and a way of responding to stressful circumstances that demonstrates stress can be valuable, and that stressful events always at least provide the opportunity to learn and grow.
- facilitates positive group sensemaking of experiences, in how tasks and missions are planned, discussed, and executed, and also how mistakes, failures, and casualties are spoken about and interpreted. While most of this "sensemaking" influence occurs through normal day-to-day interactions and communications, occasionally it can happen in the context of more formal after-action reviews or debriefings that can focus attention on events as learning opportunities and create shared positive constructions of events and responses around events.[34]
- seek out (and create if necessary) meaningful/challenging group tasks, and then capitalizes on group accomplishments by providing recognition, awards, and opportunities to reflect on and magnify positive results (such as photographs, news accounts, and other tangible mementos).

The leader who by example, discussion, and established policies communicates a positive construction or reconstruction of shared stressful experiences exerts a positive influence on the entire group in the direction of interpretation of experience toward more resilient and hardy sensemaking. And while leadership is important, other factors also may influence how individuals make sense of experiences. Policies and regulations throughout the organization not only can have the effect of increasing or decreasing stress levels, but may also directly and indirectly influence commitment, control, and challenge aspects of the hardy resilient response pattern. A better knowledge of all these factors will permit more effective approaches to build stress resilience not just in military organizations, but anywhere people are exposed to highly stressful circumstances.

This chapter is based in part on "Harnessing Hardiness: Psychological Resilience in Persons and Organizations," a paper presented by the author at the Human Sciences Research Conference on Harnessing Talent for Complex Operations, held at National Defense University on March 21, 2007, and sponsored by Assistant Secretary of the Navy for Manpower and Reserve Affairs, the Honorable William A. Navas, Jr.

NOTES

1 G.A. Bonanno, "Loss, Trauma and Human Resilience: Have We Underestimated the Human Capacity to Thrive after Extremely Aversive Events?" *American Psychologist* 59 (2004), 20–28.

2 Ibid.

3 C. Castro and A. Adler, "OPTEMPO: Effects on Soldier and Unit Readiness," *Parameters* 29 (1999), 86–95.

4 D.B. Bell et al., *USAREUR Family Support during Operation* Joint Endeavor: *Summary* Report, ARI Special Report 34 (Alexandria, VA: U.S. Army Research Institute for the Behavioral and Social Sciences, 1997).

5 P.T. Bartone, A.B. Adler, and M.A. Vaitkus, "Dimensions of Psychological Stress in Peacekeeping Operations," *Military Medicine* 163 (1998), 587–593.

6 P.T. Bartone, "Psychosocial Stressors in Future Military Operations," paper presented at the Cantigny Conference Series on Future of Armed Conflict, Wheaton, IL, June 2001.

7 Castro and Adler.

8 Bartone, Adler, and Vaitkus, 587–593.

9 Bartone, "Psychosocial Stressors in Future Military Operations."

10 Other studies have also identified powerlessness as a damaging influence for soldiers on peacekeeping operations. For example, Lars Weisaeth and Arne Sund found that in Norwegian soldiers serving in Lebanon in the United Nations peacekeeping mission, the feeling of being powerless to act or intervene when witnessing some atrocity was a main contributor to post-traumatic stress symptoms. See Weisaeth and Sund, "Psychiatric Problems in UNIFIL and the UN Soldier's Stress Syndrome," *International Review of the Army, Navy, and Air Force Medical Service* 55 (1982), 109–116.

11 S.C. Kobasa, "Stressful Life Events, Personality, and Health: An Inquiry into Hardiness," *Journal of Personality and Social Psychology* 37 (1979), 1–11.

12 Ibid.

13 S.R. Maddi and S.C. Kobasa, *The Hardy Executive* (Homewood, IL: Dow Jones-Irwin, 1984).

14 M. Heidegger, *Being and Time* (New York: Harper Collins Publishers, 1986).

15 V. Frankl, *The Doctor and the Soul* (New York: Knopf, 1960).

16 L. Binswanger, *Being in the World: Selected Papers of Ludwig Binswanger*, tr. J. Needleman (New York: Basic Books, 1963).

17 S.R. Maddi, "The Existential Neurosis," *Journal of Abnormal Psychology* 72 (1967), 311–325.

18 Kobasa, 1–11.

19 See, for example, R.J. Contrada, "Type A Behavior, Personality Hardiness, and Cardiovascular Responses to Stress, *Journal of Personality and Social Psychology* 57 (1989), 895–903; S.C. Kobasa, S.R. Maddi, and S. Kahn, "Hardiness and Health: A Prospective Study," *Journal of Personality and Social Psychology* 42 (1982), 168–177; D.L. Roth et al., "Life Events, Fitness, Hardiness, and Health: A Simultaneous Analysis of Proposed Stress-resistance Effects," *Journal of Personality and Social Psychology* 57 (1989), 136–142; and D.J. Wiebe, "Hardiness and Stress Moderation: A Test of Proposed Mechanisms," *Journal of Personality and Social Psychology* 60 (1991), 89–99.

20 P.T. Bartone, "Psychosocial Predictors of Soldier Adjustment to Combat Stress," paper presented at the Third European Conference on Traumatic Stress, Bergen, Norway, June 1993; P.T. Bartone, "Hardiness Protects against War-related Stress in Army Reserve Forces," *Consulting Psychology Journal* 51 (1999), 72–82; P.T. Bartone, "Personality Hardiness as a Predictor of Officer Cadet Leadership Performance," paper presented at the International

Military Testing Association Meeting, Monterey, CA, November 1999; and P.T. Bartone, "Hardiness as a Resiliency Factor for United States Forces in the Gulf War," in *Posttraumatic Stress Intervention: Challenges, Issues, and Perspectives*, ed. J.M. Violanti, D. Paton, and C. Dunning (Springfield, IL: C. Thomas, 2000), 115–133.

21 P.T. Bartone et al., "The Impact of a Military Air Disaster on the Health of Assistance Workers: A Prospective Study," *Journal of Nervous and Mental Disease* 177 (1989), 317–328.

22 T.W. Britt, A.B. Adler, and P.T. Bartone, "Deriving Benefits from Stressful Events: The Role of Engagement in Meaningful Work and Hardiness," *Journal of Occupational Health Psychology* 6 (2001), 53–63.

23 V. Florian, M. Mikulincer, and O. Taubman, "Does Hardiness Contribute to Mental Health during a Stressful Real Life Situation? The Role of Appraisal and Coping," *Journal of Personality and Social Psychology* 68 (1995), 687–695.

24 M. Westman, "The Relationship between Stress and Performance: The Moderating Effect of Hardiness," *Human Performance* 3 (1990), 141–155.

25 P.T. Bartone et al., "Factors Influencing Small Unit Cohesion in Norwegian Navy Officer Cadets," *Military Psychology* 14 (2002), 1–22.

26 Bartone, "Hardiness Protects against War-related Stress in Army Reserve Forces"; Bartone, "Personality Hardiness as a Predictor of Officer Cadet Leadership Performance."

27 M. Horowitz, N. Wilner, and W. Alvarez, "Impact of Events Scale: A Measure of Subjective Stress," *Psychosomatic Medicine* 41 (1979), 209–218.

28 Britt, Adler, and Bartone, 53–63.

29 I. Janis, *Groupthink*, 2d ed. (Boston: Houghton Mifflin, 1982).

30 P.L. Berger and T. Luckmann, *The Social Construction of Reality* (Garden City, NY: Doubleday, 1966); Binswanger.

31 K.E. Weik, *Sensemaking in Organizations* (Thousand Oaks, CA: Sage, 1995).

32 G.W. Allport, "The Historical Background of Social Psychology," in *Handbook of Social Psychology*, 3d ed., vol. 1, ed. G. Lindzey and E. Aronson (New York: Random House, 1985), 1–46.

33 Bartone et al., "Factors Influencing Small Unit Cohesion in Norwegian Navy Officer Cadets."

34 A 2002 National Institute of Mental Health report on best practices for early psychological interventions following mass violence events noted great confusion regarding the term *debriefing*. The authors recommend that the term be reserved for operational after-action reviews and not be applied to psychological treatment interventions such as critical incident stress debriefing. See J.T. Mitchell and G.S. Everly, *Critical Incident Stress Debriefing: An Operations Manual for CISD, Defusing and Other Group Crisis Interventions*, 3d ed. (Ellicott City, MD: Chevron Publishing, 1997. For groups such as the military, after-action group debriefings, properly timed and conducted and focused on events rather than emotions and reactions, can have great therapeutic value for many participants by helping them to place potentially traumatizing events in a broader context of positive meaning. See P.T. Bartone, "Predictors and Moderators of PTSD in American Bosnia Forces," 5th European Conference on Traumatic Stress, Maastricht, The Netherlands, June 1997.

chapter **18**

NEUROPLASTICITY, MIND FITNESS, and MILITARY EFFECTIVENESS

ELIZABETH STANLEY

> *In COIN* [counterinsurgency], *the side that learns faster and adapts more rapidly—the better learning organization—usually wins. Counterinsurgencies have been called learning competitions.*[1]
>
> —U.S. ARMY FIELD MANUAL 3–24/U.S. MARINE CORPS WARFIGHTING PUBLICATION 3–33.5, *COUNTERINSURGENCY*

The missions that the U.S. military is being called upon to execute in today's security environment require a tremendous amount of awareness—of self, others, and the wider environment. Situational awareness, mental agility, and adaptability have become buzzwords for the characteristics we want to cultivate in the military to succeed in complex, fluid, and unpredictable environments. In part, this complexity comes from the number and nature of the different missions the military must concurrently conduct. As Leonard Wong argues, "OIF [Operation *Iraqi Freedom*] requires junior leaders to be warriors, peacekeepers, and nation-builders—simultaneously."[2] Many soldiers call this state of complexity and unpredictability "the faucet," requiring adjustment to situations that could change from hot to cold instantaneously. Servicemembers are being asked to navigate morally ambiguous situations with balance and nonreactivity, while drawing on stores of cultural awareness to "win hearts and minds." Finally, these missions require that decisionmaking be pushed down to the lowest levels. As General Charles Krulak, USMC (Ret.), points out with his concept of the "strategic corporal,"[3] a corporal manning a checkpoint could make a split-second lethal decision that could be broadcast on al Jazeera and end up having strategic significance. In short, the key to these kinds of operations is awareness, which is the foundational metaskill for being able to "learn and adapt."

This chapter proposes a new way to achieve such awareness. It draws on the well-documented theory of neuroplasticity, which states that the brain changes in response to changes in environment. The first section examines some of the ways that "negative" experiences such as stress and trauma can change the brain in deleterious ways. Given that decisionmaking in the military operational environment is stressful, this research is very relevant for a military audience. In contrast, the second section examines some of the ways that "positive" experiences such as mind fitness training can change the brain in beneficial ways. The third section builds on these ideas to propose ways that mind fitness training could enhance military effectiveness. Given that the brain is constantly building new neural connections from all of our daily experiences, it makes sense to concentrate on skills that build those neural connections most beneficial for operational effectiveness and soldier well-being. The conclusion extends this research by examining potential limitations and highlighting possible applications in the national security environment.

Negative Experiences: Stress and Trauma

In the last 20 years, our understanding of how the brain regenerates itself has changed dramatically. We used to believe that when a baby was born, it had the most neurons it would have for its entire life, and that over time, these would die off and not be regenerated. We now know that the brain is plastic, fluid, and ever-changing—it generates new synapses and neurons and discards old ones in response to sensory input from the environment. It is not overly dramatic to say that every experience we have changes our brains. Every moment, waking or sleeping, our brains experience continuous fluctuation in the flow of arterial blood and the utilization of oxygen and glucose within different regions. For example, areas of the brain that control vision will get extra blood flow when we gaze at a painting or the horizon or when we imagine a visual image. Areas of the brain that control motor skills get extra blood flow when we ride a bicycle or take a walk. Areas of the brain that control our response to threat or danger receive more blood flow if a person tells a story describing an assault, rape, or battle. The prominence of these changes to brain blood flow and energy use varies from person to person, based on his or her specific pattern of life experience. As neurologist and trauma researcher Robert Scaer argues, "Life experience, based on specific sensory experiences, therefore changes the brain permanently in the way that it specifically reacts to subsequent similar experiences."[4]

There is also evidence that the physical structure of the brain also changes with life experience. New connections are formed between neurons. New neurons may also be formed and existing ones discarded, at least in the hippocampus, the region of the brain that controls conscious memory. Thus, not only might patterns of blood flow and energy use be altered in different brain regions, but areas of the brain might also shrink or expand—become more or less functional—based on experience. In other words, the brain, like the rest of the body, builds the "muscles" it most uses, sometimes at the expense of other abilities. Any time we perform any type of task (physical or mental), the region of the brain that serves that particular function shows increased neuronal activity. Experience and training can produce a functional and structural reorganization of the brain.

Given that all experience changes the brain, we might categorize experiences as negative and positive. From the negative perspective, a variety of research indicates that harmful conditions such as chronic stress, neglect, and abuse produce deleterious changes in the brain. Stress is produced by real or imagined events that are perceived to threaten an individual's physical and mental well-being. Today, stress is commonly understood to mean external events or circumstances, and as a result, we tend to think of stress as something *external* to us. However, as originally conceived, stress is a *perceived, internal* response. The right amount of stress will allow a decisionmaker to function at peak performance. However, excessive stress has biological and psychological consequences that reduce working memory and the capacity to process new information and learn. Stressors may bias decisionmaking more toward a reactive, unconscious emotional process. Stress can also trigger habits, which are recurrent, often unconscious patterns of behavior and thought acquired through frequent repetition.

For example, the brains of animals that received prolonged exposure to high levels of cortisol, the hormone researchers use to measure the stress response in the body, showed damage in their hippocampi, the brain region that controls conscious memory. Similarly, Vietnam veterans who suffered from post-traumatic stress disorder (PTSD) were found to have smaller hippocampi.[5] Imaging studies in the brains of victims of various types of trauma, especially trauma experienced in childhood, have also consistently shown decreased hippocampal volume. Douglas Bremner deduced that loss of hippocampal volume in these individuals caused deficits in memory. When he tested memory function in combat veterans with PTSD and victims of childhood trauma or sexual abuse, both groups showed deficits in a variety of types of memory function, including declarative memory, short-term memory, immediate and delayed recall, and percent retention of memory.[6]

Recent empirical research about decisionmaking in stressful military environments demonstrates that trauma and stress lead to deficits in both cognitive functioning and cognitive adaptation to the stressful circumstances. One large study of U.S. Army troops found that Soldiers who served in Iraq were highly likely to show subtle lapses in memory and an inability to focus, a deficit that often persisted more than 2 months after they arrived home.[7] In the study, 654 Soldiers who deployed to Iraq between April 2003 and May 2005 performed significantly worse in tasks that measured spatial memory, verbal ability, and their ability to focus than 307 Soldiers who had not deployed. In contrast, the Soldiers who had deployed outperformed those who had not in terms of quick reaction time (for example, how long it takes to spot a computer icon and react). This finding suggests that the Soldiers' minds had adapted to the dangerous, snap-judgment conditions of war, but in the process saw other cognitive capabilities degrade. In effect, the deployed Soldiers' brains built the capacity for quick reaction time, a function more necessary for survival in Iraq, at the expense of other mental capacities.

While this adaptation could be construed as positive, recent research shows that mind fitness training techniques can lead to more efficient mental processing, which may decrease the need for such a dramatic degradation of some mental capacities in order to build others. In other words, although the "untrained" minds of the Iraqi veterans showed a positive adaptation, with mind fitness training, more efficient mental processing may allow the brain to cultivate new capacities such as quick reaction time without degrading others.

Other studies of military environments have found substantial degradation in cognitive performance when subjects experience sleep deprivation and other environmental stressors. A recent study of sleep deprivation among Navy SEALs and Army Rangers during a field training exercise demonstrated that the lack of sleep affected soldiers so badly that after a week they performed worse on cognitive tests than if they were sedated or legally drunk. In this study, the SEALs and Rangers showed severe degradation in reaction time, vigilance, visual pattern recognition, short-term memory, learning, and grammatical reasoning skills.[8] Another group of studies examined more than 530 soldiers, sailors, and pilots during military survival training, including time in mock prisoner of war camps, to prepare them to withstand the mental and physical stresses of capture. In these studies, exposure to acute stress resulted in symptoms of disassociation (alterations of one's perception of body, environment, and the passage of time), problem-solving deficits (as measured by objectively assessed military performance), and significant inaccuracies in working memory and spatial memory (as measured by eyewitness identification tests). For example, only one-third of the volunteers could correctly identify their interrogator in photo spreads or lineups 24 hours after being released from the camp.[9] These findings corroborated those of other studies that found that multistressor environments lead to substantial degradation of cognitive skills that correspond to "higher brain" (as opposed to "limbic brain") functions.[10] Such declines in cognitive function have been linked to battlefield errors, such as friendly fire incidents and collateral damage.[11]

These findings about how stress and trauma change the brain are very salient for the military, because the national security environment is stressful. Decisionmaking in this environment is characterized by uncertainty, unpredictability stemming from interaction with adversaries and third parties, intense time pressure, and incomplete and ambiguous information. Moreover, soldiers face emotional and mental stress, including the loss of comrades, as well as physical distress and deprivation (sleep loss, hunger, physical danger, lack of privacy, harsh environmental conditions, and physical exertion). These stressors can impede operational effectiveness and soldier well-being.

Positive Experiences: Mind Fitness Training

Not all experiences are harmful to the brain. The brain of an expert such as a surgeon, taxi driver, or musician is functionally and structurally different from that of a nonexpert. For example, researchers studying London cab drivers found that they have larger hippocampi than matched controls, and the amount of time an individual worked as a cab driver predicted the size of the posterior hippocampus. These differences in hippocampus size are thought to be the result of experience and training as a cab driver, not of preexisting differences in the hippocampal structure.[12] In another study, monkeys were trained to maintain hand contact with a rotating disk in order to get bananas. The training caused a clear reorganization of the cortical region, a reorganization that lasted for a significantly long time after the training was discontinued.[13] This research demonstrates how experience and training can produce a functional and structural reorganization of the brain.

The role of training has been recognized in the military context for a long time, although there has been less focus on training the brain. A more nuanced look at the data from studies about stressful military environments leading to degradation of cognitive functioning shows that soldiers showed variation in physiological and psychological stress, and much of this variation was correlated with training. Soldiers with more training released less cortisol, which is the hormone used to measure the stress response and associated with hippocampal damage. These soldiers also produced more neuropeptide-Y, which has been associated with reducing anxiety responses. In other words, these soldiers were not only better equipped for handling stress, but they also recuperated more quickly so that they were better prepared to handle any subsequent stressors. Because of findings such as these, military training has long been designed as "stress immunization" *to habituate* soldiers to a variety of stressors and thus reduce the stress response. As we will see below, mind fitness training could complement and strengthen such training by helping soldiers *to perceive and relate to* those stressors differently.

Because of neuroplasticity, humans can create new neural pathways, often with beneficial results. This scientific fact is starting to be recognized and applied with mind fitness training, as a variety of bestselling books suggest.[14] Mind fitness training techniques have existed for thousands of years, having originated in Eastern spiritual traditions, but in recent decades they have been adapted for secular use, including in corporations, prisons, and elementary schools. Mind fitness training has also been incorporated into several interventions that are now widely available in medical and mental health settings. For example, UCLA researcher Gary Small and a team of experts in memory training, stress reduction, and other healthy longevity strategies offer a one-day "Brain Boot Camp" to teach brain-healthy lifestyles and enhance memory ability.[15] However, the most common health intervention is mindfulness-based stress reduction; more than 250 U.S. hospitals offer programs in the technique.[16] Recently, these techniques have been extended to Iraq War veterans returning with PTSD. There are at least five mindfulness-based interventions with war veterans under way at this time.[17]

Mind fitness includes the capacities of mental agility, emotional intelligence, improved attention and working memory, and awareness of self, others, and the wider environment. Like physical fitness, mind fitness can be strengthened through attention and concentration practices that literally change the brain structurally and functionally. And like physical fitness, mind fitness is protective, building resiliency and leading to faster recovery from mental and psychological stress. Mind fitness may also slow the cognitive degradation associated with aging by building cognitive reserve, the capacity to optimize performance by recruiting alternative brain networks or cognitive strategies in response to the demands of a task.

How does mind fitness training work? Each time we choose to ignore an old, maladaptive habit, we weaken the neural circuitry associated with that pattern (what we could call a "brain groove"). Each time we choose to build a new habit, we strengthen a new brain groove. This concept is something that athletes, musicians, and martial artists have known for a long time: with repetition, the body begins to develop muscle memory. Physical trainers talk about developing muscle memory with 300 repetitions and muscle mastery with 3,000 repetitions. But the same process can occur with habits of mind. For example, it is possible to change our conditioning toward anxious or angry trains of thought. To do this, when the anxious or angry thought arises, rather than follow the impulse, a person can turn his mind elsewhere, "starving" the old thought pattern, and consciously choose to see the situation in a different way, "feeding" a new thought pattern. Mind fitness allows an individual to experience even very strong emotions with greater objectivity and less reactivity, which serves to counter our habitual tendency to avoid or deny difficult emotional states. Instead, individuals can learn to use such states as information ("What is this anger trying to tell me?"), which can lead them to access a wider, more adaptive range of coping skills.

Mind fitness is cultivated through a variety of training techniques, but most of them are based on a foundational skill called *mindfulness*. Mindfulness has been described as a process of "bringing one's attention to the present experience on a moment-by-moment basis"[18] and as "paying attention in a particular way, on purpose, in the present moment and non-judgmentally."[19] Mindfulness differs from a more conceptual mode of processing information, which is often the mind's default way of perceiving and cognizing. In other words, paying attention is not the same thing as thinking, although we often confuse the two.

Numerous training techniques have been designed to develop mindfulness skills. One central practice is to teach individuals to train their attention by focusing attention on their breath and returning to the breath when they notice that their attention has wandered. Other practices focus on paying attention to other internal experiences such as emotions, thoughts, or bodily sensations. Still others focus on paying attention to aspects of the external environment, such as sounds and sights. Some techniques are practiced while the individual is sitting still, others while the individual is moving about or interacting with others.[20] Perhaps the most common way to cultivate mindfulness skills is meditation, but the martial arts, physical exercise, eating, talking, and even washing dishes can all serve as mindfulness training. For this reason, incorporating mind fitness training into existing military training would not require much additional time on the training schedule.

For example, the military already incorporates mindfulness training—although it does not call it this—into perhaps *the* most foundational soldier skill, firing a weapon. Soldiers learning how to fire the M-16 rifle are taught to pay attention to their breath and synchronize the breathing process to the trigger finger's movement, "squeezing" off the round while exhaling. This synchronization has two effects. The gross motor effect is to steady the soldier's aim, so that the round is more likely to hit its target. But this training also has the more subtle psychological effect of helping to steady the soldier's concentration and heart, to offer something concrete to focus on as bullets are whizzing past. It provides an anchor in the midst of very trying physical and psychological conditions.

There is growing empirical scientific evidence that increasingly supports the efficacy of mindfulness-based interventions. Clinical studies of civilian mindfulness-based interventions demonstrate that patients who participate in such programs see improvement in many conditions, including anxiety disorders, depression, fibromyalgia, chronic pain, substance abuse, insomnia, binge eating, cancer, smoking cessation, menopause, irritable bowel syndrome, hypertension, HIV/AIDS, and skin-related diseases. Moreover, patients report a decrease in mood disturbance from and stress related to these conditions.[21] Mindfulness training has also been shown to produce brain changes consistent with more effective handling of emotion under stress, with significant increases in left-sided activation of the anterior cortical area of the brain as well as an increase in immune functioning.[22]

Given this evidence that mind fitness training can have such positive effects in other environments, there appears to be an intriguing and timely opportunity to adapt such training techniques to the national security context. And given how stressful the contemporary security environment is, these techniques could be particularly helpful for deploying soldiers. Mind fitness, together with mission essential skills and physical fitness, is one of three components necessary for optimal soldier readiness and operational effectiveness. While all three components are crucial for equipping soldiers to handle the challenges and stressors of deployment, there is a dearth of training in the mind fitness arena.

Mind Fitness Training and Military Effectiveness

Mind fitness training could help soldiers cultivate a variety of critical competencies that are often in short supply on the modern battlefield. For example, soldiers are frequently asked to perform tasks outside their usual specialties and take on roles they never envisioned. This requires an immense capacity to adapt to new routines and learn new tasks quickly, from negotiating with the local sheik to rebuilding an electrical power grid. Moreover, they are asked to exhibit self-awareness and situational awareness and tailor decisions to these constraints. This state is similar to how the U.S. Army defines adaptability: "an individual's ability to recognize changes in the environment, identify the critical elements of the new situation, and trigger changes accordingly to meet new requirements."[23] Finally, they are asked to make decisions, allocate resources, and take actions that are internally consistent with and supportive of broader strategic and political goals. Means-ends consistency is particularly important for counterinsurgency operations: as U.S. Army Field Manual 3–24/Marine Corps Warfighting Publication 3–33.5, *Counterinsurgency*, argues, "Unity of effort must be present at every echelon of a COIN operation. Otherwise, well-intentioned but uncoordinated actions can cancel each other or provide vulnerabilities for insurgents to exploit."[24]

Given the importance of such competencies, and the fact that soldiers are simultaneously being exposed to a variety of stressors while being called upon to exhibit these competencies, mind fitness training could help in six ways: by improving self-regulation, enhancing attention skills, increasing flexibility in awareness and attention, decreasing reactivity, cultivating emotional and social intelligence, and enhancing values clarification and integrity. The rest of this section draws on recent neuroscience and clinical psychological research from civilian contexts to suggest exactly how mind fitness training could enhance these six competencies.

Improved Self-regulation

Self-regulation involves making conscious, intentional choices about how to relate to the current situation in a way that supports personal or organizational needs, values, and interests. Mindfulness can improve self-regulation by cultivating ongoing attention and sensitivity to psychological, somatic, and environmental clues.[25] Such information allows for more deliberate decisionmaking so that decisionmakers can employ a wider, more flexible, and more adaptive range of coping skills, even in highly stressful environments. In addition, self-regulation can help an individual cultivate a higher tolerance for unpleasant internal and external conditions, as well as the ability to remain clear-headed in these situations. This, in turn, can increase an individual's cognitive, emotional, and behavioral flexibility. Finally, mindfulness helps individuals to not act on impulse, but rather to access the gap between intention and action, so that all decisions and resulting behavior can be deliberate and intentional.

While performing brain surgery that required the patients to remain conscious, neurosurgeon Benjamin Libet asked them to move their finger and report the precise moment when they became aware of the urge to do so. During this process, Libet monitored the electrical activity in the patients' brains that regulated the movement, allowing him to separate the moment of *intention* to move from the moment of *awareness* of that intention and from the moment of actual *action*. He discovered that freely voluntary acts are preceded by a specific electrical change in the brain (what he called the readiness potential) that begins 550 milliseconds (ms) before the act. The patients become aware of the intention to act 350–400 ms *after* the readiness potential starts, but 200 ms *before* the actual action. He concluded that the volitional process is thus initiated unconsciously, but the conscious mind could still control the outcome—it could veto the actual act, and thus "free will" is not excluded.[26] In other words, once a person is aware of the intent to move, there is another quarter-second before the movement begins. This is the crucial window during which mindfulness is so important. With mindfulness, this window allows a conscious choice-point to break the chain of reactivity or habit and choose responses that are most aligned with our interests and needs.

For example, mindfulness training has been shown to increase tolerance of unpleasant physical states, such as pain. Recent studies about mindfulness and pain suggest that experienced mindfulness meditation practitioners required an average of 2 degrees more thermal heat applied to their legs before perceiving pain, and they showed a lower breathing rate during the heat stimulus than control subjects. Moreover, when all participants were asked to concentrate directly on the heat stimulation on their legs, control subjects reported much higher levels of subjective unpleasantness than the mindfulness practitioners did.[27]

Self-regulation also helps individuals to feel more autonomous and in charge of their decisions and actions.[28] With improved self-regulation, individuals can learn that although they cannot control *what* they are experiencing internally or externally, they can always choose consciously and control *how they relate to* that experience. Autonomy and self-esteem are important qualities for cultivating initiative, which is important when decisionmaking authority gets devolved down to the lowest levels. Self-regulation depends on an individual taking responsibility for his own thoughts, emotions, and actions. In other words, self-regulation leads to professional accountability.

In his study of distress and war termination, Stephen Rosen argues that uncontrollable, unpredictable stressors in the environment, such as those experienced by the losing side in war, can produce distress among those fighting on the losing side. Without training, these soldiers can become pessimistic about the outcomes of their actions, lack energy to perform their tasks, and perceive and understand less information from their environment that could be relevant to finding solutions for their problems. This lack of attention and energy can lead to a vicious, self-reinforcing cycle, as the distress of one soldier can quickly spread to his comrades.[29] Self-regulation can break this cycle. In other words, by increasing distress tolerance and undercutting fear-based reactivity, self-regulation can lead to social control.

Better Attention Skills

Mind fitness training can also improve different facets of attention. Cognitive psychology delineates many different aspects of attentional abilities, including the capacity to attend for long periods of time to one object (vigilance or sustained attention),[30] the ability to shift the focus of attention between objects or mental sets at will (switching),[31] and the ability to inhibit secondary processing of thoughts, feelings, and sensations (cognitive inhibition).[32] Similarly, one neuroscience model proposes that attention consists of three distinct cognitive networks that carry out the operations of alerting, orienting, and conflict monitoring. Alerting consists of achieving and maintaining a vigilant or alert state of preparedness, orienting directs and limits the attention to a subset of possible inputs, and conflict monitoring prioritizes among competing tasks and responses.[33]

How could better attention skills improve decisionmaking in military operations? First, soldiers could maintain a vigilant and alert state while on patrol or while manning checkpoints during deployments. Given that "every soldier is a collector" of intelligence, improved alerting skills could lead to more accurate and timely intelligence information for decisionmakers at higher levels. Second, decisionmakers with better orienting and conflict-monitoring skills could strengthen concentration and filter out distractions and extraneous information. Moreover, concentration enables decisionmakers at all levels to learn new tasks and assimilate new technologies. Finally, stronger attention skills can improve what Jeremy Gray and his colleagues call "fluid intelligence," the intelligence associated with *using* facts (rather than simply knowing them).[34]

Mind fitness training has been shown to be effective in increasing alerting and orienting skills and working memory, which decreases the need to rely on cognitive "shortcuts."[35] Mindfulness practitioners have also been shown to have greater cortical thickening in areas of the right prefrontal cortex and right anterior insula, brain areas that are associated with sustained attention and awareness.[36]

In addition to these grosser aspects of attention, recent research shows that mind fitness training can also affect more subtle attentional processes. One of these is called the *attentional blink* (AB), which occurs when the brain has difficulty processing two temporally close stimuli roughly a half-second apart. With AB, the brain devotes too many resources to processing the first stimulus and, as result, misses the second one. However, mindfulness training has been associated with decreasing the size of the AB window, so that individuals can see the fast-following second target more often. In one recent study, relative to controls, study participants who had just completed a 3-month meditation retreat showed a significantly smaller AB window and were able to see targets inside the typical AB window more often. The scientists argue that because detecting the second target is dependent on efficient mental processing of the first one, mindfulness training can decrease the mental resources devoted to the first target and allow the brain to process the second one more easily.[37]

Increased Flexibility in Awareness and Attention

Related to better attention skills is cognitive flexibility,[38] which allows decisionmakers to access information from a wider variety of channels and decrease filters on that information coming in. Cognitive flexibility also allows decisionmakers to gain a larger perspective, which promotes insight into self, others, and the wider environment. From this larger perspective, decisionmakers are better able to discern which new tasks they need to learn to operate most effectively in their environment. As a result, mindfulness may enhance behavioral flexibility, particularly through adaptive responding based on actual, rather than imagined, contingencies.[39] Learning to see clearly—indeed, learning in general—depends upon the ability to disengage from prior patterns and beliefs.

Most importantly, increased flexibility in awareness and attention leads decisionmakers to display greater accuracy and more objectivity in assessing situations, making decisions and allocating scarce resources.[40] U.S. Army Field Manual 6–22, *Army Leadership*, calls this trait "mental agility," defined as "a flexibility of mind, a tendency to anticipate or adapt to uncertain or changing situations."[41] Mental agility is important today because leaders need to adapt to the external environment and other actors, including potential adversaries, allies, and local populations. They also need to be able to transition quickly from fighting in one moment to relating peacefully with the local community in the next. This requires accessing information from a wider variety of channels and displaying greater accuracy and more objectivity in gathering information. These improved inputs to decisionmaking can lead to more creative brainstorming, as well as a more realistic assessment of various alternatives. In other words, increased flexibility in awareness and attention could help protect against the whiplash of the "the faucet"—the rapid cycle between hostile and friendly encounters that soldiers in COIN environments face—and offer new methods for coping with its effects.

Less Reactivity

Mind fitness training can also enhance a soldier's capacity for nonreactivity. Recent studies have demonstrated that mindfulness is strongly associated with less reactivity.[42] Nonreactive decisionmakers are better able to maintain contact with threatening emotional stimuli and make decisions from a place of receptive and realistic observation rather than emotional judgment. Nonreactive decisionmakers can also diminish impulsive or defensive reactions to unpleasant experiences and thus inhibit escalatory spirals with others. As a result, they can use improved self-awareness to *respond* instead of habitually or impulsively *reacting*.

Nonreactivity can be a powerful weapon in today's environment. As the *Counterinsurgency* field manual argues, sometimes doing nothing is the best reaction. Often insurgents carry out a terrorist act or guerrilla raid with the primary purpose of enticing the counterinsurgents to overreact, or at least react in a way that insurgents can exploit—for example, opening fire on a crowd or executing a clearing operation that creates more enemies than it takes off the streets. If an assessment of the effects of a course of action determines that more negative than positive results may result, an alternative should be considered, potentially including not acting.[43]

Obviously, nonreactivity is linked to self-regulation, because it requires self-control and emotional regulation. Nonreactivity and self-regulation work hand in hand so that individuals can better override or change inner, often habitual, responses and refrain from acting on undesired behavioral tendencies. Both capacities in tandem allow decisionmakers to diminish impulsive or defensive reactions to unpleasant experiences and inhibit escalatory spirals with others.

Mindfulness and the nonreactivity it cultivates allow decisionmakers to act from a place of realistic observation rather than emotional judgment. Thus, decisionmakers could become more accurate and less biased at gathering information and assessing situations. Mindfulness allows a clear perception of incoming sensory data, from the external environment as well as internally from the body, emotions, and thoughts. Mind fitness training allows soldiers and leaders to "get out of their heads" and access somatic information ("trust your gut"). Such an impartial, balanced view of incoming data helps the decisionmaker to see all of the information available with clarity and cut through cognitive and emotional biases. This improved input and a wider perspective could lead to more appropriate and thus more effective decisions.

Improved Emotional and Social Intelligence

Emotional intelligence refers to an individual's ability to fully comprehend his own emotions and thoughts, as well as notice and interpret the moods, temperaments, motivations, and intentions of others.[44] Emotional intelligence comprises four skills: knowing one's emotions, managing emotions, recognizing emotions in others, and handling relationships. Decisionmakers with high emotional intelligence are more in tune with their own strengths and weaknesses and open to feedback. They are also more likely to understand the moods of those around them and stay in touch with the mood of their organization. They show greater awareness of others' affective tone and nonverbal behavior, and exhibit more open and less defensive responses to challenging people and events. This can lead them to a better assessment of themselves, other people, and their environment, as others candidly share information they need to know.

In addition, emotional intelligence allows decisionmakers to more easily see situations from and respect others' perspectives. This capacity is obviously important in *cross-cultural situations*, such as during COIN operations, where building social rapport and trust are paramount to the goal of "winning hearts and minds." The capacity to see situations from and respect others' perspectives can also be important across *organizational cultural lines*, such as between the military, the State Department, and humanitarian relief nongovernmental organizations (NGOs), which must work together on these kinds of missions. Emotional intelligence allows decisionmakers to recognize how their own conditioning creates biases and filters to incoming information and see through that conditioning to assess themselves, others, and the environment more realistically. With more awareness of others' perspectives, as well as of their own conditioning, decisionmakers are more able to seat themselves appropriately within wider social, cultural, and political contexts. Finally, emotional intelligence allows decisionmakers to establish better rapport with others, improve team cohesion, and lead more effectively.

Given its ongoing operations in Afghanistan, Iraq, and other locations, the U.S. Government is placing a greater focus on awareness of the cultures of potential adversaries, the nations where U.S. forces are deployed, and their allies and partners in multinational operations. Most soldiers deploying to Iraq, for example, have the opportunity to be trained at mock Iraqi villages at the National Training Center at Fort Irwin, California. Moreover, those soldiers who will be assigned as advisors to train indigenous Iraqi military and police units participate in a rigorous 60-day training program.[45] The U.S. military is also recruiting psychologists, cultural anthropologists, and others with local cultural expertise to deploy and participate in "Human Terrain Teams" that advise Provincial Reconstruction Teams in Afghanistan and Iraq. This kind of training is crucial for having what the military calls "domain knowledge" of cultural, geographic, and political differences and sensitivities. The domain knowledge is particular *context-specific facts* about other cultures, their customs, traditions, doctrinal principles, and operational methods. However, it is important to note that emotional intelligence is a necessary *complement* to domain knowledge, because it provides the more general, foundational social skills that allow decisionmakers to *access those context-specific facts and use them* in their interactions with others.

Thus, emotional intelligence would be useful for at least four kinds of interactions in the national security context: with the indigenous population where the soldier is deployed, with allies or coalition partners from other countries that are participating in the mission, with members of humanitarian relief and other NGOs that often have very different organizational cultures, values and, goals, and with members of other U.S. Government agencies or organizations that likewise may have different organizational cultures, doctrines, and operational methods.

Given the importance of emotional intelligence for responsiveness, how can mindfulness enhance this quality? Scholars have theorized that mindfulness may promote a greater ability or willingness to take interest in others' thoughts, emotions, or welfare.[46] It may also enhance communication with others through improved listening skills and greater awareness of others' affective tone and nonverbal behavior.[47] At the same time, a person with mindfulness may be more aware of his or her own cognitive, emotional, and verbal responses in communication, which can reduce impulsive, destructive behavior in social settings.[48] Similarly, scholars have suggested that greater insight into the self, others, and human nature, along with the easing of ego-based concerns, encourages more empathy with others.[49] Self-report empirical research provides some support for these hypotheses.[50] Finally, mindfulness is associated with a more open and less defensive response to challenging social events. For example, in a study of being excluded from a group, participants with a high degree of mindfulness were less likely to perceive rejection and "take it personally."[51]

Enhanced Values Clarification and Integrity

Given the importance of means-ends consistency in the national security realm—linking tactical actions to strategic policy choices—enhanced values clarification is an important quality. It is virtually impossible for individuals to make decisions that are supportive of strategic goals until they identify what those goals are. Similarly, it is difficult for individuals to identify strategic goals until they know what they value. Norms and values, whether implicit or explicit, are crucial for helping us constitute our identities, worldviews, interests, and goals.[52] Mindfulness may help individuals to recognize what is meaningful for them and what they truly value.

As the discussion above about self-regulation suggested, automatic processing often limits individuals from considering options that would be most congruent with their interests, needs, and values. In contrast, as a study of participants using pagers over 3 weeks to record their daily activities showed, empirical evidence suggests that when subjects are "acting mindfully," they act in ways that are more congruent with their actual interests, needs, and values.[53] In other words, decisions and actions become more intentional, more internally consistent, and more likely to support broader personal and organizational goals and interests.

Correspondingly, mindfulness, and the improved attention it cultivates, makes it less likely that decisionmakers will "pretend not to know" and avoid or deny unpleasant truths; default to cognitive or emotional dissonance while gathering and processing information, assessing situations, and making decisions; and act in ways not aligned to their values or interests. Instead, mindfulness helps decisionmakers to recognize what is meaningful for them and their organizations and then act congruently with their actual values, interests, and needs. This allows them to assess the situation, respond to it, and act authentically and consistently.

These traits can lead to consistency in thought, word, and action, which is how the military profession defines integrity. At the personal level, integrity allows for personal integration, having the various aspects of ourselves come together and be expressed outwardly with consistency and authenticity. Integrity and integration at the organizational or state level are holographically similar: an organization is integrated if its actions are consistent with its strategic goals and values. At the organizational or strategic level, integrity allows decisionmakers to choose policies that are aligned with personal and organizational values, link those policies to the strategic goals of the organization or state, and choose policy tools that align with those values and goals. Thus, it is no wonder that when policies are not aligned with values and goals, we say the policies "lack integrity."

Mind Fitness Training in the National Security Environment

Winning "hearts and minds," coping with complex, fluid, and unpredictable environments, navigating moral ambiguity, and empowering leaders at lower levels require a tremendous amount of awareness of self, others, and the wider environment. This chapter has proposed a new way to achieve such awareness. New research into neuroplasticity and clinical studies of the effects of mindfulness show that humans can rewire their neural pathways and strengthen skills that allow them to make better decisions. Mind fitness training inoculates against stress by permitting more adaptive responses to and interpretation of stressors, providing a type of "mental armor." In addition, mind fitness training can help to optimize soldier performance by cultivating six competencies critical for today's security environment: improved self-regulation, better attention skills, increased flexibility in awareness and attention, less reactivity, improved emotional and social intelligence, and enhanced values clarification and integrity.

In other words, mind fitness training could not only help deploying soldiers relate more effectively to the stressors of the deployment, but it could also enhance the skills of immense flexibility and agility in thinking and behavior that these kinds of missions require. Mind fitness can ease these transitions by helping individuals to increase their tolerance of and exhibit more open, less defensive responses to challenging people and events.

Building social rapport and trust is paramount. Mind fitness training could help deploying individuals to better recognize their own conditioning and how that conditioning affects interactions with others, as well as to see situations more easily from and respect others' perspectives. These skills would be useful for both cross-cultural and cross-organizational cultural interactions.

To be sure, there are some potential limitations that need to be explicitly acknowledged. Regarding the existing clinical research, many early studies had methodological weaknesses that make it difficult to draw strong conclusions about the effects of mindfulness-based interventions, including small sample size, lack of control groups, static research protocols that did not control for preexisting conditions or brain structure, and poor self-report measures of mindfulness. Nevertheless, these preliminary data are promising, especially when considered with more recent research that has begun to correct for such weaknesses. This underscores the need for additional rigorous empirical research to further delineate the causal mechanisms at work.

More importantly, there are at least four possible issues with incorporating mind fitness training in the national security context. First, unlike instantaneous and simple solutions, mind fitness training is not a quick-fix "silver bullet." While mind fitness skills are easily taught, they require ongoing effort and commitment to develop and strengthen over time—just like physical exercise for the body. Second, although the end result is less reactivity, more self-control, and more inner strength, individuals may need to go through a destabilizing period during the training. Third, it is critical that students have access to a well-trained instructor who has a longstanding personal practice. Thus, there is some tension between providing effective instruction, on the one hand, and creating a training program that would be replicable and scalable, on the other. Finally, mind fitness training and the kinds of attributes it inculcates could be threatening to hierarchical institutions like the military. At the very least, it presents a paradox: to develop the physical and mental qualities that the military says it wants—flexible, agile leaders with initiative, creative problemsolving skills, and emotional intelligence—it is also necessary to develop skills that the bureaucratic elements may find threatening.

Despite these potential limitations and given that these training techniques have worked in other contexts, applying them to the military could provide an intriguing and timely opportunity to enhance military effectiveness. A natural first step to investigating this chapter's ideas in a concrete way is to implement a *prospective* study of the effects of mind fitness training on operational effectiveness and individual soldier well-being, as Amishi Jha, a neuroscientist at the University of Pennsylvania, and I are currently doing. This training will be provided prior to deployment. We argue that such mind fitness training is likely to lead to improvements in operational effectiveness during the deployment and reduce the rate of psychological dysfunction such as PTSD, depression, and anxiety upon return.

There are at least four other ways that mind fitness training could be effectively incorporated into civilian and military professional development. These ideas are included to suggest a longer term vision of what might be possible, ranging from conventional to radical, both in terms of the targeted population and the potential effects on national security decisionmaking.

First, mind fitness training could be offered to chaplains, ombudsmen, inspectors general, Judge Advocate General attorneys, and medical and mental health professionals who provide care for soldiers, tactical leaders, and strategic decisionmakers. It could also be offered to rear detachment commanders and other community service providers who support families at home while units are deployed. In both cases, mind fitness training could help these individuals improve their own capacity to be present with their clients' needs without suffering burnout.

Second, mind fitness training could be incorporated into the training of anyone deploying in counterinsurgency, stability operations, or postconflict reconstruction missions. Similarly, mind fitness training could be offered to teams that advise and train indigenous militaries as part of the U.S. counterterrorism strategy abroad. While the COIN training the military already conducts, such as scenarios at mock "Iraqi villages" that soldiers encounter during their Joint Readiness Training Center and National Training Center rotations—can help to reduce soldiers' stress response by *habituating* them to a variety of COIN-related stressors, mind fitness training could complement and strengthen that training by helping soldiers to *perceive and relate to* those stressors differently.

Third, mind fitness training could be incorporated into the professional development of intelligence collectors and analysts. For intelligence collectors, mind fitness training could help them to access information from a wider variety of channels and display greater accuracy and more objectivity in gathering information. For intelligence analysts, the training could help them recognize how their own conditioning creates filters to incoming information and see through that conditioning, in order to process all available information with clarity and cut through cognitive and emotional biases.

Finally, mind fitness training could be incorporated in the regular professional development of enlisted soldiers, noncommissioned officers, officers, and civilian government officials in the national security realm. The military officer corps is perhaps the most promising target audience. Being selected for these courses is a sign of future promotion potential, which suggests that the officers who attend are likely to incur even larger leadership roles with greater decisionmaking responsibilities in the future. Mind fitness training could be beneficial not only for these leaders and the quality of the decisions they make, but also for the command climate they create.

Given the frequency of these courses, initial mind fitness training could occur early in a leader's career, and refresher training could be offered in subsequent courses. While leaders attend one of these courses, they would be relieved of regular command or staff responsibilities so that they can concentrate on their professional development. Moreover, because the operational tempo at these courses is much slower, many of the stressors discussed in this paper are muted. Finally, leaders attend these courses with a group of peers who could provide a support network during new skill acquisition. Having these skills and habits cemented before they return to line duty increases the potential that they will be able to maintain the new skills, given how easy it is to fall back into habitual or automatic conditioning under stress. Thus, professional development courses are perhaps the *most* ideal time for leaders to learn mind fitness skills and new behavioral, cognitive, and emotional habits, which take some time to develop and stabilize.

The security environment today is complex and rapidly changing, and those who work in this environment need new competencies that rely on the foundational skill of awareness. Mind fitness—together with mission essential skills and physical fitness—is one of the three components necessary for optimal soldier readiness, yet there is a dearth of training in this area. This chapter has proposed ways that mind fitness training could be incorporated into the military setting. Given that the brain is constantly building new neural connections from all of our daily experiences, the question becomes: how can we build neural connections that will be most beneficial for operational effectiveness and soldier well-being, rather than neural connections that will lead to reactivity and cognitive degradation in stressful environments?

NOTES

1 U.S. Army Field Manual 3–24/U.S. Marine Corps Warfighting Publication 3–33.5, *Counterinsurgency* (Washington, DC: Headquarters Department of the Army, December 2006), ix–x.

2 Leonard Wong, *Developing Adaptive Leaders: The Crucible Experience of Operation* Iraqi Freedom (Carlisle, PA: US Army War College Strategic Studies Institute, July 2004), 4.

3 Charles C. Krulak, "The Strategic Corporal: Leadership in the Three Block War," *Marines Magazine* (1999).

4 Robert Scaer, *The Trauma Spectrum: Hidden Wounds and Human Resiliency* (New York: W.W. Norton, 2005), 17.

5 Paul J. Rosch, "Stress and Memory Loss: Some Speculations and Solutions," *Stress Medicine* 13 (1997), 1–6.

6 Douglas Bremner, *Does Stress Damage the Brain? Understanding Trauma-related Disorders from a Mind-Body Perspective* (New York: W.W. Norton, 2002).

7 Jennifer J. Vasterling et al., "Neuropsychological Outcomes of Army Personnel Following Deployment to the Iraq War," *Journal of the American Medical Association* 296, no. 5 (2006), 519–529.

8 Harris R. Lieberman et al., "Severe Decrements in Cognition Function and Mood Induced by Sleep Loss, Heat, Dehydration and Under-nutrition During Simulated Combat," *Biological Psychiatry* 57 (2005), 422–429.

9 Charles A. Morgan III et al., "Accuracy of Eyewitness Memory for Persons Encountered during Exposure to Highly Intense Stress," *International Journal of Law and Psychiatry* 27 (2004), 265–279; Charles A. Morgan III et al., "Stress-induced Deficits in Working Memory and Visuo-constructive Abilities in Special Operations Soldiers," *Biological Psychiatry* 60 (2006), 722–729; Charles A. Morgan III et al., "Accuracy of Eyewitness Identification is Significantly Associated with Performance on a Standardized Test of Face Recognition," *International Journal of Law and Psychiatry* 30 (2007), 213–223.

10 D.R. Haslam, "The Military Performance of Soldiers in Sustained Operations," *Aviation, Space, and Environmental Medicine* 55 (1984), 216–221; Harris R. Lieberman et al., "Effects of Caffeine, Sleep Loss and Stress on Cognitive Performance and Mood during U.S. Navy Seal Training," *Psychopharmacology* 164 (2002), 250–261; K. Opstad, "Circadian Rhythm of Hormones is Extinguished during Prolonged Physical Stress, Sleep, and Energy Deficiency in Young Men," *European Journal of Endocrinology* 131(1994), 56–66.

11 G. Belenky et al., "Sustaining Performance during Continuous Operations: The U.S. Army's Sleep Management System," in *Pennington Center Nutritional Series, Vol. 10, Countermeasures for Battlefield Stressors*, ed. K. Friedl, H. Lieberman, D.H. Ryan, and G.A. Bray (Baton Rouge: Louisiana State University Press, 2000).

12 E.A. Maguire et al., "Navigation Expertise and the Human Hippocampus: A Structural Brain Imaging Analysis," *Hippocampus* 13, no. 2 (2003), 250–259.

13 William M. Jenkins et al. "Functional Reorganization of Primary Somatosensory Cortex in Adult Owl Monkeys after Behaviorally Controlled Tactile Stimulation," *Journal of Neurophysiology* 63 (1990), 82–104; John J. Ratey, *A User's Guide to the Brain* (New York: Vintage Books, 2002), 167–171.

14 Sharon Begley, *Train Your Mind, Change Your Brain* (New York: Random House, 2007); Ryuta Kawashima, *Train Your Brain: 60 Days to a Better Brain* (New York: Kumon Publishing, 2005); Joel Levy, *Train Your Brain: The Complete Mental Workout for a Fit and Agile Mind* (New York: Barnes and Noble Publishing, 2007).

15 Offered by the UCLA Semel Institute Memory and Aging Center. See **www.npi.ucla.edu/memory/brain.php**.

16 Amishi P. Jha et al., "Mindfulness Training Modifies Subsystems of Attention," *Cognitive, Affective, and Behavioral Neuroscience* 7, no. 2 (2007), 109–119.

17 First, a recent pilot study with mindfulness-based group therapy at the Ann Arbor Veterans Administration (VA) PTSD clinic found that war veterans, relative to their pre-study baselines, showed decreased PTSD symptoms after the 8-week intervention. Available laboratory data in this chronic PTSD group—including neuroimaging, saliva cortisol, skin conductance, and heart rate—suggested decreased psychological and physiological reactivity to trauma recall. The Ann Arbor VA is now conducting a second pilot study with an active control group. Researchers on the study believe that an 8-week intervention is not long enough for effective PTSD treatment, and they are considering an additional 8-week course on emotional regulation to complement the mindfulness skills. Similarly, the Emory University School of Medicine and the Atlanta VA Medical Center are jointly starting a study with returning Iraq War veterans that will compare a Mindfulness-Based Stress

Reduction intervention with an active control group receiving supportive therapy, including hatha yoga and other techniques. Similar mindfulness-based training programs with war veterans suffering from PTSD are ongoing at VA hospitals in Albuquerque, Puget Sound, and Palo Alto. Interview with Anthony King, investigator on the study at the Ann Arbor VA PTSD clinic and University of Michigan, June 7, 2007; interview with Sandra DiVitale, investigator on the study at Emory University, June 8, 2007; email correspondence with Sarah Bower, University of Washington and Puget Sound VA medical center, July 22, 2007.

18 G.A. Marlatt and J.L. Kristeller, "Mindfulness and Meditation," in *Integrating Spirituality into Treatment: Resources for Practitioners*, ed. William R. Miller (Washington, DC: American Psychological Association, 1999), 67–84, at 68.

19 Jon Kabat-Zinn, *Wherever You Go, There You Are: Mindfulness Meditation in Daily Life* (New York: Hyperion, 1994), 4.

20 Thich Nhat Hanh, *The Miracle of Mindfulness* (Boston: Beacon Press, 1976); Bhante H. Gunaratana, *Mindfulness in Plain English* (Boston: Wisdom Publications, 2002); Jon Kabat-Zinn, *Wherever You Go* and *Full Catastrophe Living: Using the Wisdom of Your Body and Mind to Face Stress, Pain and Illness* (New York: Delacorte, 1990); Joan Borysenko, *Minding the Body, Mending the Mind* (New York: Bantam Books, 1987).

21 For a good overview of mindfulness-based clinical interventions and the empirical results of clinical studies of these interventions, see Ruth A. Baer, "Mindfulness Training as a Clinical Intervention: A Conceptual and Empirical Review," *Clinical Psychology: Science and Practice* 10, no. 2 (Summer 2003), 125–143; and K.W. Brown et al., "Mindfulness: Theoretical Foundations and Evidence for its Salutary Effects," *Journal of Clinical Psychology* 18, no. 4 (2007), 211–237. For studies that focused especially on stress symptoms and mood disturbance, see M. Speca et al., "A Randomized, Wait-list Controlled Clinical Trial: The Effect of a Mindfulness Meditation-Based Stress Reduction Program on Mood and Symptoms of Stress in Cancer Outpatients," *Psychosomatic Medicine* 62 (2000), 613–622; and Kirk Warren Brown and Richard M. Ryan, "The Benefits of Being Present: Mindfulness and Its Role in Psychological Well-Being," *Journal of Personality and Social Psychology* 84, no. 4 (2003), 822–848, especially the fifth study.

22 Richard J. Davidson et al., "Alterations in Brain and Immune Function Produced by Mindfulness Meditation," *Psychosomatic Medicine* 65, no. 4 (2003), 564–570.

23 Field Manual 6–22, *Army Leadership* (Washington, DC: Headquarters Department of the Army, October 2006), section 10–48.

24 Field Manual 3–24, *Counterinsurgency* (Washington, DC: Headquarters Department of the Army, December 2006), 1–22.

25 R.M. Ryan and E.L. Deci, "Self-determination Theory and the Facilitation of Intrinsic Motivation, Social Development and Well-being," *American Psychologist* 55 (2000), 68–78; Brown and Ryan, 822–848.

26 Benjamin Libet, "Do We Have Free Will?" *Journal of Consciousness Studies* 6, no. 8–9 (1999), 47–57.

27 Joshua Grant, "Pain Perception, Pain Tolerance, Pain Control, and Zen Meditation," presentation at the Mind and Life Summer Research Institute, June 5, 2007.

28 Brown and Ryan, 822–848.

29 Stephen Peter Rosen, *War and Human Nature* (Princeton: Princeton University Press, 2005), 127.

30 Raja Parasuraman, *The Attentive Brain* (Cambridge: MIT Press, 1998).

31 M.I. Posner, "Orienting of Attention," *Quarterly Journal of Experimental Psychology* 32, no. 1 (1980), 3–25.

32 J.M.G. Williams, A. Mathews, and C. MacLeod, "The Emotional Stroop Task and Psychopathology," *Psychological Bulletin* 120, no. 1 (1996), 3–24.

33 M.I. Posner and S.E. Peterson, "The Attention System of the Human Brain," *Annual Review of Neuroscience* 13 (1990), 25–42.

34 Jeremy R. Gray et al., "Neural Mechanisms of General Fluid Intelligence," *Nature Neuroscience* 6 (2003), 316–322.

35 E.R. Valentine and P.L.G. Sweet, "Meditation and Attention: A Comparison of the Effects of Concentrative and Mindfulness Meditation on Sustained Attention," *Mental Health, Religion, and Culture* 2 (1999), 59–70; Jha et al.

36 Sara Lazar et al., "Meditation Experience Is Associated with Increased Cortical Thickness," *NeuroReport* 16 (2005), 1893–1897. Because this study was not longitudinal in design, it could not control for preexisting differences in brain structure; nonetheless, its suggestive findings argue for the need for further research.

37 Heleen Slater et al., "Mental Training Affects Distribution of Limited Brain Resources," *PLoS Biology* 5, no. 6 (2007), e138.

38 Brown et al., "Mindfulness: Theoretical Foundations," 9, in unpublished manuscript; J.M.G Williams et al, "Mindfulness-based Cognitive Therapy Reduces Over-general Autobiographical Memory in Formerly Depressed Patients," *Journal of Abnormal Psychology* 109 (2000), 150–155; C.N. Alexander et al., "Transcendental Meditation, Mindfulness, and Longevity: An Experimental Study with the Elderly," *Journal of Personality and Social Psychology* 57 (1989), 950–964.

39 T.D. Borkovec, "Life in the Future Versus Life in the Present," *Clinical Psychology: Science and Practice* 9 (2002), 76–80.

40 C.E. Lakey et al., "The Role of Mindfulness in Psychological Gambling," unpublished manuscript (2006), cited in Brown et al., "Mindfulness: Theoretical Foundations."

41 *Army Leadership*, section 6–3.

42 Brown et al., "Mindfulness: Theoretical Foundations," 29; Lisbeth Nielsen and Alfred W. Kasniak, "Awareness of Subtle Emotional Feelings: A Comparison of Long-Term Meditators and Non-meditators," *Emotion* 6, no. 3 (2006), 392–405; J.J. Arch and M.G. Craske, "Mechanisms of Mindfulness: Emotion Regulation Following a Focused Breathing Induction," *Behavior Research and Therapy* 44 (2006), 1849–1858; J. David Cresswell et al., "Neural Correlates of Mindfulness during Affect Labeling," draft manuscript, 2006.

43 *Counterinsurgency*, 1–27.

44 Daniel Goleman, *Emotional Intelligence* (New York: Bantam Books, 1996); David Abrams, "Emotional Intelligence and Army Leadership: Give It to Me Straight!" *Military Review* (March-April 2007), 86–93.

45 Michael R. Gordon, "Army Expands Training for Advisors Who Will Try to Improve Iraq's Security Forces," *The New York Times*, November 25, 2006; email communication with advisory team leader MAJ J.R. Deimel, USA, January 3, 2007.

46 John Welwood, *Love and Awakening* (New York: HarperCollins, 1996).

47 It is important to note that even when a high degree of mindfulness has been developed, actual training in communication, listening, and perspective taking skills is often helpful (or depending upon the person, even necessary) for improved social interaction abilities. However, mindfulness can dramatically increase the retention and utilization of such skills.

48 Sylvia Boorstein, *Transpersonal Psychotherapy* (Albany: State University of New York Press, 1996).

49 Richard J. Davidson and Anne Harrington, eds., *Visions of Compassion: Western Scientists and Tibetan Buddhists Examine Human Nature* (New York: Oxford University Press, 2002).

50 L. Tickle-Degnan and R. Rosenthal, "The Nature of Rapport and its Nonverbal Correlates," *Psychological Inquiry* 1 (1990), 285–293; Baer et al.; Brown and Ryan, 822–848. Because this body of research is largely built on self-report questionnaires, taking these studies at face

value may be problematic. To some degree, asking individuals without mindfulness training to assess their level of mindfulness is asking them to do something they cannot do.

51 J. D. Cresswell, N.I. Eisenberger, and M.D. Lieberman, "Neurobehavioral Correlates of Mindfulness during Social Exclusion," unpublished manuscript (2006), cited in Brown et al., "Mindfulness: Theoretical Foundations."

52 There is a large constructivist literature that elaborates on this point. See, for example, James A. Aho, *The Things of the World: A Social Phenomenology* (Westport, CT: Praeger, 1998), and Peter L. Berger and Thomas Luckmann, *The Social Construction of Reality* (New York: Anchor Books, 1966).

53 Brown and Ryan, 822–848.

chapter 19

BIO-INSPIRED NETWORK SCIENCE

João Hespanha and Francis Doyle III

At the beginning of the 21st century, we find ourselves in a technological revolution that is changing our lives as much as the Industrial Revolution did in the late 18th century. Our revolution started in the late 1980s as personal computers started to become household items and is now in full force as broadband Internet is also widely available. However, a next qualitative step is about to occur and it is being triggered by advances in very-large-scale integration fabrication and design, which resulted in the availability of low-cost, low-power, small computational elements that are able to communicate wirelessly. In addition, advances in micro-electromechanical systems are producing solid-state sensors and actuators. The net result is inexpensive computation and ubiquitous embedded sensing, actuation, and communication, which provide tremendous opportunities for societal impact, but also present great challenges in the design of networked systems. Although both the scientific community and industry have recognized the possibilities enabled by these advances and are starting to plan and build truly large-scale systems, appropriate design methods are lacking.

In broader terms, for the large-scale networked systems that are appearing to fully realize their potential, a *logical and systematic approach* to the analysis and design of such systems needs to be developed. Moreover, this approach should strive for laws that are *generalizable* to broad classes of systems and thus permit transfer of knowledge across domains. What we just described constitutes a science, specifically a network science.

In this chapter, we discuss how large-scale networks that arise in biological systems can make significant contributions to the development of a network science, not only by serving as a subject matter to exercise and validate new analytical tools, but also by providing inspiration for the development of new methodologies and architectures to design artificial networks.

We start by focusing our attention on biochemical networks that express complex interactions between the chemical species that regulate cell behavior. While such networks are characterized by great uncertainty at the level of individual components, they typically exhibit levels of aggregate robustness that are unmatched by comparable engineered systems. In addition to enabling fundamental contributions for the development of a network science, the study of biochemical networks has the potential for a tremendous impact in medicine.

We also discuss the biological networks formed by groups of individuals engaged in activities that benefit from a group effort. We focus our discussion on activities that involve motion coordination, falling under the generic term of "swarming." Biological swarms typically achieve tight motion coordination by relying only on "noisy" local interactions between individuals. Understanding the design principles behind this feat will enable the design of large-scale artificial networked systems.

Biochemical Networks

As with engineering networks, biophysical networks are remarkably diverse, cover a wide spectrum of scales, and inevitably are characterized by a range of complex behaviors. These networks have attracted a great deal of attention at the level of gene regulation, where dozens of input connections may characterize the regulatory domain of a single gene in a eukaryote, as well as the protein level where literally thousands of interactions have been mapped in so-called protein interactome diagrams that illustrate the potential coupling of pairs of proteins.[1, 2] However, these networks also exist at higher levels, including the coupling of individual cells via signaling molecules, the coupling of organs via endocrine signaling, and ultimately the coupling of organisms in ecosystems. To elucidate the mechanisms employed by these networks, biological experimentation and intuition are by themselves insufficient. The field of systems biology has laid claim to this class of problems, and engineers, biologists, physicists, chemists, mathematicians, and many others have united to embrace these problems with interdisciplinary approaches.[3] In this field, investigators characterize dynamics via mathematical models and apply systems theory with the goal of guiding further experimentation to better understand the biological network that gives rise to robust performance.[4]

An ideal example of such networked biological complexity is the circadian clock, which coordinates daily physiological behaviors of most organisms. The mammalian circadian master clock resides in the suprachiasmatic nucleus (SCN), located in the hypothalamus.[5] It is a network of multiple autonomous noisy oscillators, which communicate via neuropeptides to synchronize and form a coherent oscillator.[6, 7] At the core of the clock is a gene regulatory network in which approximately six key genes are regulated through an elegant array of time-delayed negative feedback circuits. The activity states of the proteins in this network are modulated (activated/inactivated) through a series of chemical reactions including phosphorylation and dimerization. These networks exist at the subcellular level. Above this layer is the signaling that leads to a synchronized response from the population of thousands of clock neurons in the SCN. Ultimately, this coherent oscillator then coordinates the timing of daily behaviors, such as the sleep/wake cycle. Left in constant conditions, the clock will free-run with a period of only approximately 24 hours such that its internal time, or phase, drifts away from that of its environment. Thus, vital to a circadian clock is its ability to entrain to external time through environmental factors.[8, 9, 10]

A second example is the apoptosis network in which an extracellular input "controls" the response of the cell as a result of this information processing network. (Apoptosis is the process of "programmed cell death," machinery that nature uses to strategically kill off unneeded cells. However, this mechanism becomes impaired in cancer, leading to unchecked proliferation.) The specific example employed here details the response of kidney cells to exposure to staphylococcal enterotoxin B (SEB), a problem of particular interest to the military. Understanding apoptosis in a broader sense will lead to a better knowledge of the common platform for emergence of cancer cells, and perhaps point to possible cures for certain types of cancer. The complexity of apoptosis, however, makes the understanding very difficult without a systems level approach using a mathematical representation of the pathway. Further, analysis of an apoptosis model can reveal the fragility points in the mechanism of programmed cell death that can have physiological implications not only for explaining the emergence of cancer cells but also for designing drugs or treatment for reinstating apoptosis in these cells.

Biological Inspiration for Engineered Networks

Biological networks offer a number of opportunities for inspired design of engineering networks. Aside from the overlapping computational toolkit (for example, simulation methods for high-dimensional, stochastic, stiff, multiscale systems), there are numerous behaviors in biological networks that offer promise for improved communications and sensors networks. Given the space constraints, we highlight only a few of them here.[11]

The coexistence of extreme robustness and fragility constitutes one of the most salient features of highly evolved or designed complexity. Optimally, robust systems are those that balance their robustness to frequent environmental variations with their coexisting sensitivity to rare events. As a result, robustness and sensitivity analysis are key measures in understanding and controlling system performance. Robust performance reflects a relative insensitivity to perturbations; it is the persistence of a system's characteristic behavior under perturbations or conditions of uncertainty. Measuring the robustness of a system determines the behavior (the output or performance) as a function of the input (the disturbance). Formal sensitivity analysis allows the investigation of robustness and fragility properties of mathematical models, yielding local properties with respect to a particular choice of parameter values. Natural systems have honed the robust performance characteristics over millennia of evolutionary pressure, and consequently have achieved levels of performance unrivaled in engineering networks.

Synchronization is manifested in natural circuits in, among other ways, the coherent response of an ensemble of otherwise noisy components. There are a number of examples of this behavior in neuronal systems, most notably the coherence in circadian timekeeping achieved by cells in the suprachiasmatic nucleus.[12] The remarkable feature of these cells is that their individual characteristics are remarkably diverse, with individual periods ranging from 21 to 26 hours; moreover, a given cell may exhibit fluctuations in the period of its firing rate from 2 to 10 hours between consecutive cycles. Despite this huge range of "component tolerances," the intact signaling network is able to reduce this dispersion by over an order of magnitude in timekeeping precision. By contrast, engineering networks are built from high precision components, yet often struggle with network performance characteristics such as time synchrony.[13]

The Role of Networks for Medical Applications

Systems biology is driving the interdisciplinary approach to unraveling design principles in complex biological networks. From a basic research perspective, the field has revolutionized thinking about networks as opposed to components, and how the coordinated actions in a network must be understood in order to unravel the principles involved in a particular phenotype, such as a disease state.

The medical driving forces include the identification of "targets" in the network for therapeutic intervention. A "systems" analysis reveals that a single point perturbation is often less effective than a vectoral perturbation, and at the same time, a point perturbation will often propagate beyond the intended action, leading to undesirable side effects. Systems methods are also applied to these networks to determine "signatures," or markers, of the propagation of a disease state. For example, the temporal progression of the SEB-induced response in the network could be tracked by gene expression profiling, and one can determine the time course of the response and apply appropriate therapy at the optimal point in the disease progression.

The size and complexity of cellular networks make intuition inadequate for deducing cellular behavior from the underlying gene and protein interactions. Systems analysis can help to unravel this complexity. One such method is sensitivity analysis,[14] in which linear sensitivities quantify how much the system behavior changes as the parameters are varied. In cellular networks, high sensitivities point to the weakest links in the system that cellular behavior strongly depends on. Perturbations on these links can potentially lead to a large disruption in the network behavior—that is, the network is not robust to the uncertainty in these pathways. By mapping critical pathways back to the genome, one can point to the set of genes and interactions that controls the cellular behavior. Such information can be used for guiding data-fitting and model refinement in the reverse-engineering of cellular networks.

These *hot spots* (or fragile nodes) have several implications. First, further model refinements on the hot spots in the network may be necessary. Second, when the model is sufficiently accurate, the differences between the analysis with and without SEB treatment can have physiological relevance and provide the candidate pathways in the network for targeted detection and/or therapy of the SEB toxin response.

Swarms

Coordinated motion is commonly observed in flocks of geese, schools of sardines, herds of wildebeests, and packs of wolfs. The emergence of these complex behaviors indicates that coordinated motion can provide a significant evolutionary advantage that outweighs the costs incurred by the individuals in developing a swarming behavior.

Migratory birds such as ducks, geese, and pelicans provide the best example of motion coordination. Especially in long-range migrations, these birds fly in a V-shaped formation with a bird in the lead and the remaining birds trailing behind in two straight lines. Several hypotheses have been proposed to explain what benefits can be drawn from this flying formation. There is widespread consensus that it can improve aerodynamic efficiency, but it has also been noted that this formation can optimize the visual contact between the individual members. Energy savings arise as a trailing bird uses the wingtip vortices of the bird ahead to gain lift. These types of vortices generally increase drag for the wing that produces them, resulting in less efficient flight. However, they also create upwash regions that essentially provide free lift to a second wing flying behind. This extra lift allows a trailing bird to flap its wings less often and less hard, meaning that its muscles do not work as hard and its heart rate drops. It has been theoretically shown that the increased aerodynamic efficiency of formation flight can allow 25 birds to increase their range by about 70 percent as compared to a bird flying alone.[15] These predictions have been experimentally validated on several bird species. For example, great white pelicans show an energy savings of as much as 11 to 14 percent when flying in the vortex wake generated by another bird.[16] In addition to energy savings, V-shaped formations also improve visual contact between individuals, and evidence exists of some agreement between the formation structure and the visual anatomy of several bird species.[17] Although manned aircraft can also exploit the drag-reduction benefits of formation flight, this is generally not done because it is difficult to maintain the required close formation with current technology. Nevertheless, military aircraft squadrons still operate in V formations to maintain visibility in the same way that birds do. However, these aircraft formations are generally spaced too far apart to enjoy energy savings.

Fish schooling refers to the synchronized motion of individuals of similar size, generally moving in the same direction. It is believed that this complex behavior can be observed in over 10,000 species,[18] which raises the question of what evolutionary mechanisms could be responsible for schooling. The most widely accepted hypotheses are related to antipredatory adaption. Schooling may have evolved as a "cover-seeking" response, as each individual tries to reduce its chance of being caught by a predator.[19, 20] According to this explanation, schooling may not necessarily benefit the overall population by reducing a predator's success rate, but it may simply arise because natural selection may favor individuals that are prone to schooling.[21, 22] It has also been proposed that schooling evolved as a mechanism to simultaneously improve the probability of detecting predators and confuse an attacking predator.[23, 24] The hypothesis that schooling evolved as an antipredatory response is supported by experimental evidence. For example, guppy populations subject to intense predation exhibit a more pronounced schooling behavior, and even within the same species, schooling can be more pronounced in environments that lack predators.[25] Schooling has also been proposed as a mechanism by which social individuals gain hydrodynamic advantage over solitary ones. This hypothesis has been tested experimentally and supporting evidence was found mostly for large predatory species,[26] whereas for prey fish, antipredatory mechanisms seemed to dominate the formation of schools.[27, 28] This dichotomy can be explained by the fact that predatory schools are free from predation and therefore free to adopt a hydrodynamically efficient structure, whereas prey schools may not be.

Challenges Posed by the Network Structure

One of the key challenges in swarming behavior is the need to coordinate the motion of a large number of individuals. In natural swarms this coordination typically relies on visual queues and on sensory organs through which an individual can perceive vortices created by its neighbors.[29] Regardless of the sensing mechanism used, individuals are generally only able to monitor the motion of a few close neighbors and the overall motion of the swarm must emerge from local interactions between neighboring individuals.

The requirement that the motion of a swarm must emerge from distributed control laws that only use local measurements imposes limitations on the agility of the swarm and also on how large the swarm can grow. This can be understood by analyzing simple distributed control laws by which an individual adjusts its heading based on the positions of its neighbors so as to achieve synchronized motion. While many such laws can be shown to lead to stable flocking behavior,[30] a small amount of noise in the local measurement can be amplified by the network structure and result in large stochastic fluctuations of the individuals.[31,32] This effect is especially noticeable in swarms for which the graph that characterizes the interactions between individuals resembles a one-dimensional lattice, such as the V-shaped formations observed in many species of migratory birds, which never grow very large when compared, for example, to three-dimensional fish schools.

Biological Inspiration from Natural Swarms

Synchronized motion has numerous applications in engineered systems. Formations of spacecraft can provide distributed sensing for gravitational field mapping or contemporaneous spatial sampling of atmospheric data, whereas formations of aircraft can provide near-simultaneous observations of the same science target by instruments on multiple platforms or synthetic aperture imagery. Formation control of aircraft also has the potential to reduce energy consumption. In addition, many of the algorithms discovered/constructed to achieve synchronized motion can also be used to accomplish a multitude of other tasks ranging from time synchronization to sensor fusion.[33]

One of the most well known biologically inspired models for swarming was introduced for the creation of computer animations of flocking behavior.[34] This work was eventually granted the Scientific and Engineering Award presented by The Academy of Motion Picture Arts and Sciences for pioneering contributions to the development of three-dimensional computer animation for motion picture production. The heuristic model was based on three rules to govern flocking behavior: cohesion, which causes an individual to stay close to nearby flockmates; separation, which prevents collision with nearby flockmates; and alignment, which is responsible for matching velocity with nearby flockmates. Although a complete analysis of this type of algorithm is still unavailable, some of its simplest forms are sufficiently well understood to be applied to distributed estimation and control problems.[35, 36, 37]

The observation that several animal species were able to achieve tight motion coordination for very large swarms prompted the search for network structures that could be used to overcome stochastic fluctuations caused by noise in local measurements. This led to the discovery that in formations with interaction graphs that resemble one-dimensional lattices, noise has an additive effect and, as the number of individuals grows, distributed control laws result in stochastic fluctuations whose variance grows linearly with the size of the formation. On the other hand, swarms such as fish schools for which the interaction graphs form three-dimensional structures exhibit much more favorable noise attenuation properties and simple distributed control laws can achieve tight formation control even in the presence of significant noise in the local measurements.[38, 39] These theoretical results are consistent with observations of natural swarms as, although birds in V-shaped formations could benefit from ever increasing formation sizes, they never grow as much as three-dimensional fish schools.

NOTES

1 A.L. Barabasi, "Network Biology: Understanding the Cell's Functional Organization," *Nature Reviews Genetics* 5 (2004), 101–113.

2 A.M. Malcolm and L.J. Heyer, *Discovering Genomics, Proteomics, and Bioinformatics* (San Francisco: Benjamin Cummings, 2003).

3 H. Kitano, "Systems Biology: A Brief Overview," *Science* 295 (2002), 1662–1664.

4 Ibid.

5 S.M. Reppert and D.R. Weaver, "Coordination of Circadian Timing in Mammals," *Nature* 418 (2002), 935–941.

6 E.D. Herzog et al., "Temporal Precision in the Mammalian Circadian System: A Reliable Clock from Less Reliable Neurons," *Journal of Biological Rhythms* 19 (2004), 35–46.

7 A.C. Liu et al., "Intercellular Coupling Confers Robustness against Mutations in the SCN Circadian Clock Network," *Cell* 129 (2007), 605–616.

8 Z. Boulos et al., "Light Visor Treatment for Jet Lag after Westward Travel across Six Time Zones," *Aviation, Space, and Environmental Medicine* 73 (2002), 953–963.

9 J.C. Dunlap, J.J. Loros, and P.J. DeCoursey, eds., *Chronobiology: Biological Timekeeping* (Sunderland, MA: Sinauer Associates, Inc., 2004).

10 S. Daan and C.S. Pittendrigh, "A Functional Analysis of Circadian Pacemakers in Nocturnal Rodents. II. The Variability of Phase Response Curves," *Journal of Comparative Physiology* 106 (1976), 253–266.

11 For additional details, see the National Research Council, *Network Science* (Washington, DC: National Academies Press, 2005).

12 Herzog et al., 35–46.

13 J. Elson and D. Estrin, "Time Synchronization for Wireless Sensor Networks," *Parallel and Distributed Processing Symposium, Proceedings 15th International* (2001), 1965–1970.

14 Dunlap, Loros, and DeCoursey, eds.

15 P.B.S. Lissaman and Carl A. Shollenberger, "Formation Flight of Birds," *Science* 168, no. 3934 (1970), 1003–1005.

16 H. Weimerskirch et al., "Energy Saving in Flight Formation," *Nature* 413, no. 6857 (2001), 697–698.

17 John P. Badgerow, "An Analysis of Function in the Formation Flight of Canada Geese," *The Auk* 105, no. 4 (1988), 749–755.

18 E. Shaw, "Schooling Fish," *American Scientist* 66 (1978), 166–175.

19 G.C. Williams, *Adaptation and Natural Selection* (Princeton: Princeton University Press, 1966).

20 W.D. Hamilton, "Geometry for the Selfish Herd," *Journal of Theoretical Biology* 31, no. 2 (1971), 295–311.

21 Ibid.

22 D. Bumann, J. Krause, and D.I. Rubenstein, "Mortality Risk of Spatial Positions in Animal Groups: The Danger of Being in the Front," *Behaviour* 134, no. 13–14 (1997), 1063–1076.

23 H.R. Pulliam, "On the Advantages of Flocking," *Journal of Theoretical Biology* 38 (1973), 419–422.

24 Douglass H. Morse, "Feeding Behavior and Predator Avoidance in Heterospecific Groups," *BioScience* 27, no. 5 (1977), 332–339.

25 Benoni H. Seghers, "Schooling Behavior in the Guppy (Poecilia icrosyste): An Evolutionary Response to Predation," *Evolution* 28, no. 3 (1974), 486–489.

26 Brian L. Partridge, Jonas Johansson, and John Kalish, "The Structure of Schools of Giant Bluefin Tuna in Cape Cod Bay," *Environmental Biology of Fishes* 9, no. 3–4 (1983), 253–262.

27 B.L. Partridge and T.J. Pitcher, "Evidence against a Hydrodynamic Function for Fish Schools," *Nature* 279 (1979), 418–419.

28 Mark V. Abrahams and Patrick W. Colgan, "Risk of Predation, Hydrodynamic Efficiency and their Influence on School Structure," *Environmental Biology of Fishes* 13, no. 3 (1985), 195–202.

29 T.J. Pitcher, B.L. Partridge, and C.S. Wardle, "A Blind Fish Can School," *Science* 194, no. 4268 (1976), 963–965.

30 R. Olfati-Saber, "Flocking for Multi-Agent Dynamic Systems: Algorithms and Theory," *IEEE Transactions on Automatic Control* 51, no. 3 (2006), 401–420.

31 P. Barooah and J. Hespanha, "Estimation on Graphs from Relative Measurements: Distributed Algorithms and Fundamental Limits," *IEEE Control Systems Magazine* 27, no. 4 (2007).

32 P. Barooah and J. Hespanha, "Graph Effective Resistances and Distributed Control: Spectral Properties and Applications," in *Proceedings of the 45th Conference on Decision and Control*, December 2006.

33 Barooah and Hespanha, "Estimation on Graphs from Relative Measurements."

34 C. Reynolds, "Flocks, Herds, and Schools: A Distributed Behavioral Model," ACM SIGGRAPH Computer Graphics, published in *Computer Graphics* 21, no. 4 (1987), 25–34.

35 Olfati-Saber, "Flocking for Multi-Agent Dynamic Systems," 401–420.

36 Barooah and Hespanha, "Estimation on Graphs from Relative Measurements."

37 Barooah and Hespanha, "Graph Effective Resistances and Distributed Control."

38 Barooah and Hespanha, "Estimation on Graphs from Relative Measurements."

39 Barooah and Hespanha, "Graph Effective Resistances and Distributed Control."

part five

IMPLICATIONS for the DEPARTMENT OF DEFENSE

chapter 20

ETHICS and the BIOLOGIZED BATTLEFIELD: MORAL ISSUES in 21ST-CENTURY CONFLICT

William D. Casebeer

Imagine that the nation-state of which you are a citizen is at war with another. The conflict has gone badly for your country; at times, it seems as though your nation's very survival as a political entity is at stake. Happily, your scientists engineer a virus that can be released into the opposing country that will temporarily disable anyone it infects. It is short lived, and should not spread beyond the borders of that nation, perhaps barely even beyond the capital region where it will be released. You fully expect the virus to completely decimate the command and control infrastructure of your adversary for several days and destroy almost all the biological neural nets responsible for decisionmaking in the enemy's autonomous tank brigades. One downside: it causes permanent blindness in 40 percent of the people it infects. Do you release the virus into the capital city?

Picture a second scenario. Your nation is combating a deadly insurgency. The guerrillas are as indiscriminate as terrorists come, killing innocents at a whim. While you expect this behavior will eventually cause their movement to collapse, their charismatic leader is holding things together in the short term, and the guerrillas have holed up for the winter in a series of valleys. You know that, in anticipation of the bad weather, the insurgents have stockpiled the wheat they will need to survive the season. That grain is the favorite food of a kind of weevil, and your ecologists recommend you seed the weevil—a species not native to this biome—into the valley in order to decimate the wheat stockpiles and force the insurgents into the cities, where you will better be able to confront them. You will take some risks, among them the possibility of not being able to control the spread of the weevil. On the other hand, if you do not act, you are fairly certain that the insurgency will kill thousands more innocent people, as you have intelligence indicating that, owing to their apocalyptic beliefs, they may have a dirty bomb under development that they would be willing to use in case of imminent defeat. Do you introduce the weevils into the valley's ecosystem?

A third case. Biological augmentation has become more common. Soldiers of other countries routinely use designer drugs and genetic enhancements to increase their combat efficiency. Your nation has decided to heavily regulate the use of everything from "super pep pills" that obviate the need for sleep for weeks, to genetically tailored enhancements that boost memory and recall and can produce mental giants capable of thinking clearly in times of combat stress. You hope that by regulating these developments you can set an example for other countries that will prevent a biological augmentation arms race. However, your soldiers—involved in stabilization and reconstruction operations in another country far away, and stretched to the limit with a theater-wide humanitarian intervention in another region—ask (actually, *demand*) that the government release some of these drugs and therapies for use in contingency circumstances. To do so will short-circuit standard safety testing and present multiple risks such as addiction and the need for post-military treatment, but you know that giving in will help out dramatically in the short term, even if it would mean violating a combat pharmacology protocol ratified by Congress 3 years ago that prohibits the use of many of the enhancements. Do you give in to the demands of your soldiers?

These are difficult cases posing challenging questions. At first glance, they may seem unanswerable. But even as the biological sciences—widely thought to be the king of 21st-century science in the wake of physics' domination of the 20th—simultaneously enhance and undermine our ability to answer normative questions, we have on hand at least some of the ethical tools required to reason well about what we ought to do in these circumstances. While the biologized battlefields of this century will present moral and ethical complications, they are not ones for which we are "at sea" when it comes to general guidelines. Exactness in ethics may be too much to ask, but by reasoning well together we will be able to rate some answers to these pressing moral questions as "beyond the pale" or "poor," others as at least "acceptable," and some as "very good." The introduction of these technologies and techniques complicates our decisionmaking, but will drive home one general trend in the morality of the professionalized use of violence: it is becoming more important that we equip individual soldiers at the lowest ranks with the cognitive kit they will need to be able to think comprehensively about what they can and cannot do, about what kinds of people they ought to be, and what kinds of outcomes they ought strive to produce.

In this chapter, I will discuss some traditional moral reasoning tools—ethical theories—that produce general guidance for questions about what we ought to do. After reviewing how some of the big-picture moral distinctions such as that between combatant and noncombatant will be blurred by the biologized battlefield, I will briefly discuss how the questions posed and tentatively answered by these theories interact with the circumstances of the brave new world of biowarfare to make it even more important that we produce individual soldiers with sound moral reasoning. New circumstances may also require that we modify or reject previous answers to questions on the regulation of violence, such as those supplied by the traditional just war theory. The rapid pace of evolution in the biologized battlefield heightens the importance of pushing down responsibility for sensible moral judgment to the line soldier. This is where it has always ultimately resided, and biowarfare is merely clarifying.

Basics of Ethical Theory

Ethical questions are normative questions. They deal with what we ought to do, what is permitted in good and right thought and conduct, and what kind of people we ought to be. Humans have struggled with these questions for millennia. A useful beginning to any question of morality might then be to look at the three grand traditions of ethical theory for assistance: virtue theory, deontology, and utilitarianism. These theories deal with three different aspects of moral questions. The first tackles the person taking an action; the second focuses on the nature of the action being taken; the third highlights the consequences of the action. All but the most adamant partisans of a particular approach to moral theory can agree that, at least for heuristic value, these three traditions have thrived because they focus attention on ethical aspects of a situation that we might otherwise be prone to ignore. Here, then, are thumbnail sketches of each approach. It is important to note that these are secular ethical theories that assume no particular position about religious metaphysical questions. Given the plurality of religious views in the world, and the prevalence of natural law reasoning in the Western tradition (which points out that the use of reason can lead you to the same moral truths revealed by the normative aspects of the great faith traditions), contemporary applied ethicists generally set aside what are called "divine-command theories" of morality.

Virtue theorists, such as the Greek philosophers Plato (427–347 BCE)[1] and Aristotle (384–322 BCE),[2] make paramount the concept of "human flourishing": to be maximally moral is simply to function as well as one can given one's nature. This is a practical affair that involves the cultivation of virtues such as wisdom, and the avoidance of vices such as intemperance. Deontologists, exemplified by the Prussian philosopher Immanuel Kant (1724–1804),[3] do not place emphasis upon the consequences of actions, as utilitarians would, or on the character of people, as a virtue theorist would. Instead, they focus on the maxim of the action, the intent-based principle that plays itself out in an agent's mind. We must do our duty as derived from the dictates of pure reason and the "categorical imperative" for duty's sake alone. Deontologists are particularly concerned to highlight the duties that free and reasonable creatures (paradigmatically, human beings, but potentially any creature that can be shown to possess a free will that can be "conditioned" by the dictates of reason) owe to one another. Maximizing happiness or cultivating character is not the primary goal of this scheme; instead, ensuring that we do not violate another's rights is paramount. The typical utilitarian, such as British philosopher John Stuart Mill (1806–1873),[4] thinks one ought to take that action or follow that "rule" that, if taken (or followed) would produce the greatest amount of happiness for the largest number of sentient beings (where, by *happiness*, Mill means the presence of pleasure or the absence of pain). The second flavor of utility we just described, "rule utilitarianism," is probably the most popular.

For any given biowarfare development, we can ask: Would developing and implementing this technology corrupt our character or somehow make us dysfunctional as human beings? Would it constitute a rights violation or somehow lead to disrespect for human dignity and autonomy? Would it produce bad results in terms of increasing net unhappiness among the world's sentient creatures in the long run? These questions of virtue, duty, and consequence lay at the heart of moral reasoning.

One brief aside: there is insufficient space in this chapter to discuss and dismiss two alternatives to the moral worldview. One says that attempting to answer questions like these is silly, as every answer is equally good. This "moral relativism" is not a useful approach to decisionmaking, as it offers no advice regarding what one ought to do since any decision is as good as any other. That you are bothering to read this chapter indicates that you are concerned about making good decisions in the biowarfare realm and is good enough reason to dismiss moral relativism as a serious approach to normative questions. The other alternative is a radical realism, which says that, especially in warfare, moral questions have no place and nations are free to do what is in their unenlightened interest. If this means killing innocents in wartime, not worrying about the long-term ecological impact of a bioweapon, or freely using your own citizen-soldiers as mere tools by engineering their genome even against their wishes, then so be it. All is permitted in this view of nature. This other nihilist alternative is equally implausible. To consider moral prohibitions and obligations on the conduct of both war and peace, then, is a reasonable approach that allows us to steer between these extremes of relativism and nihilism.

These moral theories themselves underpin several distinctions that have proven useful in the history of warfare, such as that between combatant and noncombatant. For utilitarian reasons, it is useful to distinguish between those contributing to the war effort and those who are not. At the very least, harming those not involved in actually inflicting violence upon others would be a waste of resources that could otherwise be channeled toward actually winning the war. More robustly, ensuring that the war is ended in a manner appropriate to reestablishing a long-term peace requires discrimination—that is, telling the difference between combatants and noncombatants. Otherwise, you set yourself up for a continuation of the violence later. From a deontic view, it is easy to see why the combatant/noncombatant distinction is justified. People have a right, and we have a corresponding duty to respect that right, not to be treated merely as a means to an end. Killing an innocent person is a canonical example of that violation of the "categorical imperative," the demands that duty makes upon us as a result of our natures as free and rational creatures. In this view, innocent people such as noncombatants have a right to not be killed even if doing so would end the war sooner. Finally, from a virtue-theoretic perspective, the just and virtuous person would harm only those who deserve to be harmed—the guilty or, in the context of war, combatants. So, one important wartime distinction has its roots in all three moral theories. They each justify a combatant/noncombatant distinction in their own way.

The three major Western approaches to morality support and undergird other aspects of the justice of war as well. While it is beyond the scope of this chapter to explore just war theory and its connections to morality in any detail, a brief primer on the theory will help set the stage for the next section of the paper.

Jus ad bellum ("the justice of the war") considerations govern when one can begin to wage a war in a morally permissible manner. Traditionally, there are seven tenets of *jus ad bellum*:[5]

- *Just cause*: The purpose must be the preservation and protection of value, such as defending the innocent, punishing evil, or reclaiming wrongly taken property.
- *Right authority*: The agent authorizing the use of force must be the representative of a sovereign body.
- *Right intention*: The intent must accord with the just cause; the war must not be fought for territorial aggrandizement or out of bloodlust.
- *Proportionality of ends*: The net good achieved by the war must outweigh the net harm caused by waging it.
- *Last resort*: No other means are available to achieve the ends sought.

- *Reasonable hope of success*: There must be a good chance that the means used will actually achieve the ends sought.
- *Aim of peace*: We should strive to end the war in a fashion that will restore security and the chance for peaceful interaction.

Of course, we should not think of these considerations as an absolute checklist, as there may very well be wars that fail to meet a few of the tenets but that are nonetheless just. These guidelines are distillations of the practical experience of those concerned about beginning wars in a morally permissible and hopefully morally praiseworthy fashion and are not intended to be an unquestionable list of necessary and sufficient conditions for the morality of war.

There are two traditional tenets of *jus in bello* ("justice in the war") that govern the conduct of operations that take place while the war is being waged:

- *Proportionality of means*: Acts causing gratuitous or unnecessary harm are to be avoided; the good achieved by a particular means should outweigh the harm done by employing it.
- *Noncombatant protection ("discrimination")*: Direct harm to noncombatants should be avoided; efforts should be taken to protect noncombatants.

Generally, considerations of *jus ad bellum* and *jus in bello* do not interact. That is, if you are waging an unjust war from the *ad bellum* perspective, this does not mean that you are "cleared hot" to violate the *in bello* tenets (indeed, what excuse could you have for *not* following them given that the war you are waging is itself *unjust*?). Conversely, the fact that you are waging a just war does not give you permission to wage it in an unjust manner (although see Michael Walzer's interesting identification of a possible exception to this rule that he captured in 1977 with his doctrine of "supreme emergency,"[6] the name of which was originally articulated by Winston Churchill in World War II).

These just war considerations do more than summarize the thoughts and practices of those involved in waging war in an ethical manner. In many respects, they have close connections to the *general* moral theories we all use at one point or another to help us reason about what we ought to do or how we ought to behave.

There are other moral distinctions in warfare that are driven by recognition of the seeds of moral truth contained in all three of these traditional approaches to ethics. The next section of this chapter discusses how the evolving biological battlefield might influence just war theory and these critical distinctions.

The Evolving Biological Battlefield

As the scope of the volume you hold in your hands reveals, the use of biotechnology in the controlled application of violence will further complicate what is an already complicated battlefield environment. Biotechnology offers the promise of spurring technological advances in everything ranging from efficient use of energy, self-repairing armor, genetically enhanced soldiers, bio-engineered weapons, adaptive camouflage, autonomous and semi-autonomous battlefield agents, combat medical advances, and the like. Reviewing the moral upshot of each of these technologies in turn would take far too long. Instead, in this section, I highlight some general features of these developments that have moral and ethical upshots.

Many of these technologies will be perfectly orthogonal to moral considerations: that is, their use and development will have no particularly unique ethical implications. For example, the development of a new kind of pistol that has a slightly longer range probably offers no unique moral issues. However, the development of a weapon that could kill instantly from halfway around the world would spur unique moral questions (for example, on the battlefield, soldiers are thought of as being moral equals in the sense that they expose themselves to the same risks and dangers while waging war. If soldiers do not have to expose themselves to these same risks, do they then lose some of the unique moral protections offered to soldiers in wartime?). In general, the technologies discussed in this book break down or make more problematic the distinction between combatant and noncombatant, accelerate the observe-orient-decide-act (OODA) loop for moral decisionmaking, boost the chances of unintended consequences, offer a new suite of tools for warfighters that lowers the moral obligation to not use force in some circumstances but raises it in others, and pose fundamental questions about changing the nature of what it means to be human. I will discuss each of these in turn.

The combatant/noncombatant distinction is justified by the three major moral theories discussed earlier. Observing it on the battlefield is conducive to the greater good (utilitarianism), respects the obligations we have to bearers of rights (deontology), and makes us better human beings (virtue theory). It is probably the most important of the *in bello* aspects of just war theory. Biologized battlefields complicate this distinction both by breaking it down and by making it more difficult to distinguish between combatants and noncombatants when there is a distinction to be made. The combatant/noncombatant distinction is supported at least in part by considerations regarding what it means for someone to be an agent: the kind of creature that is autonomous and able to structure its decisions according to the dictates of reason. Being free and rational creatures is what gives us "moral standing" in a Kantian deontic worldview, gives us a human dignity that makes us bearers of rights, and makes us worthy of a respect. As we invent and deploy autonomous agents—for example, a highly autonomous combat tank that is driven by a biological neural network—questions of agency will arise. Is a tank that can make decisions of its own accord and in conformance with its simple logic an agent worthy of some moral consideration, or is it just another tool like a pistol? For now, this question may seem merely academic, but in the future it will be of critical moral importance. Let us stipulate that at some point in the relatively near future, autonomous agents will be given some sort of deontic respect—we will think of them in the same way we think, perhaps, of children, as being bearers of rights but not possessing full agency (in which case we hold the parents fiducially responsible for the actions of the child). The presence of children on the battlefield in places such as Africa already complicates moral decisionmaking for the line soldier. Imagine what will happen if we introduce a third class of agents on the battlefield, one for which we have little day-to-day experience making moral judgments about how we ought to treat them and who ought to be held responsible for their behavior. If we eventually settle on some sort of mechanism for dealing with this new class of autonomous creatures, we have the problem of *identifying* them on the battlefield. For human beings, we use uniforms and other markers of combatancy status. What will be the equivalent for agents who come in all shapes and sizes on the late 21st-century battlefield?

A second complication is related to this first development. Biotechnology threatens to accelerate the rate at which we must make important moral choices. The literature on how technology, especially the biological sciences, accelerates decisionmaking in general is rich and varied. When advances in biowarfare are factored in, the problem space enlarges yet again. Is this swarm of autonomous agents approaching me merely a flock of birds or a deadly assortment of networked micro–unmanned aircraft systems? Does the use of a genetically engineered virus with a fast burn-time justify quickly unleashing a counter-virus? Battlefield decisions have always been complex and time-sensitive—biologized battlefields even more so. Our OODA loops will be more compressed, amplifying the challenge of getting inside our adversaries' OODA loop while still making morally defensible decisions.

This highlights a third moral danger for the biologized battlefield: the chances of unintended harmful moral consequences resulting from actions is far higher in the ecologically complex battlespaces of the 21st century. The danger presented by bioweapons in this regard is obvious, and is one reason why precursors and analogues to robust biological weapons such as pathogenic bacteria have been highly regulated. When dealing with ecosystems, especially ecosystems that contain agents (human beings, animals, artificial life forms, sapient autonomous technologies), the chance that the introduction of a new technology will have a radically different effect than intended increases. For example, it is very difficult to predict the consequence that an introduced species will have on the environment. While the kudzu vine might have provided nice shade initially, its introduction has had a dramatic and unforeseen negative economic impact on southern areas of the United States.

A fourth challenge presented by these developments deals with what justifies using *military* force, as opposed to another kind of force, to redress wrongdoing. This is not a trivial distinction, as members of the military are allowed to use force in a very different manner than are members of a police force. In general, military professionals are obligated to use as *much* force as is *permitted* to achieve their objectives within the general limits set by just war theory. Members of a police force, on the other hand, use as *little* force as is *required* to keep the peace. This is because the contexts in which the two different kinds of force are applied is very different, in one case against other human beings who are not necessarily citizens of your own nation-state, and in another case in a situation where the chances of great harm coming from not acting are very different. These facets of what amount of force is required will generate conflicting obligations on the biologized battlefield. On the one hand, the stakes could be very high—a failure to act quickly might produce devastating consequences for humanity, in which case you might be morally permitted to use more force than would otherwise be the case to forestall the great harm (think of Walzer's case of supreme emergency). On the other hand, biologized battlefields will be *extended* battlefields in that the arena in which the struggle takes place will be fluid and will often overlap with traditionally "civilian" areas of concern, in which case you might be obligated to act more like a policeman using only minimal nondeadly force than like a soldier looking down the barrel of a gun. Of course, these complications are already with us even in more traditional forms of warfare; the biologized battlefield merely has the potential to greatly complicate the interaction between the harm that is to be prevented and the harm allowed to be done in order to prevent it.

On the whole, biological developments will force us to face a final existential moral challenge. As our tools and weapons become more like us and we more like them, at what point do we refuse to use biotechnology out of fear that embracing it will cause us to become something other than we are? Questions about the use of genetic enhancement and biological (especially neurobiological) augmentation confront these issues explicitly. At what point do soldiers become mere tools and, if they are mere tools, what moral obligations do we have to them over and above the obligations we owe to, say, a screwdriver? When one has finished using a screwdriver, one can put it back into the toolbox. Will engineering, augmentation, and specialization prevent me from demobilizing my soldiers at the end of conflict so they can return to an otherwise normal human life or an approximation thereof? Worries about post-traumatic stress disorder may seem trivial compared to these big-picture existential questions that radical biotechnologies will force us to confront. This is a question that traditional just war theory has not grappled with in any meaningful way.

Whither Ethics and Warfare?

One response to these developments might be to suggest that we start anew. Traditional just war theory and the three major moral theories that underpin it are not useful for thinking about these issues (the critic might say). However, this reaction would be too quick. While the questions that just war theory has us ask traditionally might be too limiting in many circumstances (for example, what does the requirement to declare a war mean in a world where the primary opponent is a nonstate actor?), the moral theories that give shape to *jus ad bellum* and *jus in bello* considerations are still very useful and applicable. They have been around for hundreds or, in the case of virtue theory, thousands of years, and have proven their usefulness in the fields of practical action: what ought I to do here, now? Consider the three cases introduced at the start of this chapter. We can usefully interrogate all three using the major moral theories, drawing upon the resources of just war theory where appropriate. What are the utilitarian, deontic, and virtue-theoretic aspects of targeted killing? What complications present themselves when we start to target the entire system of an adversary and not just fielded military forces? What does it mean to be a good human being, let alone a good soldier, in the genetically enhanced and cognitively augmented extended battlefield? I will not discuss these cases in any detail, as this chapter is intended to serve as an invitation to start a moral conversation about the use of these technologies. I merely want to point out that it is entirely sensible to dissect these cases using traditional tools of moral inquiry, keeping in mind the complications and provisos discussed in the last section.

The Moral Big Picture and the Importance of Individual Judgment

The development of robust biotechnologies will have a disruptive effect on some traditional moral issues in warfare. The lines between war and peace and the distinction between combatant and noncombatant will become blurred. We will have to make decisions more quickly and with more of a chance that the stakes will be very high than in the past. Unintended consequences will be at the forefront of our minds. The moral bar that must be cleared before we can apply lethal force will change more often, and more rapidly, than in the past, complicating our moral decisionmaking. But the flip side of the biologized battlefield also offers moral promise and ethical hope: we may be able to prevent grave harms more easily. We might be able to attack problems using a police model rather than a military model, preventing the harm that accompanies outright war. And at the end of the day, we might be able to confront challenges such as cognitive augmentation and genetic enhancement in a manner that reaffirms, rather than attacks, our fundamental humanity.

The key to dealing sensibly with the moral challenges of the biologized battlefield will be cultivation of good moral judgment at the individual level. Perhaps more so than in the past, the strategic corporal, the ethical airman, and the sensible seaman will be critically important. Confronting these developments in an intelligent and humane way means developing individuals of good judgment, with thoughts and sensibilities tutored by our best ethical theories, as applied to the new circumstances of war.

NOTES

1 Plato, *The Republic*, trans. Robin Waterfield (New York: Oxford University Press, 1998).

2 Aristotle, *Nichomachean Ethics*, trans. Terence Irwin (Indianapolis: Hackett Publishing Company, Inc., 1985).

3 Immanuel Kant *Groundwork of the Metaphysics of Morals*, trans. H.J. Paton (San Francisco: Harper Torchbooks, 1964).

4 John Stuart Mill, *On Liberty and Other Essays* (New York: Oxford University Press, 1998).

5 See James Turner Johnson, *Morality and Contemporary Warfare* (New Haven, CT: Yale University Press, 1999), for an enlightening discussion of the nature and origin of these tenets.

6 Michael Walzer, *Just and Unjust Wars* (Basic Books: New York, 1977).

chapter 21

LEGAL ISSUES AFFECTING BIOTECHNOLOGY

JOSEPH RUTIGLIANO, JR.

The Biological Weapons Convention (BWC) of 1972[1] prohibits the development, production, stockpiling, acquisition, or retention of "microbial or other biological agents, or toxins whatever their origin or method of production, of types and in quantities that have no justification for prophylactic, protective or other peaceful purposes."[2] This, in combination with the Geneva Protocol of 1925,[3] which bans the use of "bacteriological methods of warfare," effectively eliminates an entire class of weapons. The United States officially renounced its offensive biological weapons program in 1969, but the existence of restrictions and prohibitions in the law can be traced back thousands of years.

In this chapter, I will briefly explore early uses of biological agents in warfare, and what, if any, the "legal" response was to those uses. I will then examine the present legal regime to help clarify the parameters. Finally, I will look at some of the advances in biotechnology and comment on how we may be able to move forward without violating our domestic law and international treaty obligations.

Historical Overview

There are indications dating as far back as 1885 BCE that biological agents may have been used for nefarious purposes. The Bible indicates that during the time of Joseph, the Egyptian Pharaoh employed a cupbearer whose sole purpose was to ensure the Pharaoh would not be poisoned.[4] Solomon, who began his rule in 971 BCE, also employed cupbearers.[5] And Nehemiah, who returned to Jerusalem in 444 BCE[6] to rebuild the city, was cupbearer to the Persian king, Artaxerxes.[7] While it is not clear whether a chemical or biological substance was the poison of choice, it is not unreasonable to assume a natural or biological substance was being used as a poison.

In approximately 1500 BCE, on the Indian subcontinent, the Laws of Manu specifically prohibited the use of poison in warfare: "When he fights with his foes in battle, let him not strike with weapons concealed (in wood), nor with (such as are) barbed, poisoned, or the points of which are blazing with fire."[8] This is the earliest known proscription against the use of biological agents that this author could find. The existence of the proscription suggests biological warfare, at least in the form of poisons, had been resorted to or, at the very least, threatened.

Around 600 BCE, the Athenians are reported to have used biological agents by contaminating a stream with a poisonous type of lily root. The stream supplied the water for the adversary's city, and when the inhabitants drank the contaminated water, many became ill.[9] The Carthaginians were also known to use poisons against their enemies. In fact, Hannibal is alleged to have catapulted pots of poisonous snakes into enemy ships, and other Carthaginian generals poisoned wine before retreating so when their adversaries took control of the abandoned property, they would become ill or die from drinking the wine.[10]

In the Middle Ages, use of biological agents continued, sometimes with catastrophic effect. For instance, in the 1346 siege of Kaffa on the Crimea, Mongol hordes catapulted corpses infected with plague into the city. The infected residents fled to Italy as the city fell to the Mongols, and the plague subsequently spread throughout Europe, killing an estimated one-quarter of the European population.[11]

In 1763, British troops traded smallpox-infected blankets with American Indians. The following year, the Mingoe, Delaware, and Shawanoe tribes suffered smallpox epidemics. It has been reported that the British troops were under orders to "do well to try to inoculate the Indians by means of the blankets, as well as to try every other method that can serve to extirpate this execrable race."[12]

During the Civil War, Confederate forces forced farm animals into ponds where they were shot and left to contaminate the water.[13] On the Union side, the 20th Maine, commanded by Lieutenant Colonel Joshua L. Chamberlain, was quarantined during the battle of Chancellorsville because of an epidemic of smallpox. Apparently, poorly prepared serum used to inoculate the 20th Maine resulted in the outbreak. Frustrated when General Daniel Butterfield refused his request for the 20th Maine to get into the fight, Chamberlain said, "If we couldn't do anything else we would give the rebels the smallpox!" General Butterfield, "not too impressed with such a modern and devastating conception of warfare," detailed the 20th Maine to guard a telegraph line.[14]

During World War I, the Germans infected livestock of Allied nations, as well as that of neutral nations in an attempt to prevent its sale to the Allies. They used anthrax and glanders, an infectious horse bacterium. In addition to attacking livestock, the Germans also targeted French cavalry horses and mules in Mesopotamia.[15]

Between World War I and II, Belgium, Canada, France, Great Britain, Italy, the Netherlands, Poland, and the Soviet Union reportedly conducted experiments with biological weapons.[16] During World War II, the infamous Japanese Unit 731 developed plague, cholera, and typhoid agents and experimented on prisoners of war and Chinese civilians. Although it was reported that Japanese use of biological weapons caused approximately 10,000 casualties among its own troops, including 1,700 dead from cholera,[17] Chinese survivors claim that as many as 94,000 people were killed by bubonic plague between 1937 and 1945.[18] The Japanese also made preparations to use biological weapons against U.S. forces and the U.S. mainland. They built a bacterial bomb, a defoliation bacilli bomb, weather balloons to reach across the Pacific (even sending test balloons as far as South Dakota), and submarine-launched weapons. Moreover, they made plans to use germ-infected suicide troops who would charge advancing U.S. forces in an attempt to infect them.[19]

While the United States and the Allies never used biological weapons during World War II, they did experiment with them. In fact, the United States planned to produce 500,000 anthrax bombs per month, and experimented with biotoxins for potential use against Japanese crops.[20] The United Kingdom conducted tests on sheep with anthrax on the Scottish island of Gruinard. Because of the stability of the anthrax bacillus, and the fact it can remain dormant for up to 100 years, the island remained uninhabitable after the experiments.[21] In 1986, the British attempted to decontaminate it.[22]

After World War II, the United States experimented with, and developed, biological weapons until 1969, when it formally rejected them.[23] In fact, the Pentagon admitted to conducting experiments over San Francisco, Minneapolis, St. Louis, Key West, and Panama City, Florida, and more specific tests in the New York City subway, at National Airport in Washington, DC, and on the Pennsylvania Turnpike. Chemicals designed to simulate the spread of the agents rather than actual biological weapons were used. These chemicals themselves caused an outbreak of urinary tract infections in San Francisco, resulting in one death.[24] Although the Soviet Union and other communist countries alleged that the United States engaged in biological warfare, none of the charges could be substantiated.[25]

Ironically, the Soviets did not have the same view toward biological agents that the United States developed in 1969, and may have begun a biological warfare program as early as the mid-1920s.[26] A German assessment done during World War II stated that Joseph Stalin officially began a biological and chemical weapons program in 1939.[27] In the late 1970s, the vice president of the Soviet Academy of Sciences began a program to develop biological agents using modern biotechnology, including genetic engineering, in order to persuade the Ministry of Defense to fund research into genetics and molecular biology.[28] This program grew to employ over 25,000 scientists and technical workers at more than 30 scientific institutes and 5 production plants during its peak in the late 1980s. The existence of this particular program was kept secret through civilian cover until Boris Yeltsin admitted its existence in 1992.[29] This program existed even after the Russians became a party to the BWC.

The Russians continued to pursue development of biological weapons until 1992,[30] when they agreed to terminate offensive weapons research, dismantle experimental technological lines for the production of biological agents, and close biological weapons testing facilities. They also agreed to cut the number of personnel involved in military biological programs by 50 percent, reduce military biological research funding by 30 percent, and dissolve the department in the Ministry of Defense responsible for the offensive biological program, creating in its stead a new department for radiological, biological, and chemical defense.[31]

During the first Gulf War, Iraq deployed but did not use anthrax.[32] Iraq did possess Scud missiles armed with biological agents but did not use any against Israel.[33] After the war, the United Nations Special Commission (UNSCOM) documented Iraq's efforts to acquire weapons of mass destruction (WMD), including biological weapons and long-range ballistic missiles developed to serve as delivery systems. In fact, UNSCOM reported that Iraq had produced large quantities of microbial agents and toxins for use as biological weapons.[34] The reports made it clear that the Iraqis had developed and produced biological weapons systems, including artillery shells, bombs, ballistic missiles, rockets, and aerosol spray devices.[35] Because of the secrecy of Iraq's WMD program, UNSCOM's understanding of it is incomplete. Most of Iraq's biological weapons facilities were undamaged during the conflict.[36]

States are not the only players in the biowarfare realm. Terrorist organizations and various cults have been known to be in pursuit of biological weapons, or to have actually used them. In 1984, the Rajneeshee cult, in an attempt to disrupt local elections, poisoned salad bars in The Dalles, Oregon, with salmonella, resulting in over 700 illnesses.[37] The Japanese cult Aum Shinrikyo, in addition to building its own sarin gas factory capable of producing 70 tons,[38] developed a biological weapons program that progressed to the testing stage for anthrax and botulism.[39] Even though they attempted to release anthrax on more than one occasion, the cult was unsuccessful for a number of reasons. In 1995, U.S. law enforcement authorities caught and convicted two members of a Minnesota militia group of planning an attack on government officials with ricin.[40] Perhaps most worrisome of all is Osama bin Laden's statement that obtaining WMD is a "religious duty."[41] Subsequent to the attacks on 9/11, anthrax was used in a number of attacks through the U.S. mail, killing 5 people and infecting 17 others.

Proscriptions of Biowarfare

The previous discussion demonstrates a consistent pattern throughout the ages of a resort to some level of biological warfare. These uses did not go unanswered by the international community. For example, the prohibition against using poison in the Laws of Manu was part of the code regarding the duties and responsibilities of kings.[42]

The general rule against the use of biological agents existed only in the form of customary international law until very recently, when it was codified in treaties of the 19th and 20th centuries. Customary international law is based on two prongs: the practice of states, and the sense of legal obligation on which state practice is based. A continuous line of development can be traced from Italian jurist Alberico Gentili (1552–1608) and Dutch jurist Hugo Grotius (1583–1645), whom many consider to be the father of modern international law, through The Hague Conventions of 1899 and 1907, the Geneva Protocol of 1925, and the 1972 BWC.[43] This is not to say prohibitions were not codified before the Hague Conventions. For instance, during the Civil War, President Abraham Lincoln commissioned Dr. Francis Lieber to promulgate instructions governing the Union armies.[44] The Lieber Code, as it has came to be known, stated, "Military necessity does not admit of cruelty—that is, the infliction of suffering for the sake of suffering or for revenge, nor of maiming or wounding except in fight, nor of torture to extort confessions. It does not admit of the use of poison in any way, nor of the wanton devastation of a district."[45] In Article 70, the Lieber Code makes a strong statement against the use of poisons, and for the universal condemnation of such use. It states, "The use of poison in any manner, be it to poison wells, or food, or arms, is wholly excluded from modern warfare. He that uses it puts himself out of the pale of the law and the usages of war."[46]

Shortly after the Civil War, there were several treaties and draft treaties that specifically addressed the issue of biological warfare. The Declaration of St. Petersburg of 1868 stated that the "contracting Parties engage mutually to renounce, in case of war among themselves, the employment by their military or naval troops of any projectile of a weight below 400 grammes, which is either explosive or charged with fulminating or inflammable substances."[47] The treaty stated that "the employment of such arms would therefore be contrary to the laws of humanity."[48] This is evidence that there existed as of 1868 a longstanding custom of abhorrence toward chemical and, presumably, biological weapons.

In 1874, a conference in Brussels resulted in a declaration concerning the law of war, in which it was stated that it was "especially *forbidden*" to employ "poison or poisoned weapons."[49] The Brussels Declaration also reiterated the prohibition of using projectiles banned under the St. Petersburg Declaration and prohibited the "employment of arms, projectiles or material calculated to cause superfluous injury."[50] While the Brussels Declaration was never ratified by the countries involved in the negotiation, it laid the groundwork for the First International Peace Conference, held at The Hague, in 1899.[51] The Hague Convention of 1899 provided for the same prohibition against the employment of poison or poisoned weapons, as well as forbidding the employment of "arms, projectiles, or material calculated to cause unnecessary suffering."[52]

Following the Russo-Japanese War of 1905, and at the behest of President Theodore Roosevelt, among others, the international community once again met at The Hague to address the laws and customs of war. In 1907, one of the resulting conventions again set forth the prohibition to employ poison or poisoned weapons, as well as limiting the rights of belligerents to adopt means of injuring the enemy by prohibiting weaponry "calculated to cause unnecessary suffering."[53] These obligations, however, only applied to nations that became a party to the convention. In fact, it specifically stated that the convention and its annexed regulations "do not apply except between Contracting Powers, and then only if all the belligerents are parties to the Convention."[54]

Although there had been a string of treaties addressing the use of poisons, and containing prohibitions against causing superfluous injury or unnecessary suffering, during World War I, Germany used both chemical and biological weapons. This resulted in the issue being directly addressed in the Treaty of Versailles, which ended World War I. Article 171 of the treaty states: "The use of asphyxiating, poisonous or other gases and all analogous liquids, materials or devices being prohibited, their manufacture and importation are strictly forbidden in Germany." The same applies to materials specially intended for the manufacture, storage, and use of the said products or devices.[55]

Subsequent to World War I, and before the Geneva Protocol of 1925 was signed, there were two more documents that addressed the issue of biological warfare. In 1922, the United States signed the Treaty of Washington regarding the use of submarines and noxious gases. Article V of this treaty states the following:

> The use in war of asphyxiating, poisonous or other gases, and all analogous liquids, materials or devices, having been justly condemned by the general opinion of the civilized world and a prohibition of such use having been declared in Treaties to which a majority of the civilized Powers are parties,
>
> The Signatory Powers, to the end that this prohibition shall be universally accepted as a part of international law binding alike the conscience and practice of nations, declare their assent to such a prohibition, agree to be bound thereby as between themselves and invite all other civilized nations to adhere thereto.[56]

This treaty was concluded among the United States, the United Kingdom, Italy, France, and Japan, but never entered into effect because France failed to ratify it. France's reasons for not ratifying the treaty had nothing to do with Article V quoted above.[57] The existence of this treaty, however, among these major powers, is a clear indication that the use of chemical and biological agents was considered beyond the pale. While *asphyxiating, poisonous or other gases* could be interpreted to mean strictly chemical weapons, *all analogous liquids, materials or devices* could equally well be interpreted to include bacteriological agents.

One year later, five Central American states asserted their agreement with this principle by signing a convention wherein the contracting parties agreed that "the use in warfare of asphyxiating gases, poisons, or similar substances as well as analogous liquids, materials or devices, is contrary to humanitarian principles and to international law, and obligate themselves . . . not to use said substances in time of war."[58]

It must be stated that in the Treaties of Versailles and Washington, as well as in the Geneva Protocol of 1925, there is an argument that the prohibitions only applied to chemical warfare unless the term *all analogous liquids, materials or devices* could be interpreted to include biological agents. Some of the delegates who helped draft the Geneva Protocol thought that was indeed the interpretation, although the protocol itself includes the clause, "agree to extend this prohibition to the use of bacteriological methods of warfare."[59] The additional language could be interpreted to be a simple restatement of the longstanding customary prohibition against biological warfare. It appears to this author that use of the word *poison* would include the use of biological agents as well as chemical agents. The bottom line is that whether a biological or chemical method was used to poison, say, a water source, it would have been considered an insidious and treacherous means of warfare, and a violation of all of the treaties and conventions mentioned above.[60]

Geneva Protocol of 1925

The pertinent part of the protocol states:

> *Whereas* the use in war of asphyxiating, poisonous or other gases, and of all analogous liquids, materials or devices, has been justly condemned by the general opinion of the civilized world; and
>
> *Whereas* the prohibition of such use has been declared in Treaties to which the majority of Powers of the world are Parties; and
>
> *To the end* that this prohibition shall be universally accepted as a part of International Law, binding alike the conscience and the practice of nations;
>
> *Declare:*
> That the High Contracting Parties, so far as they are not already Parties to Treaties prohibiting such use, accept this prohibition, agree to extend this prohibition to the use of bacteriological methods of warfare and agree to be bound as between themselves according to the terms of this declaration.[61]

Some interpretations of this language are in order. First, the prohibition against the use of chemical and biological agents applies only during war. This raises the issue of what *war* actually means. It would appear that a commonsense reading of the language of the protocol indicates that the term would mean armed conflict between two or more of the High Contracting parties. The language indicates this when it states, "the High Contracting Parties . . . agree to be bound as between themselves." Subsequent treaties also indicate that the term *war* refers to actually declared wars between nations, or any other armed conflict that may exist between two or more nations.[62] This would mean that the Geneva Protocol would not apply to those conflicts not of an international character.[63]

The term "between themselves" also makes clear the prohibition does not extend to nations that did not ratify the protocol. While this may be true, there is a very strong argument to be made that the prohibition existed in the form of customary international law before it was codified in the Geneva Protocol. For instance, the protocol itself clearly states that this method of warfare "has been justly condemned by the general opinion of the civilized world." This is important, because unlike treaty law, which is generally binding only on those nations that are party to the treaty, customary international law is binding on all nations.

This language also is different from that contained in Hague IV, which stated it applied only with respect to other High Contracting Parties, and as long as all the belligerents to a conflict were parties to the convention. Here, the parties were restricted from use against other parties to the conflict regardless of whether all belligerents to a conflict were parties to the Geneva Protocol. This, coupled with the fact that the Geneva Protocol only prohibits the *use* of chemical and biological agents, led to many nations filing reservations when they became a party. Most notable were France, Great Britain, and the Soviet Union, which became parties in 1926, 1930, and 1928, respectively. The reservations were virtually identical. The reservation of France read: "The said Protocol shall *ipso facto* cease to be binding on the Government of the French Republic in regard to any enemy State whose armed forces or whose allies fail to respect the prohibitions laid down in the Protocol."

In essence, the Geneva Protocol forced the parties to it to adopt a no-first-use policy. This essentially was the U.S. position. When the United States ratified the Geneva Protocol in 1975, it filed the following reservation: "With reservations to Protocol as follows: to cease to be binding as regards use of chemical agents with respect to any enemy State whose armed forces or allies do not observe provisions." Note the United States limited its reservation to "chemical agents," reserving a right to retaliate with chemical weapons should it be attacked with chemical weapons. The United States limited its reservation to chemical agents only because, simultaneously with ratifying the Geneva Protocol, it ratified the BWC, which banned biological agents for offensive purposes.

This overview has demonstrated that the prohibition against the use of biological agents in warfare has been with us for some time. In the form of treaty law, it has existed for the past 150 years. It clearly existed beforehand in the form of customary international law as nearly all the treaties noted above refer to condemnation by the "civilized world" of use of biological weapons. Now, we will explore the current legal regime.

The Biological Weapons Convention of 1972

Understanding what the BWC was intended to cover will allow for determinations of what biotechnologies may be permissible. While this may appear axiomatic, there are cases that at first blush are not clear cut. For instance, several years ago, the Judge Advocate General (JAG) of the Air Force was preparing a legal review of a nonlethal weapon system referred to as pepper foam. It was so-called because it contained oleoresin capsicum (OC), or pepper spray. The United States has declared OC to be a riot control agent pursuant to the Chemical Weapons Convention (CWC) of 1993.[64] Overlooked in previous legal reviews of OC was the fact that it is derived from a living organism and had never been analyzed under the lens of the BWC. This author conducted research into the applicability of the BWC to OC, which was incorporated into the Air Force JAG legal review of pepper foam.[65] The following discussion is taken from that research. Keep in mind that while the discussion specifically relates to OC, it is clear that many biotech capabilities could easily be inserted into the analysis.

The BWC prohibits the development, production, stockpiling, or acquisition by other means, or retention of microbial or other biological agents or toxins, as well as weapons, equipment, or means of delivery designed to use such agents or toxins for hostile purposes or in armed conflict.[66] While the BWC does not define what biological agents or toxins are, the World Health Organization (WHO) has formulated a definition of biological agents considered to be authoritative.[67] In a 1970 report, the WHO defined biological agents "as those that depend for their effects on multiplication within the target organism and are intended for use in war to cause disease or death in man, animals or plants; they may be transmissible or non-transmissible."[68]

Clearly, OC is excluded from this definition of a biological agent because it does not depend for its effect on multiplication within the target organism. Neither is it intended for use in war to cause disease or death in man, animals, or plants. This was the prevailing definition at the time the BWC was negotiated; therefore, it is clear the parties to the BWC were not contemplating prohibiting substances such as OC, but rather were more concerned with weapons of mass destruction. The same can be said for toxins, which are "poisonous products of organisms; unlike biological agents, they are inanimate and not capable of reproducing themselves."[69] OC would be excluded from the definition of toxin because it is not intended for use in war to cause disease or death in man, animals, or plants. "Since toxins are chemicals by nature, their inclusion in the [BWC] was a step towards the projected ban on chemical weapons."[70] Therefore, the U.S. position that OC is a riot control agent and is governed by the CWC and Executive Order 11850[71] and not by the BWC is correct.

This conclusion is supported further by the phrase in the BWC, "hostile purposes or in armed conflict" which, based on the U.S. position at the time, is consistent with the CWC term "method of warfare." The Department of State, in its letter to President Nixon recommending the BWC be transmitted to the Senate for its advice and consent to ratification, pointed out that on November 25, 1969, after receiving the results of a comprehensive study of U.S. policies and programs relating to chemical and biological warfare, Nixon "announced that the United States was unilaterally renouncing biological methods of warfare, and would destroy its existing stocks of these weapons." It further pointed out that on February 14, 1970, "The United States also renounced the use of toxins as a method of warfare." Included in the November 25 statement was express support by the United States for the principles and objectives of the draft Convention on Biological Weapons, which had been proffered by the United Kingdom in July 1969. Therefore, the final language of the BWC, "hostile purposes or in armed conflict," must be read to be consistent with the U.S. position of banning biological agents and toxins as a method of warfare. As a riot control agent, OC may not be used as a method of warfare. Therefore, the BWC does not prohibit the acquisition and use of OC, either as a spray or foam.

Furthermore, in 1969, when the United States renounced the use of biological weapons, it was clear that the President was talking about the horrors of biological agents that have "massive, unpredictable, and potentially uncontrollable consequences," with the potential to produce "global epidemics and profoundly affect the health of future generations." OC does not fit that description. Also, consider what President Nixon actually said in banning biological weapons: "I have decided that the USA will renounce the use of any form of deadly biological weapons that either kill or incapacitate."

Therefore, it is only those deadly agents that can either kill or incapacitate that are banned. For example, smallpox is a deadly disease that does not always kill. It can, however, certainly incapacitate for lengthy periods those it fails to kill. No reasonable person can consider OC to be a deadly agent.

In a statement made to the Senate during the BWC ratification process in 1974, General George Brown, then Chairman of the Joint Chiefs of Staff, reinforced this view when he explained that:

> Biological warfare could produce massive, unpredictable, and potentially uncontrollable consequences. The specter of global epidemics, with their profound impact on the health of present or future generations, is not a happy one. This Convention was designed to reduce this specter and to enhance the security of the world community.

Thus, it is clear that the BWC was designed to prohibit biological weapons of mass destruction, not personal protection substances such as OC.

The Biological Weapons Anti-Terrorism Act

The act, which implements the BWC, defines the term *biological agent* as:

> any microorganism (including, but not limited to, bacteria, viruses, fungi, rickettsiae, or protozoa) or infectious substance, or any naturally occurring, bioengineered, or synthesized component of any such microorganism or infectious substance, capable

> of causing . . . death, disease, or other biological malfunction in a human, an animal, a plant, or another living organism; . . . deterioration of food, water, equipment, supplies, or material of any kind; or . . . deleterious alteration of the environment.[72]

The act defines *toxin* as:

> The toxic material or product of plants, animals, microorganisms (including, but not limited to, bacteria, viruses, fungi, rickettsiae or protozoa), or infectious substances, or a recombinant or synthesized molecule, whatever their origin and method of production, and includes . . . any poisonous substance or biological product that may be engineered as a result of biotechnology produced by a living organism; or . . . any poisonous isomer or biological product, homolog, or derivative of such a substance.[73]

The United States has not taken a position on whether OC and capsaicin constitute biological agents or toxins. While it appears OC would fall outside the statutory definition of biological agent, the statutory definition of toxin is broad enough to seemingly cover substances such as OC and capsaicin. In my opinion, OC, capsaicin, and pepper sprays were not intended to, and do not, fall within the purview of the BWC or the statutory prohibitions on biological and toxin weapons. The legislative history to the act bears this out.

In Senate Report 101–210,[74] it is made clear that the act was meant to address the threat posed by weapons of mass destruction, not self-defense devices such as OC. For instance, the report noted:

> Although they are often associated in the public mind, biological weapons are different than chemical weapons. Most biological weapons embody living organisms that multiply once inside the victim, whereas chemical weapons—nerve gas, for example—are inanimate. Biological weapons usually consist of some form of bacteria, virus, or fungus. Examples would be typhus, cholera, anthrax, and yellow fever. But there are also biological weapons that contain dead substances made from living organisms. These are known as toxins, and examples would include the botulism toxin and shellfish poison. Toxins are substances that fall between biological agents and chemicals, because while inanimate, they are produced by biological or microbic processes.

In 1925, the Geneva Protocol prohibited the first use in war of chemical and biological weapons, but it did not prohibit their development, production, possession, or transfer. Consequently, the United States and other nations sought to outlaw the manufacture and stockpiling of these heinous weapons of mass destruction.[75]

It is clear from this statement that the Congress viewed the BWC as a step toward banning weapons of mass destruction. OC is not a weapon of mass destruction. Hence, Congress did not view either the BWC or the act as preventing the acquisition and use of OC. The statement of President George Bush upon signing the act further supports this position. The President noted on May 22, 1990, that:

> This Act will impose new criminal penalties against those who would employ or contribute to the dangerous proliferation of biological weapons, and it will add teeth to our efforts to eradicate such horrible weapons.... The United States has renounced these weapons, as have all civilized countries, by joining the Biological Weapons Convention of 1972. . . . I call upon the leaders of all nations to join us in our drive to rid the world of biological and chemical weapons and to do everything in their power to stop the proliferation of these weapons of mass destruction.[76]

There can be no doubt that the purpose of the BWC and the act is to address the problem of weapons of mass destruction, not to eliminate OC for use in self-defense devices.

Biological Weapons

Even if OC itself could be considered a toxin under the definition of the act or the WHO definition, this device would not be banned by either the act or the BWC because it is not a prohibited weapon. Article I of the BWC clearly states that it applies to "weapons . . . designed to use such agents or toxins for hostile purposes or in armed conflict." As has been demonstrated above, the term "hostile purposes or in armed conflict" is synonymous with the CWC term "method of warfare." Since pepper spray and foam devices will not be used as a method of warfare—that is, with hostile purposes or in armed conflict—they are not the type of weapon governed by the BWC or the act. These systems are properly governed under the riot control agents regime of the CWC and Executive Order 11850.

The custom and practice of nations since the signing of the BWC further support this conclusion. Article VI of the BWC affords any party "which finds that any other State Party is acting in breach of obligations deriving from the provisions of the [BWC] may lodge a complaint with the Security Council of the United Nations." While there have been several allegations of noncompliance with the BWC, we are unaware of any involving the use of OC.[77]

Moreover, additional evidence that the international community does not consider OC as prohibited under the BWC is that OC, capsaicin, and pepper sprays are not controlled items on the Australia Group's[78] List of Dual-Use Biological Agents for Export Control. This document does not include OC or capsaicin on the list of toxins or other biological agents that are subject to export or transfer controls such as those imposed by the BWC. We also note that OC and capsaicin are not listed in the Australia Group's Control List of Dual-Use Chemicals.

Conclusion

It is apparent from the discussion of the BWC and the Biological Weapons Anti-Terrorism Act using OC as an example that most biotechnologies will not run afoul of our international and domestic legal obligations. Biotechnologies that result in stronger material or more protective clothing or are designed for other peaceful purposes do not violate the BWC or the act, because both provide exceptions for "prophylactic, protective or other peaceful purposes."[79] Biotechnologies that are being considered for hostile purposes or use in armed conflict will have to undergo stringent review. Any weaponization of a biotechnology will be required to undergo a legal review[80] as well as an arms control treaty review.[81] Many of the factors considered in the analysis of OC will need to be considered.

Biotechnology is a promising area that offers capabilities that can better protect our Servicemembers. It also can provide alternatives to deadly force, which under certain circumstances may not be the best manner of responding. The BWC and the act are primarily geared to address those biological agents that can cause mass casualties and result in the spread of deadly disease. Clearly, there is a lot of room to maneuver under the current legal regime to take advantage of these promising biotechnologies.

NOTES

1 Convention on the Prohibition of the Development, Production, and Stockpiling of Bacteriological (Biological) and Toxin Weapons and on their Destruction, April 10, 1972 (26 UST 583; TIAS 8062; 1015 UNTS 163) (hereinafter, the "BWC"), available at **www.opbw.org/**. The BWC entered into force on March 26, 1975.

2 BWC, Article I (1), available at **www.opbw.org/**.

3 Protocol for the Prohibition of the Use in War of Asphyxiating, Poisonous or Other Gases, and of Bacteriological Methods of Warfare, Geneva, June 17, 1925 (26 UST 571; TIAS 8061; 94 LNTS 65) (hereinafter, the "Geneva Protocol"). This treaty entered into force on February 8, 1928, but not for the United States, which became a party to this treaty on April 10, 1975.

4 See Genesis 40; also for dates, see *Nelson's Complete Book of Bible Maps and Charts* (Nashville: Thomas Nelson, 1996), at first chart. The King James Version refers to Pharaoh's butler; but the Hebrew word used for "butler" is the same word used for "cupbearer" in 1 Kings, 2 Chronicles, and Nehemiah. See James Strong et al., *The Strongest Strong's—Exhaustive Concordance of the Bible* (Grand Rapids, MI: Zondervan, 2001).

5 See 1 Kings 10:15, and 2 Chronicles 9:4; for dates, see Nelson's, 112–113.

6 See Nelson's, 160.

7 See Nehemiah 1:11.

8 The Laws of Manu, Chapter VII (translated by George Bühler, *Sacred Books of the East*, volume 25), available at **www.sacred-texts.com/hin/manu.htm**.

9 Paul Christopher, *The Ethics of War and Peace: An Introduction to Legal and Moral Issues* (New York: Prentice Hall, 1994), 202, cited in Scott Keefer, "International Control of Biological Weapons," *ILSA Journal of International and Comparative Law* 107 (Fall 1999), 112–113.

10 Paul Christopher, note 6, 202–203; cited in Keefer, note 6, 113.

11 Keefer, note 6, 113, citing Paul Christopher, note 6, 204; also citing John D. Steinbruner, "Biological Weapons: A Plague Upon All Houses," *Foreign Policy* 109 (1997–1998) 85, 86; George W. Christopher, "Biological Warfare: A Historical Perspective," *Journal of the American Medical Association* 412 (1997) (which states this account of infected corpses may oversimplify the spread of the plague in Europe, as other carriers may have contributed).

12 Francis Parkman, *The Conspiracy of Pontiac*, cited in Paul Christopher, note 6, 204.

13 Paul Christopher, note 6, 205.

14 Willard M. Wallace, *Soul of the Lion—A Biography of General Joshua L. Chamberlain* (New York: T. Nelson, 1960), 66–67.

15 Keefer, note 6, 113–114, citing George W. Christopher, note 8, 413.

16 Ibid.

17 Keefer, note 6, 114, citing Paul Christopher, note 6, 205, and George W. Christopher, note 8, 413.

18 Jonathan Watts, "Japan Taken to Court over Germ Warfare Allegations," *Lancet* 657 (1998), 351, cited in Keefer, note 6, 39.

19 Keefer, note 6, 114, citing Thomas B. Allen and Norman Polmar, *Code-Name Downfall: The Secret Plan to Invade Japan and Why Truman Dropped the Bomb* (New York: Simon and Schuster, 1993), 187–189, 256–257.

20 Keefer, note 6, 114, citing Allen and Polmar, note 16, 178–183.

21 Keefer, note 6, 114, citing Robin Clarke and J. Perry Robinson, "Research Policy: United Kingdom," in *CBW: Chemical and Biological Warfare*, ed. Steven Rose (Boston: Beacon Press, 1969), 105, 108–109.

22 Keefer, note 6, note 43, citing George W. Christopher, note 8, 413.

23 Keefer, note 6, 115, citing George W. Christopher, note 8, 414.

24 Leonard A. Cole, *The Eleventh Plague: The Politics of Biological and Chemical Warfare* (New York: Henry Holt, 1997), 18; and George W. Christopher, note 8, 414.

25 George W. Christopher, note 8, 415.

26 Rita R. Colwell and Raymond A. Zilinskas, "Bioethics and the Prevention of Biological Warfare," in *Biological Warfare: Modern Offense and Defense*, ed. Raymond A. Zilinskas (Boulder, CO: Lynne Rienner, 2000), 225, 227.

27 Ibid.

28 Ibid.

29 Ibid.

30 Richard Boucher, Office of Public Affairs, U.S. Department of State, Press Release, September 14, 1992, "Joint US/UK/Russian Statement on Biological Weapons."

31 Ibid.

32 Keefer, note 6, 117, citing Steinbruner, note 8, 87.

33 Keefer, note 6, 117, citing Raymond A. Zilinskas, "Iraq's Biological Weapons: The Past as Future?" *Journal of the American Medical Association* 278 (1997), 418, 422.

34 Colwell and Zilinskas, note 23, 225.

35 Ibid.

36 Ibid.

37 Keefer, note 6, 118, citing George W. Christopher, note 8, 416. There were 751 cases of enteritis reported, with 45 hospitalizations.

38 Presentation of Jonathan B. Tucker, Ph.D., Senior Fellow, Center for Nonproliferation Studies, Monterey Institute of International Studies, March 14, 2006.

39 Keefer, note 6, 118, citing Robert P. Kadlec et al., "Biological Weapons Control: Prospects and Implications for the Future," *Journal of the American Medical Association* 278 (1997), 351, 354.

40 Keefer, note 6, 118, citing Richard Danzig and Pamela B. Berkowsky, "Why Should We Be Concerned About Biological Warfare?" *Journal of the American Medical Association* 278 (1997), 431, 432, and George W. Christopher, note 8, 416.

41 Tucker, note 35.

42 Ibid., see chapter VII, line 1.

43 Stockholm International Peace Research Institute (SIPRI), *The Problem of Chemical and Biological Warfare*, Volume III, CBW and the Law of War, 1973, 23.

44 "Instructions for the Government of Armies of the United States in the Field," promulgated as General Order No. 100, by President Abraham Lincoln, on April 24, 1864, available at **http://fletcher.tufts.edu/multi/texts/historical/LIEBER-CODE.txt**.

45 Ibid., article 16.

46 Ibid., article 70.

47 Declaration of St. Petersburg of 1868 to the Effect of Prohibiting the Use of Certain Projectiles in Wartime, November 29–December 11, 1868. The United States is not a party.

48 Ibid.

49 Article XIII, Brussels Conference of 1874, International Declaration Concerning the Laws and Customs of War, signed at Brussels, August 27, 1874 (hereinafter, the "Brussels Declaration").

50 Ibid.

51 See SIPRI, note 41, 96.

52 Article XXIII, Annex of Regulations Respecting the Laws and Customs of War on Land, to the Convention with Respect to the Laws and Customs of War on Land, First International Peace Conference, The Hague, 1899, Acts signed at The Hague, July 29, 1899 (32 Stat. 1803; TS 403; 1 Bevans 247) (hereinafter, the "Hague Convention of 1899"), and entered into force September 4, 1900; for the United States, on April 9, 1902. Note the use of the term "superfluous injury" in the Brussels Declaration, and the use of the term "unnecessary suffering," found in the Hague Convention of 1899. This is the codification of the Law of War principle of Humanity, which is also known as "Unnecessary Suffering." Both terms

are found in more modern treaties. See Article 35(2), Protocol Additional to the Geneva Conventions of August 12, 1949, and Relating to the Protection of Victims of International Armed Conflicts (Protocol I), of June 8, 1977.

53 See Hague Convention No. IV Respecting the Laws and Customs of War on Land ("Hague IV"), and Annex Thereto Embodying Regulations Respecting the Laws and Customs of War on Land ("Hague Regulations IV"), October 18, 1907 (36 Stat. 2277; TS 539; 1 Bevans 631), in particular, Articles XXII and XXIII, to the Annexed Regulations. This convention entered into force January 26, 1910. The United States is a party.

54 Article 2, Hague IV.

55 Treaty of Peace with Germany, concluded at Versailles, June 28, 1919. The United States is not a party because the U.S. Senate refused to provide its advice and consent to ratification.

56 Treaty of Washington of 1922 Relating to the Use of Submarines and Noxious Gases in Warfare, signed at Washington, February 6, 1922.

57 See SIPRI, note 41, 22.

58 See Article V, Convention for the Limitation of Armaments of Central American States, signed at Washington, February 7, 1923. The convention entered into force on November 24, 1924. The five Central American countries were Guatemala, El Salvador, Costa Rica, Honduras, and Nicaragua.

59 See SIPRI, note 41, 23.

60 Ibid., 93–96. "As regards *biological* weapons . . . it is indisputable, first, that toxins, evidently, come under the definition of poison. . . . Second, it is certain that no rule of customary law has emerged which could have the effect of excluding biological agents from the prohibition of poison." In fact, the Hague Regulations appear to have implicitly included the spreading of contagious diseases under the heading of 'poison and poisoned weapons,' for as the minutes of the two Hague Conferences show, the expression was taken over without discussion from the Declaration of the Brussels Conference of 1874, which had never been ratified by the governments but formed the basis for the Hague Conference of 1899. The records of the Brussels Conference show that in 1874 the reference to poison and poisoned weapons was meant to include the spreading of disease on enemy territory. In line with this, the U.S. Army manual from 1914 (as well as the manual from 1940) stated that the prohibition of poison expressed in Article 23(a) of the Hague Regulations applied 'to the use of means calculated to spread contagious diseases.'"

61 Geneva Protocol, note 3.

62 See Common Article 2, Geneva Conventions of August 12, 1949, which states, "The present Convention shall apply to all cases of declared war or of any other armed conflict which may arise between two or more of the High Contracting Parties, even if the state of war is not recognized by one of them."

63 There are those who argue that the mere fact an insurgent party is not bound to the terms of an international treaty because it is not an official party to that treaty raises only a formal problem. "The obligation of an insurgent party within a state to observe humanitarian rules accepted by that state is no more paradoxical than is the obligation of a new state, born into a community of states, to observe the customary rules of that community." See SIPRI, note 41, 29.

64 The Convention on the Prohibition of Development, Production, Stockpiling and Use of Chemical Weapons and on Their Destruction, dated January 13, 1993, entered into force April 29, 1997 (hereinafter, the "CWC"). The United States became a party on April 29, 1997.

65 See "AF/JA Memorandum for HQ AMC/JA dated 17 April 2003, Subject: Legal Review of MK3 First Defense/Pepper Foam" (on file with author).

66 BWC, article I.

67 Jozef Goldblat, "The Biological Weapons Convention: An Overview," *International Review of the Red Cross* 318 (June 30, 1997), 251–265.

68 Ibid., 253–254 (citing World Health Organization, *Health Aspects of the Use of Chemical and Biological Weapons*, Geneva, 1970).

69 Goldblat, 254.

70 Ibid.

71 Executive Order 11850, "Renunciation of Certain Uses in War of Chemical Herbicides and Riot Control Agents," April 8, 1975. This Executive Order establishes Presidential approval for use of riot control agents (RCA) in time of war, and then only for defensive military modes to save lives. The Executive Order gives four examples of defensive modes where the use of RCA may be permitted. They are the use of RCA in riot control situations in areas under direct and distinct U.S. military control, to include rioting prisoners of war; use of RCA in which civilians are used to mask or screen attacks and civilian casualties can be reduced or avoided; use of RCA in rescue missions in remotely isolated areas, of downed aircrews and passengers, and escaping prisoners; and use of RCA in rear echelon areas outside the zone of immediate combat to protect convoys from civil disturbances, terrorists, and paramilitary organizations.

72 18 U.S.C. §178(1).

73 18 U.S.C. §178(2).

74 See 1990 U.S. Code Congressional and Administrative News, 186.

75 Ibid., 188.

76 Ibid., 195.

77 The most notorious incidents, the "Sverdlovsk" and "Yellow Rain" incidents, led to U.S. accusations of noncompliance by the Soviet Union. Sverdlovsk involved the airborne release of anthrax, which caused an outbreak in the city of Sverdlovsk, in April and May 1979. The Yellow Rain incident involved an allegation made by the United States that the Soviet Union had been involved in the production, transfer, and use of trichothecene mycotoxins in Laos, Kampuchea, and Afghanistan in the late 1970s. See Goldblat, 262–263.

78 The Australia Group was instituted in 1984 by states dedicated to impeding the proliferation of chemical and biological weapons and agents by encouraging countries to adopt and enforce export controls on these weapons and agents, and encouraging compliance with the 1925 Geneva Protocol, the Chemical Weapons Convention, and the Biological Weapons Convention. The Australia Group currently has 30 members, including the United States. It maintains a list of controlled biological and chemical substances.

79 See BWC, Article I, section (1); 18 U.S.C. §175.

80 See paragraph E1.1.15, Department of Defense (DOD) Directive 5000.1, "The DOD Acquisition System," May 12, 2003; for nonlethal weapons, *see* paragraph E.6.b, DOD Directive 3000.3, "Policy for Non-Lethal Weapons," July 9, 1996.

81 See generally paragraph 4.6, DOD Directive 2060.1, "Implementation of, and Compliance with, Arms Control Agreements," of January 9, 2001, certified current as of November 24, 2003.

chapter 22

BUILDING the NONMEDICAL BIO WORKFORCE for 2040

Timothy Coffey and Joel Schnur

In the 21st century, biology is becoming a major factor in the development of emerging technologies. A number of critical breakthroughs in areas such as molecular biology, genetics, and bio-informatics have led to the emergence of the rapidly growing biotechnology sector of the economy. At present, this sector accounts for about 0.003 percent of the Nation's gross domestic product (GDP). However, the sector has been showing about a 14 percent[1] annual growth rate since 1992. If this growth rate were to persist through 2040 and if the GDP maintained its average growth rate of 3.3 percent, then the biotech sector would account for about 16 percent of the Nation's GDP in 2040. While this seems to be a very high percentage, it should be noted that the information technology (IT)–producing sector was in a similar situation 40 years ago. This sector now accounts for about 8 percent of U.S. GDP.[2]

If biotechnology were to follow a path similar to that followed by IT over the past 40 years, then a 2040 estimate of 16 percent of GDP may be high but could be in the ballpark. If this were to happen, then the matter of providing for the scientific and engineering workforce to support that sector becomes an undertaking of significant consequence. Today, the biotechnology sector employs about 66,000 scientific and technical personnel, about half of whom are scientists.[3] The biotechnology industry today is principally focused on medical and health issues and markets. However, it is expected that nonmedical biotechnology will become increasingly important. Areas of particular interest for future development in nonmedical biotechnology include energy, sensors, materials, and agriculture. As these new areas develop, we expect that the makeup of the biotech science and engineering (S&E) workforce will change considerably.

The number of scientists and engineers that will need to be educated to provide this workforce will be in the millions.

As the field of biotechnology develops and matures, it is vital for the Department of Defense (DOD) to have a knowledgeable S&E workforce in order to recognize and provide the internal advocacy needed to take rapid advantage of relevant breakthroughs from this new area as they emerge. Historically, most of the internal scientific and technical competence of DOD has resided in the military laboratories and the military research and development centers. This is likely to remain the case, and the vitality and competence of the S&E workforces of these organizations will play an important role in determining whether DOD will effectively exploit the science and technology (S&T) emerging from new fields, such as nonmedical biotechnology. There are many issues now confronting these organizations that will impact the outcome regarding DOD's ability to exploit new S&T developments in the coming years.[4] This chapter will focus primarily on one issue: namely, the requirements confronting the Nation and DOD regarding the development of the 2040 biotechnology workforce. The chapter will attempt to identify special problems that will likely confront DOD over the intervening time.

It is, of course, impossible to accurately project 30 years into the future for a fledgling field like biotechnology. There is no "first principles" basis from which to make such predictions. There are, however, considerable data regarding the evolution of S&T, the S&E workforce, and the DOD program since World War II. The trends in this data can be very helpful in making rough order of magnitude (ROM) estimates of future states. This is the approach that will be used in this chapter. The underlying analytical technique utilized will be that of separation of time scales. This allows the data to be expressed as a sum of a base function that changes slowly with time upon which is superimposed a more rapidly varying function that oscillates about the base function. The rapidly varying function is usually associated with a specific transient event (such as World War II, the Korean War, or the Vietnam conflict). Future transients are not predictable. However, the base functions appear to have some robust characteristics and are the basis upon which our ROM estimates will be made.

Trends in the DOD Program

While the purpose of this chapter is to address future DOD manpower requirements in the area of nonmedical biotechnology, such predictions are always best cast in the reality of the past. In that regard, it is helpful to examine the post–World War II history of the DOD program and DOD civilian employment therein. Figure 22–1 provides a summary of DOD expenditures from 1929 through 2005. The funds are expressed in fiscal year (FY) 2000 dollars. The huge impact of World War II is evident. It can be viewed as a discontinuity followed by a train of oscillations with an average periodicity of about 17 years. The major peaks in these oscillations represent the Korean War, the Vietnam conflict, the Ronald Reagan–era buildup, and the Iraq War. The line represents the base about which the oscillations occur during the period from 1940 through 2005. The trend line (base function) is linear in this case and shows that the program has, on average, experienced slow but real growth of about FY 2000 $0.87 billion per year. The backward extrapolation from 1940 shows the large permanent change of state in the Defense program caused by World War II.

figure 22–1. DEFENSE SPENDING, 1929–2005 (IN FISCAL YEAR 2000 DOLLARS)

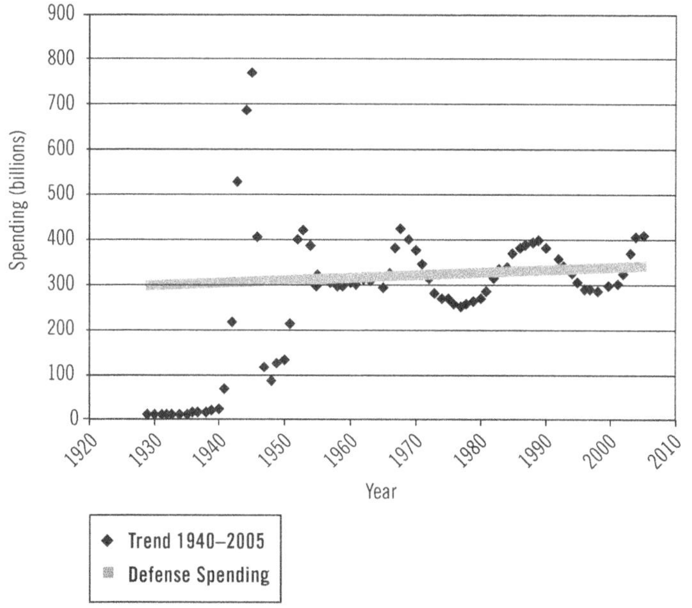

Source: Office of Management and Budget, *Historical Tables, Budget of the U.S. Government, Fiscal Year 2005*, available at **www.whitehouse.gov/omb/budget/Fy2005/pdf/hist.pdf**.

It is helpful to put the DOD program into perspective regarding the economy that it supports (see figure 22–2).

Prior to World War II, defense spending appears to have been in equilibrium with GDP at about 1.5 percent of GDP. It is clear that the war figure 22–2. DEFENSE OUTLAYS AS PERCENT OF GROSS DOMESTIC PRODUCT, 1929–2003

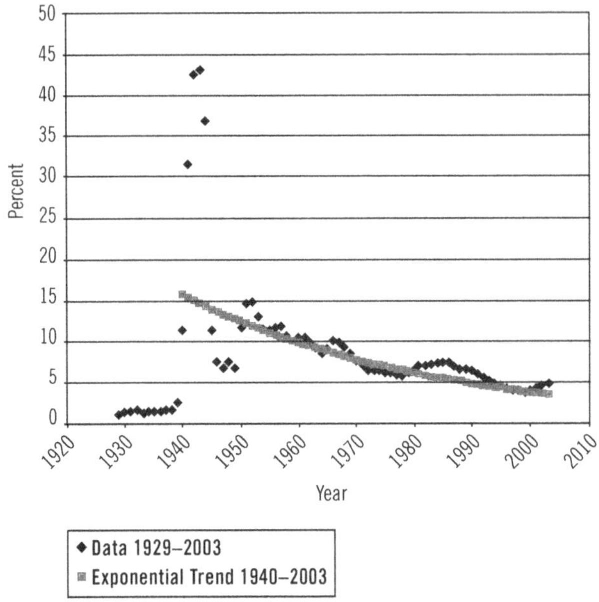

Source: Office of Management and Budget, *Historical Tables, Budget of the U.S. Government, Fiscal Year 2005,* available at **www.whitehouse.gov/omb/budget/fy2005/pdf/hist.pdf**.

created a large transient that drove the Defense program far out of equilibrium and that, since the war, defense spending has been searching for a new equilibrium. While this search has been interrupted by transients such as the Korean War, the Cold War, the Vietnam conflict, the Reagan buildup, and now the Iraq and Afghanistan wars, the search has led, on average, to steadily reducing defense programs when measured relative to GDP. Fortunately, the growth in GDP has allowed some small but real growth in the defense program even as it fell relative to GDP. If the trend continues, by 2050 the defense program will be about 1 percent of GDP.

As was seen in figure 22–1, a decline relative to GDP does not imply a decline in absolute terms. Indeed, since World War II, the absolute size of the defense program has, on average, been maintained even though defense expenditures have fallen relative to GDP. It is conceivable that this could go on indefinitely if the economy continues to grow. While this is comforting, it is also troubling that the defense program should approach zero when measured relative to the nation's economy that it exists to protect. At some point, this trend may become a threat to national security in that adequate preparedness may not be maintained relative to the scale of the nation's economic interests.

Defense Workforce Trends

The previous section provided a brief summary of historical trends in the defense program and trends relative to the economy at large. This section will attempt a similar review of the defense S&E workforce. Figure 22–3 summarizes the total DOD civilian S&E workforce that supported the DOD program from 1958 to 2005.

figure 22–3. DEFENSE SCIENCE AND ENGINEERING WORKFORCE, 1958–2005

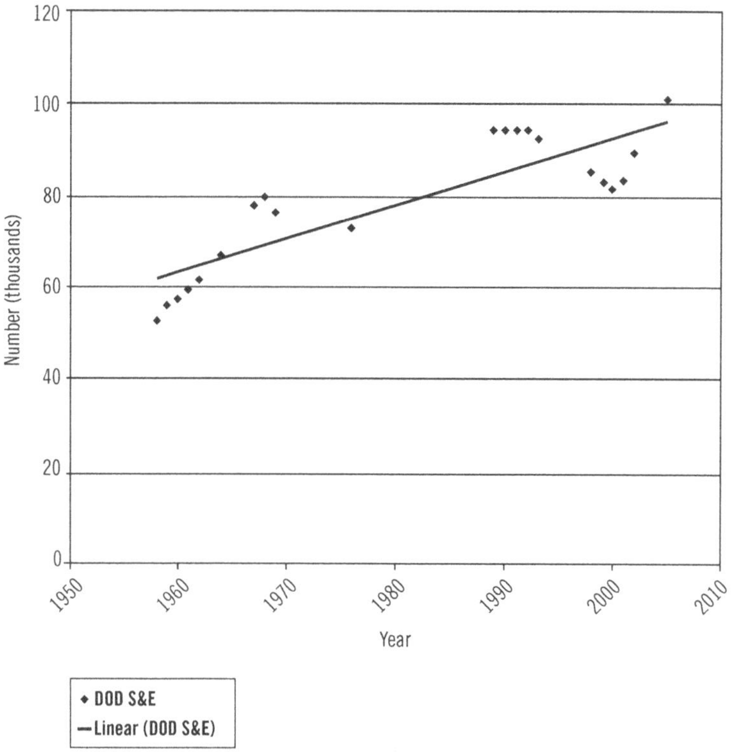

The defense S&E workforce has, on average, tracked the defense program. This presents the defense S&E workforce with long-term problems similar to those confronting the defense program. These S&E workforce issues relate largely to DOD's ability to maintain expertise in the developments occurring in the national and international S&E workforces. This becomes evident when one compares the evolution of the national S&E workforce with the defense S&E workforce. Figure 22–4 summarizes the evolution of the

U.S. S&E workforce from 1950 to 2000 and compares that evolution with the defense S&E workforce between 1960 and 2005.

In 1960, the DOD S&E workforce was representative of the national S&E workforce. However, it can be seen from figure 22–4 that the DOD 2005 S&E workforce is more representative of the 1960 national workforce than it is of today's national workforce. The difference is especially noteworthy when one compares the engineering components and the math/IT components of DOD and the Nation. It should be recalled that IT was fledgling in 1960. From 1960 to 2000, the Nation roughly tripled its engineering workforce and increased its IT workforce by a factor of 100, resulting in about as many IT professionals as engineers by 2000. The Nation was able to exploit a growing economy to continue to expand its engineering workforce while radically increasing its IT workforce. The total S&E workforce over this period grew by a factor of 5.5. DOD was not able to do this since its S&E workforce, on average, increased by a factor of 1.4 during this period. Since DOD is a hardware-/

figure 22–4. SCIENTISTS AND ENGINEERS BY OCCUPATIONAL GROUP (AS PERCENT OF TOTAL) NATIONWIDE, 1950–2000, AND IN DEPARTMENT OF DEFENSE, 1960 AND 2005

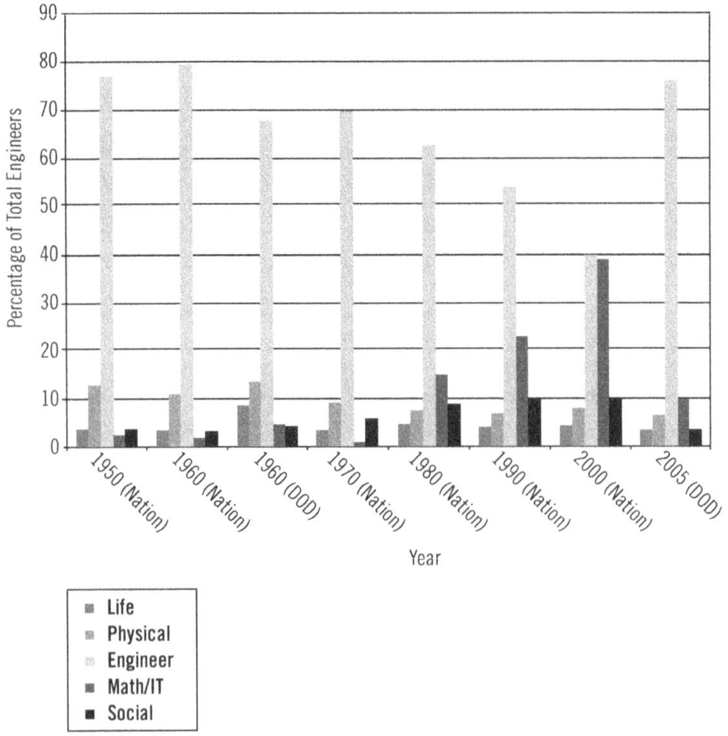

engineering-focused organization, decisions were made to accommodate the workforce constraints by predominantly outsourcing its growing IT workforce needs. This resulted in an imbalance between engineers and IT professionals in DOD compared to the Nation. DOD now finds itself poorly positioned to understand the "art of the possible" in IT, an area that is having a rapidly growing impact on DOD programs. A growing number of major DOD acquisition program problems are now IT-related. For example, it is estimated that 55 percent of military large software projects are terminated prior to completion.[5] Some of this may be related to the imbalance regarding IT professionals in the DOD civilian workforce. In support of a possible connection to the lack of adequate IT expertise in the DOD S&E workforce, we note that the General Accounting Office (GAO) reported in 2004 that IT programs with unrealistic objectives are established, requirements are poorly specified, and management and oversight of the programs are deficient.[6] While the GAO attributes the problem to management deficiencies, it seems more likely that it is due to lack of adequate DOD expertise in the appropriate scientific and technical disciplines that form the foundation of IT.

One of the important roles that the DOD civilian S&E workforce plays is that of maintaining awareness of and currency regarding developments in the national and international S&E workforces. The national S&E workforce grew at an annual rate of about 3.8 percent from 1970 to 2000. This is slightly faster than the GDP growth during this period. However, the DOD S&E workforce followed the trend shown in figure 22–3, meaning it has been declining exponentially, with a negative annual growth rate of about 3.8 percent relative to the national S&E workforce. It is informative to extrapolate this trend to the 2040 time frame considered in this chapter. To do this, we assume that the national S&E workforce will continue its past trend (growing slightly faster than the expected GDP) and the DOD S&E workforce continue its historical trend. The result of this extrapolation is shown in figure 22–5, which projects the DOD S&E workforce as a percentage of the national S&E workforce.

Figure 22–5 shows that if past DOD practices continue, its S&E workforce will exponentially approach zero relative to the national S&E workforce. By 2040, the ratio will be about 0.5 percent. This has profound and disturbing implications regarding the DOD ability to remain current in S&T developments (such as nonmedical biotechnology) and thereby to be able to predict and prepare for the impact of those developments on national security. It also has significant implications regarding the DOD ability to judge the merit of proposed scientific and technical work and to properly oversee work once it is funded. Furthermore, it has serious implications regarding DOD's ability to build the nonmedical biotechnology government workforce it will need as this new field develops. It is likely that the development of nonmedical

figure 22–5. DEPARTMENT OF DEFENSE SCIENTISTS AND ENGINEERS AS PERCENT OF NATIONAL SCIENTISTS AND ENGINEERS

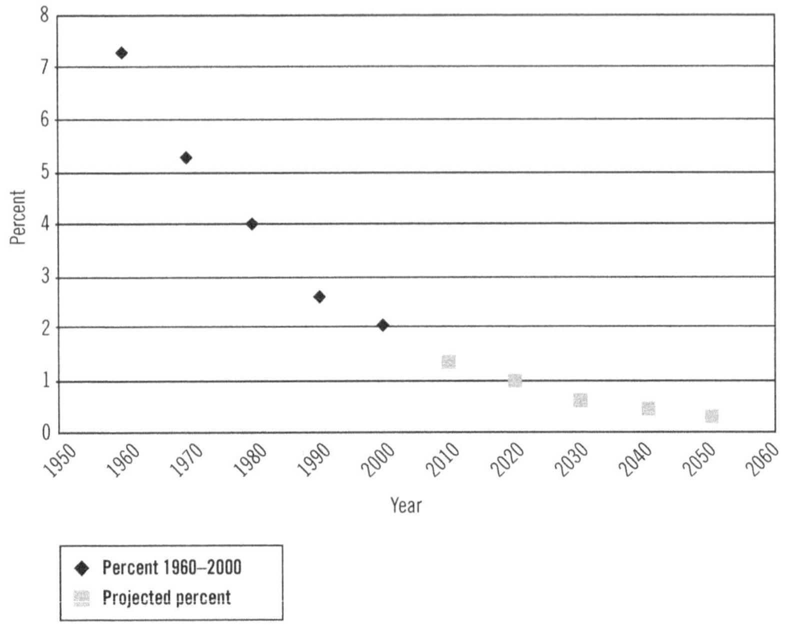

biotechnology will be rapid over the next 30 years, as was IT development over the past 30 years. It is also likely that nonmedical biotechnology will significantly impact defense systems in the coming years as did IT. If DOD manages its future S&E workforce as it did over the past 30 years, it quite likely will find itself in 2040 once again poorly positioned to exercise stewardship over an area of increasing importance to DOD missions.

The 2040 Biotech S&E Workforce

The previous sections of this chapter have provided some background on the evolution of the national and defense S&E workforces. This section will attempt to provide estimates of the 2040 biotechnology S&E workforce and issues associated with developing it. In this regard it is helpful to examine the past evolution of this workforce. From 1970 to 2000, the total number of scientists and engineers supporting the U.S. GDP underwent an annual growth rate of about 3.85 percent—slightly faster than the typical GDP growth rate. In a mature economy, it seems reasonable that the S&E workforce would roughly track the GDP since that workforce exists to support it. If one assumes that the S&E workforce will roughly follow the GDP, then the S&E workforce of 2040 will employ about 18 million scientists and engineers. Figure 22–4 indicates that, during the 1970 to 2000 period, the total number of life plus physical scientists remained relatively stable at about 13 percent of the S&E workforce. Based on this data, we will assume that this trend will continue.

The major change that occurred during this period was the rapid growth in the Math/IT category. This growth occurred mainly at the expense of the engineering category. It seems unlikely that this trend can continue much longer. Indeed, if the Math/IT category continues to grow as it has over the past 30 years, then it will employ about 52 million people by 2040. This exceeds by nearly a factor of three the projected number for all S&Es in 2040. This suggests that the current growth rate for the Math/IT category is not sustainable. The mix of engineers and math/IT professionals must come into equilibrium soon. Therefore, for the purpose of this projection at the S&E category level of description, it is assumed that there will be no radical changes beyond the 2000 relative percentages of scientists and engineers employed in the workforce at large over the next 30 years. Using this assumption, table 22–1 estimates the makeup of the 2040 S&E workforce for the Nation.

We expect that the 18 million total is reasonable based on historical GDP growth. The split among the S&E categories is likely to be much less accurate but, nevertheless, does give some indication of the number of individuals that must be educated over the next 30 years. The next step is to construct a similar table for the 2040 biotechnology workforce. One approach to this is to extrapolate from today's biotechnology S&E workforce that consists of 73.5 percent scientists, 11 percent engineers, 8.4 percent math/IT professionals, and 6.9 percent social scientists. If biotechnology continues to grow at 14 percent per year, by 2040 it will account for about 16 percent of the GDP (assuming that GDP grows at 3.3 percent per year). This biotech growth rate seems very aggressive but is assumed

table 22–1. PROJECTED SCIENTISTS AND ENGINEERS EMPLOYED IN SCIENTIFIC AND ENGINEERING WORK, 2040*

S&E Category	2000 Number (millions)	% of Total	2040 Projection (millions)
Scientists (Life & Physical)	.61	13	2.34
Engineers	1.92	40	7.2
Math/Information	1.88	40	7.2
Social Scientists	.347	7	1.26
Total	**4.757**	**100**	**18**

*Using the 2000 percentages shown in figure 22–4

table 22–2. PROJECTED SCIENTISTS AND ENGINEERS EMPLOYED IN 2040 BIOTECH WORKFORCE, SCENARIO 1

Category	2002 Number (millions)	Percent of Total	2040 Projections (millions)*
Scientists (life and physical)	.0365	73.5	2.114
Engineers	.0055	11	.321
Math/information technology	.0041	8.4	.241
Social scientists	.0037	6.9	.20
Total	**.0498**	**100**	**2.9**

*Based on assumption that 2000 biotech workforce scientist and engineer category breakdown persists and if biotech is 16 percent of U.S. gross domestic product.

to be the case for the purpose of these projections. If we assign a pro rata share of the 2040 S&E workforce to biotechnology, the ROM estimate for the 2040 biotechnology S&E workforce is 2.9 million. Table 22–2 shows the results if today's biotechnology category breakout persists through 2040.

A comparison of table 22–1 and table 22–2 indicates that, under this projection, nearly all scientists in the 2040 total S&E workforce would be in the biotechnology workforce. This seems highly, unlikely since biotechnology accounts for only 16 percent of the 2040 GDP. This suggests that the current biotechnology workforce percentages by S&E category are unlikely to persist through 2040. It seems more likely that, as biotechnology becomes a major part of the economy, its workforce will develop a different mix of S&Es and become more like the economy at large. Table 22–3 indicates the limit where a mature biotechnology S&E workforce supporting 16 percent of GDP in 2040 reflects the S&E workforce at large.

Table 22–3 is likely to be a better representation of the 2040 biotechnology workforce than is table 22–2. However, in order to achieve the table 22–3 projections, the S&T biotechnology workforce will need to grow at an annual rate of 10 percent for the next 30 years. This is three times the expected GDP growth rate. The engineering and math/IT components of the biotechnology workforce will need to grow at about 14 percent per year over this period. Meeting these requirements will place significant demands on the Nation's engineering and IT education systems and curricula. A large number of engineers and math/IT professionals will require training in the biotechnology fields they will be employed to support. This will be a daunting task and could become a limiting factor regarding the growth of biotechnology.

The DOD civilian 2040 biotechnology S&E workforce will depend upon how DOD manages its S&T workforce over the next 30 years. We will

table 22–3. PROJECTED SCIENTISTS AND ENGINEERS EMPLOYED IN 2040 BIOTECH WORKFORCE, SCENARIO 2

Category	2002 Number (millions)	2040 Projections (millions)*
Scientists	.0365	.374
Engineers	.0055	1.152
Math/information technology	.0041	1.152
Social scientists	.0037	.20
Total	**.0498**	**2.9**

*Based on assumption that current national workforce scientist and engineer category breakdown percentages apply and if biotech is 16 percent of U.S. gross domestic product.

consider two of the several scenarios could develop. In the first scenario, DOD and Congress allow the S&E trend shown in figure 22–3 to continue over this period. The total defense base S&E workforce in 2040 will be about 107,000—about 0.5 percent of the expected national S&E workforce at that time. We will call this the "Status Quo" projection. The actual number would be greater or less depending on what the transient contribution was at that time. In the second scenario, the "2 Percent" projection, DOD and Congress maintain the ratio of the DOD civilian base S&E workforce to the national S&E workforce at its present value of 2 percent. In this scenario, the 2040 DOD civilian base S&E workforce would be about 321,000. Again, the actual value would depend on the transients in effect at the time. In both scenarios, we will assume that DOD will strive to achieve an S&E category breakdown that is representative of the national breakdown shown in table 22–1. Under

table 22–4. PROJECTED SCIENTISTS AND ENGINEERS IN 2040 DEPARTMENT OF DEFENSE CIVILIAN BASE WORKFORCE UNDER STATUS QUO AND 2 PERCENT SCENARIOS*

Category	2005 Number (thousands)	2040 "Status Quo Projection" (thousands)	2040 2 Percent Projection (thousands)
Scientists	10.6	14	41.7
Engineers	77	42.8	128
Math/information technology	10	42.8	128
Social scientists	3.7	7.4	22.5
Total	**101**	**107**	**321**

*Assuming that Department of Defense scientist and engineer workforce is representative of 2040 national scientist and engineer workforce breakdown projected in figure 22–3.

these assumptions, the 2040 DOD civilian base S&E workforce category breakdown for both scenarios is shown in table 22–4.

The next matter is to project the 2040 biotechnology component of the DOD civilian base S&E workforce. The nature of biotechnology developments over the next 30 years will determine this. Since these are not predictable, there is no rigorous way to make the projection. We will simply assume that the 2040 biotechnology fraction of the Defense civilian base S&E workforce will be the same as biotechnology's projected fraction of the national workforce, namely 16 percent. This results in table 22–5.

Of course, all of the above projections may not be correct to the smallest details, but they provide a reasonable sense of possible outcomes. The 2040 breakdown between medical and nonmedical biotechnology is also difficult to estimate. The majority of biotechnology efforts today are related to medical and health technologies. This probably explains the high percentage of scientists in the current biotechnology workforce. It is likely that substantial growth will need to occur in nonmedical/health areas if biotechnology is to become a significant factor in GDP generation. The marketplace will ultimately determine the 2040 split between medical/health biotechnology and nonmedical/health biotechnology. For the purposes of this analysis, it will simply be assumed that a 50/50 split will develop by 2040. Furthermore, we will assume that table 22–3 will be representative of the biotechnology workforce in 2040. While these assumptions are somewhat arbitrary, they do provide a means to quantify, for the purpose of discussion, the requirements that will be placed on the education of S&Es in various disciplines. These assumptions imply that, at the national level, the 2040 nonmedical national biotechnology workforce will employ 187,000 scientists, 576,000 engineers, and 576,000 math/IT professionals. The number employed by DOD will depend upon the workforce strategy that DOD and Congress decide to follow. Past

table 22–5. PROJECTED BIOTECH SCIENTISTS AND ENGINEERS IN 2040 DEPARTMENT OF DEFENSE CIVILIAN BASE WORKFORCE UNDER STATUS QUO AND 2 PERCENT SCENARIOS

Category	2040 "Status Quo" Projection (thousands)	2040 2 Percent Projection (Thousands)
Scientists	2.24	6.7
Engineers	6.8	20.5
Math/information technology	6.8	20.5
Social scientists	1.18	3.6
Total	**17**	**51.3**

experience suggests that it will be very difficult for DOD to grow the required nonmedical biotechnology S&E workforce if the "Status Quo" or a lesser strategy is chosen.

The fields of science and engineering that are believed to be of special consequence to nonmedical/health biotechnology are shown in table 22–6.

table 22–6. SCIENCE AND ENGINEERING FIELDS RELEVANT TO NONMEDICAL/HEALTH BIOTECHNOLOGY

Bureau of Labor Statistics Designation	Field
	Science
19–1011	Animal Scientists
19–1012	Food Scientists and Technologists
19–1013	Soil and Plant Scientists
19–1014	Biochemists and Biophysicists
19–1015	Microbiologists
19–1023	Zoologists and Wildlife Biologists
19–1029	Biological Scientists, all other
19–1099	Life Scientists, all other
19–2012	Physicists
19–2031	Chemists
19–2032	Materials Scientists
19–2041	Environmental Scientists and Specialists, including Health
	Engineering
17–2021	Agricultural Engineers
17–2022	Biomedical Engineers
17–2023	Chemical Engineers
17–2024	Computer Hardware Engineers
17–2025	Electrical Engineers
17–2026	Electronics Engineers, except Computer
17–2081	Environmental Engineers
	Math/Information Sciences
15–1011	Computer and Information Scientists, Research
15–1031	Computer Software Engineers, Applications
15–1032	Computer Software Engineers, Systems Software
15–1061	Database Administrators
15–2041	Statisticians
15–2042	Mathematical Science Occupations, all other

The fields that will make up the DOD nonmedical biotechnology S&E workforce will likely be the same as those that make up the national nonmedical biotechnology S&E workforce. The majority of the DOD nonmedical biotechnology S&E workforce would be employed in the DOD laboratories and R&D centers. If the system were orchestrated properly, the labs and R&D centers would serve as training grounds for technically competent and experienced program managers for system and material acquisitions that have a large biotechnology component.

Conclusion

Since World War II, the U.S. defense program, on average, has shown slow but real growth. During the same period, the DOD civilian S&E workforce, on average, has shown real growth similar to the real growth in the average defense program. However, an examination of the DOD S&E workforce indicates that its makeup in terms of scientific and technical disciplines is not representative of today's national S&T workforce. This raises concerns about DOD's ability to judge the "art of the possible" regarding the new technologies that have emerged over the past few decades.

There are also several additional and disturbing trends evident at this time. For example, on average, the defense program is approaching zero exponentially when measured relative to the GDP. Furthermore, the DOD civilian S&E workforce is approaching zero exponentially relative to the national S&E workforce. As a result of this, there has emerged a significant shadow S&E workforce that is providing the DOD "brain cells" regarding new developments in science and technology and thereby new directions for defense. There are serious long-term public policy issues associated with this development.

Most of these trends are simply the result of the various tradeoffs needed in order to get the government's business done. However, there are a number of significant potential negative consequences of allowing these trends to continue indefinitely. If the defense civilian S&E workforce continues to decline relative to the national workforce, a point will be reached where it becomes irrelevant. It will not be able to renew itself and it will not be able to maintain competence in the newly developing fields of science and technology while at the same time maintaining competence in the traditional fields that will continue to be important to DOD. This could result in the government not being able to distinguish a good S&T proposal from a bad one or to competently oversee S&T work that has been funded. In addition, the defense S&E workforce will not be able to provide compelling advocacy within the government for important new S&T initiatives. Of special concern to the subject of this chapter is that DOD will find it difficult to grow internal competence in the important new area of nonmedical biotechnology. As a result, there will be no community internal to DOD that is advocating the importance of and the potential impact of this emerging field.

NOTES

1 Biotechnology Industry Organization, "Biotechnology Industry Facts," available at **http://bio.org/speeches/pubs/er/statistics.asp**.

2 D. Henry and D. Dalton, "Information Technology Producing Industries—Hopeful Signs in 2003," available at **https://www.esa.doc.gov/reports/DE-Chap1.pdf**.

3 U.S. Department of Commerce, "A Survey of the Use of Biotechnology in U.S. Industry," available at **www.technology.gov/reports/Biotechnology/CD120a_0310.pdf**

4 See, for example, Don De Young, "Silence of the Labs," *Defense Horizons* 21 (Washington, DC: National Defense University Press, January 2003); Timothy Coffey, Kenneth W. Lackie, and Michael L. Marshall, "Alternative Governance: A Tool for Military Laboratory Reform," *Defense Horizons* 34 (Washington, DC: National Defense University Press, November 2003); R. Kavetsky, M. Marshall, and D. Anand, *From Science to Seapower: A Roadmap for S&T Revitalization* (EPSC Press, 2006).

5 C. Jones, "Project Management Tools and Software Failures and Successes," available at **www.stsc.hill.af.mil/crosstalk/1998/07/tools.asp**.

6 General Accounting Office, "Defense Acquisitions: Stronger Management Practices Are Needed to Improve DOD's Software-Intensive Management Acquisitions," report to the Committee on Armed Services, 2004.

CONCLUSION

Advances in the biological sciences, as has been true in other fields, will most likely be a case of the haves and the have-nots when it comes to a nation's capacity to research, develop, and produce new products. However, dual-use and possibly seemingly innocuous biological products within an increasingly digital world and global economy can be acquired more easily by countries that are lagging behind scientifically.

This situation poses relatively new challenges for strategic defense planning. More generally, as noted in a recent RAND report, "If countries are to stay ahead in their capacity to implement applications, they will need to make continuing efforts to ensure that laws, public opinion, investment in R&D, and education and literacy are drivers for, and not barriers to, technology implementation."[1]

Of course, having biological technology does not necessarily mean adopting it or deploying it. Lack of knowledge about its importance or the existence of other economic or defense priorities could cause the technology to languish, leading different countries to have different levels of biological adoption that could have downstream consequences.

On the one hand, the United States is struggling to maintain its predominance in science and technology. Many other countries have become scientifically advanced in general or in specific areas. Mike Moodie has pointed out that even Cuba has a relatively sophisticated biotechnology (and medical) operation occurring within a relatively poor and isolated island country—the Cubans see this ability as a strategic advantage. On the other hand, biological products are increasingly easy to buy, transport, and hide. In both cases, increased education in the biological sciences and how they impact the national security of the United States is prudent.

Some advancements are perhaps more likely than others. The aforementioned RAND report maps the "implementation feasibility" (whether something fills a niche versus broad market and the level of policy difficulty it might face) and the "technical feasibility" (whether it is feasible, uncertain, or relatively unlikely to develop in the first place) of certain developments. Biological developments such as rapid biosensors/bioassays, biological sources related to hybrid vehicles and solar energy, targeted drug delivery, and tissue engineering are squarely in the middle of being both highly feasible in both categories. On the other end of the spectrum (doubly low feasibility) are biological technologies such as quantum computers, "super soldiers," drugs tailored to individual genetic makeups, and memory enhancement. This is not to say that these technologies are impossible (people are working on them right now); they are just relatively unlikely to be implemented by 2020. Of course, more scientifically advanced countries will have a greater chance of implementing less feasible technologies.

This biotechnological development also should be viewed within the framework of an increasing trend toward "sustainable development"—meeting the needs of the present without compromising the ability to meet these same needs in the future. The "triple bottom line" for sustainable industry has been described[2] as *economically viable* (creating value and profits), *environmentally compatible* (using eco-efficient products/processes to prevent pollution and depletion of natural resources), and *socially responsible* (behaving in an ethical manner). It has been posited that in order to achieve such a grand improvement, a paradigm shift to using biomimicry in industrial biotechnology is necessary. It is clear that many of the topics discussed in this volume will most likely have implications to the intertwined worlds of military research and industrial production.

Where will the interface between the natural world, biological scientists, industrial engineers, and military strategists lie? Storehouses of information like that maintained by the Biomimicry Institute are one idea.[3] In the future, engineers will be able to query a biomimicry digital library to map engineering concepts to biomimetic concepts that have evolved over 3.85 billion years in nature, to find inspirations and solutions.

According to Janine Benyus of the Biomimicry Institute,[4] four types of knowledge about the natural world need to be a part of such a database. Using an owl as an example, the four types are *semantic* (for example, the names of an owl's wing parts), *procedural* (how an owl flies), *structural* (where feathers connect to muscle), and *functional* (ecological relationship with a prey species). With the recorded number of worldwide species about 1.6 million, with estimates of the total number between 5 million and 30 million, and a huge amount of information to be catalogued for any given species, there is a lot of work to do in this area that will consume biologists for years to come. It is worth pointing out for educational purposes that this kind of descriptive work will be done by evolutionists, ecologists, animal behaviorists, physiologists, functional morphologists, and the like—people who tend to study the *outsides* of organisms rather than life at the molecular level.

Utilizing naturally occurring enzymes, evolving new ones through artificial evolution, using cells as mini-reactors in a process of metabolic engineering, and engaging the concept of industrial ecology in which companies work with materials in a faux ecosystem to avoid anything going to waste may all come into play. This also harkens the notion of a "bio-based economy"[5] in which genes replace petroleum as the basic unit of global commerce. Whereas the past 150 years or so might be termed the "Age of Geology," we may be moving toward an "Age of Biology" in which bio-based sustainable practices are the norm. Things such as genetically modified foods are a start but are only the beginning of this trend, and there are vast implications of this for the military: If the raw material of diverse genes is more important to America than the raw material of petroleum, this may influence where and whom the military fights for resources, and will also pose new challenges to foreign policy, international stability, and law. Already there are controversies over who rightfully owns intellectual property rights to discoveries of drugs that stem from rare, indigenous organisms. The host country? The biotech company? The explorer/discoverer? Maybe all, maybe none.

NOTES

1 R. Silbergritt et al., *The Global Technology Revolution 2020*, RAND report prepared for the National Intelligence Council, 2006.

2 Organisation for Economic Co-operation and Development, *The Application of Biotechnology to Industrial Sustainability—A Primer*, Working Party on Biotechnology, Directorate for Science, Technology, and Industry, Committee for Scientific and Technological Policy, 2002.

3 The mission of The Biomimicry Institute is to nurture and grow a global community of people who are learning from, emulating, and conserving life's genius to create a healthier, more sustainable planet. See **www.biomimicryinstitute.org/**.

4 See also Janine Benyus, *Biomimicry: Innovation Inspired by Nature* (New York: Harper Collins, 1998).

5 Robert E. Armstrong, "From Petro to Agro: Seeds of a New Economy," *Defense Horizons* 20 (Washington, DC: National Defense University Press, 2002). There is also the interesting question posed in this paper of whether agriculture (genes) is a "critical infrastructure" of the United States.

ABOUT the CONTRIBUTORS

Volume Editors

Robert E. Armstrong was a Senior Research Fellow in the Center for Technology and National Security Policy (CTNSP) at the National Defense University. He received a B.A. in psychology from Wabash College, an M.A. in experimental psychology from Oxford University, and an M.S. in biology and a Ph.D. in plant breeding and genetics from Purdue University. A Vietnam veteran and lifelong Army Reservist, COL Armstrong was also a graduate of the U.S. Army War College.

Mark D. Drapeau was the 2006–2008 American Association for the Advancement of Science (AAAS) Science and Technology Policy Fellow in CTNSP. Dr. Drapeau earned a B.S. in biology from the University of Rochester and a Ph.D. in ecology and evolutionary biology from the University of California–Irvine.

Cheryl A. Loeb is a Research Associate in CTNSP, where she works on health security and biological weapons proliferation threats. Previously, she worked at the Monterey Institute of International Studies as a chemical and biological weapons nonproliferation analyst. Ms. Loeb is a Ph.D. candidate in the Public and International Affairs program at George Mason University.

James J. Valdes is a Senior Research Fellow in CTNSP and the U.S. Army's Scientific Advisor for Biotechnology. He received his doctorate in neuroscience from Texas Christian University and was a postdoctoral fellow in neurotoxicology at The Johns Hopkins University. Dr. Valdes is the author of more than 120 scientific papers and a winner of a 2009 Presidential Rank Award.

Contributing Authors

COL John B. Alexander, USA (Ret.), served in key positions in Special Forces, intelligence, and research and development. Upon retirement, he joined Los Alamos National Laboratory, where he developed the concept of nonlethal defense. More recently, he served with the Army Science Board. He is author of several books and many articles on international security issues.

Paul Bartone is a Senior Research Fellow in the CTNSP Life Sciences Directorate. He has conducted numerous field studies of stress, health, and adaptation among military personnel and their families. Dr. Bartone is a charter member of the Association for Psychological Science and a Fellow of the American Psychological Association. He earned a B.A. in psychology from the University of Massachusetts and an M.S. and a Ph.D. in psychology and human development from the University of Chicago.

Yaakov Benenson is a Principal Investigator at Harvard University's Laboratory for Molecular Automata. He has a Ph.D. in computer science and biological chemistry from the Weizmann Institute of Science in Israel.

William E. Bentley is the Robert E. Fischell Distinguished Professor and Chair of the Fischell Department of Bioengineering at the University of Maryland. He holds joint appointments with the Maryland Technology Enterprise Institute and the University of Maryland Biotechnology Institute. He received B.S. and M.S. degrees from Cornell University and a Ph.D. from the University of Colorado, all in chemical engineering.

Lt. Col. William D. Casebeer, USAF, is the Chief of Eurasia Analysis at the North Atlantic Treaty Organization Military Headquarters. His previous positions include associate professor of philosophy at the U.S. Air Force Academy, Middle East analyst with the Ninth Air Force, and project fellow at the Carr Center for Human Rights Policy at Harvard University. Lieutenant Colonel Casebeer has a Ph.D. in cognitive science and philosophy from the University of California at San Diego.

James P. Chambers is Professor of Biochemistry in the Department of Biology and an Adjunct Professor of Biochemistry in the Department of Biochemistry at the University of Texas at San Antonio. Dr. Chambers has worked on Department of Defense–sponsored research projects over the past 25 years, is a recipient of numerous research awards, and has published extensively in peer-reviewed scientific journals. He holds a Ph.D. in biochemistry from the University of Texas.

Timothy Coffey has been a Distinguished Research Fellow at the National Defense University since 2001. He held a joint appointment as Senior Research

Scientist at the University of Maryland and as Edison Chair for Technology at the National Defense University from 2001 to 2007. Previously, he served as the Director of Research of the U.S. Naval Research Laboratory from 1982 to 2001.

Francis Doyle III holds the Duncan and Suzanne Mellichamp Chair in Process Control in the Department of Chemical Engineering at the University of California, Santa Barbara, as well as appointments in the electrical engineering department and the biomolecular science and engineering program. He received a B.S. from Princeton University, a Certificate of Postgraduate Study from Cambridge, and a Ph.D. from California Polytechnic State University, all in chemical engineering.

Eleanore Edson is a Life Sciences and Public Health Analyst at Booz Allen Hamilton. Previously, she was an AAAS Science and Technology Policy Fellow at the Office of Naval Research. Dr. Edson received her Ph.D. in neurobiology from Harvard University.

Andrew Ellington is the Fraser Professor of Biochemistry in the Department of Chemistry and Biochemistry at the University of Texas at Austin. Previously, Dr. Ellington was an Associate Professor in the Department of Chemistry at Indiana University and a Research Fellow in the Department of Genetics at Harvard University and the Department of Molecular Biology at Massachusetts General Hospital. He received a B.S. in biochemistry from Michigan State University and a Ph.D. in biochemistry and molecular biology from Harvard University.

Richard Genik III is Director of the Emergent Technology Research Division and Interim Director of the Transportation Imaging Laboratory at the Wayne State University School of Medicine. His expertise is in magnetic resonance imaging sequence design, functional magnetic resonance imaging data analysis, and integration of emerging technologies.

Christopher Green leads the Emergent Technology Research Division in the Department of Psychiatry at the Wayne State University Medical School. He holds a faculty position in both psychiatry and diagnostic radiology. Dr. Green was previously the Executive Director for Global Science and Technology Policy for General Motors and Chief Technology Officer for its Asia-Pacific operations.

Justin Grubich is an Assistant Professor of Biology at the American University in Cairo. Previously, he was a foreign affairs advisor on environmental and scientific issues in the Office of the Science and Technology Advisor to Secretary of State Hillary Clinton. Dr. Grubich received his doctorate in evolution and ecology from Florida State University.

Thomas X. Hammes is a Senior Research Fellow in the Institute for National Strategic Studies (INSS) at the National Defense University. Colonel Hammes' final tour prior to retiring from the U.S. Marine Corps was as Senior Military Fellow in INSS. He has an M.A. in historical research and a Ph.D. in modern history from Oxford University. He is the author of *The Sling and the Stone: On War in the Twenty-First Century* (Zenith Press, 2006).

João Hespanha is a Professor and Department Vice Chair in the Electrical and Computer Engineering Department at the University of California, Santa Barbara.

Edmund G. (Randy) Howe is a Professor in the Department of Psychiatry, Director of Programs in Medical Ethics, and Senior Scientist at The Center for the Study of Traumatic Stress at the Uniformed Services University of the Health Sciences, where he has been on the faculty since 1977. His research has focused on medical ethics with an emphasis on ethics in military medicine and clinical care at the end of life. Dr. Howe holds an undergraduate degree from Yale University, an M.D. from Columbia University, and a law degree from Catholic University.

Adrienne Huston is a Staff Scientist at the Verenium Corporation. She is a biochemist and microbiologist with over 10 years of laboratory research experience, in addition to science policy and program management experience at the National Science Foundation.

Eric W. Justh is with the Tactical Electronic Warfare Division of the Naval Research Laboratory. He received a Ph.D. in electrical engineering from the University of Maryland and held a half-time appointment as an Assistant Research Scientist in the Institute for Systems Research at the University of Maryland from 2001 to 2006.

Jeffrey Krichmar is an Associate Professor in the Department of Cognitive Sciences and the Department of Computer Science at the University of California, Irvine. Previously, he was a Senior Fellow in Theoretical Neurobiology at The Neurosciences Institute. He received a B.S. and an M.S. in computer science from the University of Massachusetts at Amherst and The George Washington University, respectively, and a Ph.D. in computational sciences and informatics from George Mason University.

Michael Ladisch is a Distinguished Professor of Agricultural and Biological Engineering in the College of Engineering and the Director, Laboratory of Renewable Resources Engineering, at Purdue University.

Judith Lytle is the Director of Science and Technology Division at Avian Engineering, LLC. Previously, she was a Science and Technology Policy Fellow at AAAS.

Thomas McKenna works on neuroscience and biorobotics in the Life Sciences Research Division at the Office of Naval Research.

Michael Moodie is a consultant with CTNSP who has worked for more than 15 years in chemical and biological weapons issues in government and the policy research community. He headed the Chemical and Biological Arms Control Institute and served as assistant director for multilateral affairs at the U.S. Arms Control and Disarmament Agency.

David C. Peters II is on the faculty of the Wayne State University School of Medicine Department of Psychiatry and Behavioral Neuroscience. He is a Ph.D. candidate in the Department of Sociology at Wayne State University and is also an attorney.

Joseph Rutigliano, Jr., a former Marine Corps officer, has been an international law specialist at Headquarters, U.S. Marine Corps, since 1998.

Joel Schnur is at the College of Science at George Mason University. In 1983, he founded the Center for Bio/Molecular Science and Engineering at the Naval Research Laboratory, where he was director until 2008. Dr. Schnur has a B.A. from Rutgers College and an M.S. in physics and chemistry and a Ph.D. in physical chemistry from Georgetown University.

John Socha is an Assistant Professor in Biomechanics in the College of Engineering at Virginia Technical University. He is an organismal biomechanist studying the relationship between form and function in animals, with a broad range of projects involving locomotion, breathing, and feeding.

Elizabeth Stanley is an Assistant Professor of Security Studies in the Edmund A. Walsh School of Foreign Service and the Department of Government at Georgetown University. Previously, she served as Associate Director of Georgetown's Security Studies Program and the Center for Peace and Security Studies. She served in Bosnia, Germany, Macedonia, Italy, and Korea as a U.S. Army military intelligence officer, leaving service with the rank of captain.

Dimitra Stratis-Cullum is a Research Chemist at the U.S. Army Research Laboratory in Adelphi, Maryland. Previously, she was a postdoctoral research fellow at Oak Ridge National Laboratory. Dr. Stratis-Cullum earned a B.S. in chemistry from Marist College and a Ph.D. in chemistry from the University of South Carolina.

James Sumner is a Research Chemist at the U.S. Army Research Laboratory, where he serves as the Team Leader for the Bio-Inspired Devices and Sensors Team in the Sensors and Electron Devices Directorate. Dr. Sumner earned a

B.S. in chemistry from High Point University and a Ph.D. in chemistry from Clemson University.

Roy G. Thompson is Chief of Molecular Engineering at the U.S. Army Edgewood Chemical Biological Research Center, and has formerly held positions as Assistant Professor at Texas Christian University, Research Associate at The Johns Hopkins University, Associate Member of the Vanderbilt Institute for Integrative Biological Research and Education, and Assistant Program Manager to the Deputy Assistant Undersecretary of Defense for Environmental Security. He received a B.A. in chemistry and psychology from Youngstown State University and has doctoral training in neuroscience at Texas Christian University.

Erica R. Valdes is a Research Chemist in the Chemical/Biological Forensic Analysis Center at the U.S. Army Edgewood Chemical Biological Research Center, where she has 27 years of research experience ranging from ultrafine particle studies and biologically inspired materials to forensic investigation of hazardous materials. Dr. Valdes holds an M.S. in chemistry and a Ph.D. in materials science from The Johns Hopkins University.

www.ingramcontent.com/pod-product-compliance
Lightning Source LLC
Chambersburg PA
CBHW071235160426
43196CB00009B/1068
9781780390406